THE BORANE, CARBORANE, CARBOCATION CONTINUUM

THE BORANE, CARBORANE, CARBOCATION CONTINUUM

Edited by

JOSEPH CASANOVA
Loker Hydrocarbon Research Institute
University of Southern California
Los Angeles, CA 90080–1661

A Wiley-Interscience Publication
JOHN WILEY & SONS, INC.

New York • Chichester • Weinheim • Brisbane • Singapore • Toronto

This book is printed on acid-free paper. ∞

Library of Congress Catologing-in-Publication Data:

The borane, carborane, carbocation continuum / edited by Joseph
   Casanova.
      p.   cm.
    "July 15, 1997."
    "A Wiley-Interscience publication."
    Includes bibliographical references and index.
    ISBN 0-471-18075-0 (alk. paper)
    1. Boranes—Congresses.   2. Carboranes—Congresses.
  3. Carbonium ions—Congresses.   I. Casanova, Joseph, 1931–.
QD181.B1B64   1998
547′.056710442—dc21
                                       97–31255
                                         CIP

Printed in the United States of America
10 9 8 7 6 5 4 3 2 1

Dr. Robert E. and Barbara S. Williams

CONTENTS

PART II   THE CARBORANE–CARBOCATION CONTINUUM

PART III   UNTANGLING MOLECULAR STRUCTURES   215

PART IV   NEW SPECIES OF BORANES AND CARBORANES

FOREWORD: THE PATTERNMAKER

During the past 45 years, Bob Williams has made numerous seminal contributions to at least three aspects of borane chemistry: (1) the discovery and subsequent characterization of the first carboranes; (2) the nomenclature of cluster species; and (3) the structral patterns and significant relationships between the various types of carborane clusters. As is clear from his contribution in this volume, he continues to make subtle and penetrating observations that invariably stimulate us to further thought and experimentation.

His original work on the carboranes (1953–1961) was classified by the U.S. government and the first publication only appeared in 1962.[1] It opened up an enormous field of novel cluster chemistry that has been of great theoretical importance and industrial significance. This has been fully reviewed many times and need not be further elaborated here.

Bob was one of the first to deprecate the continuing and constricting use of what he termed "the shibboleth of the icosahedron"[2] as the main structural basis for borane cluster geometry and was the first to point out the unifying relations involving triangulated polyhedra.[2,3] Before that (and even afterward), as is clear from various ACS and IUPAC Nomenclature Committee Reports,[4] classification had been based on the idea of closed and open networks of boron atoms distinguished by the prefixes *closo* and *nido*, respectively. Perhaps Bob disliked the emphasis on the idea of a *network of boron atoms* divorced from the structural influence of the hydrogen atoms, and he may even have recalled Samuel Johnson's famous definition in the very first Dictionary of the English Language: "*Network*: anything reticulated or decussated, at equal distances, with interstices between the intersections." In any event, he pointed out that, "for bookkeeping purposes" all carborane–borane chemistry can be divided into three categories: *closo, nido*, and *arachno*, though these last were at first considered as a subset of *nido*.[2,5]

His choice of the prefix *arachno*, in allusion to a spider's web, reveals that Bob was a much better boron chemist than zoologist because, as any observant naturalist knows, spiders' webs never feature triangular units, only quadrilaterals and occasional pentagons (see **a**). Indeed, the class *Arachnida* includes not only the 32,000 species of spiders, but also the scorpions (see **b**), and this reminds us of another of Bob's endearing traits. For, though he is one of the most generous and charming of people and a delightful colleague, he also sees it as his role to sting us into action by his provocative thoughts, and this without doubt has stimulated all of us to think more deeply about our subject.

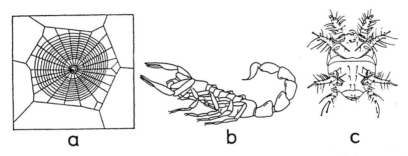

a   b   c

It might be thought less than seemly to remind Bob that, in addition to spiders and scorpions, the arachnids also include some 30,000 species of mites and ticks. These minute creatures are rarely more than 0.5–0.9 mm long, but many have bizarre patterns of behavior. Perhaps the most remarkable is the mite *Acarophenax tribolii*. The female (see **c**) produces 15 eggs, including but a single male, which all develop within the mother's body. The male emerges within his mother's shell, copulates with all his sisters and dies before birth.[16] It may not sound like much of a life, even for a boron chemist, but is suggests that there are still innumerable developments possible in our field. I am confident that Bob will continue to goad us into searching for them.

REFERENCES

1. Shapiro, I., Good, C. D., and Williams, R. E., *J. Am. Chem. Soc.*, **84**, 3837–40, 1962.
2. Williams, R. E., *Inorg. Chem.*, **10**, 210–214, 1971.
3. Williams, R. E., *Adv. Inorg. Chem. Radiochem.*, **18**, 67–142, 1976.
4. Adams, R. M. and Loening, K. L., *Inorg. Chem.*, **7**, 1945–64, 1968.; see also: *Pure Appl. Chem.*, **30**, 683–710, 1972.
5. Williams, R. E., *Progress in Boron Chemistry*, Vol. 2 (Brotherton, R. J. and Steinberg, H., eds), Pergamon Press, Oxford, 1970, pp. 37–118.
6. Gould, S. J., *The Panda's Thumb; More Reflections in Natural History*, Penguin Books, Harmondsworth, Middlesex, 1980, p. 64.

NORMAN N. GREENWOOD

FOREWORD: A BRIEF OVERVIEW

Bob Williams was the codiscoverer, along with C. D. Good, of the smaller carboranes $C_2B_3H_5$, $C_2B_5H_7$, and two isomers of $C_2B_4H_6$.[1] Good isolated the pure samples, following preliminary very-low-yield mixtures obtained by H. Landesman (in 1953) and B. Keilin. Bob established the number of hydrogens using mass spectra and deduced the polyhedral structures from the ^{11}B and ^1H NMR spectra. These structures became public knowledge on September 8, 1961.[2] Bob comments [3] that only 17 days later, I included them in a paper then submitted to the *Proceedings of the National Academy of Sciences*.[4]

At about the same time, several industrial research groups were beginning to prepare the icosahedral $C_2B_{10}H_{12}$ and its derivatives. For example, an application was filed on May 13, 1959 by J. W. Ager Jr. [5] At least two other patents filed in 1959 were issued in 1962.[6,7] Although an X-ray diffraction structure indicated [8] that $C_2B_{10}H_{12}$ consisted of a —CH=CH— unit bridging the 6,9 positions of a decarborane-like B_{10} unit, the icosahedral structure was clearly proven in two other studies.[9,10] These discoveries in the smaller carboranes and $C_2B_{10}H_{12}$ demonstrated that carbon could show the high coordination (5,6) that boron was known to have in "electron-deficient" bonding situations.

The rapture attendant upon these discoveries was tempered by slow acceptance of the ideas, reduction in funding of the rocket propellant program, and termination of support for most of main-group inorganic chemistry (including my grant) by the National Science Foundation, which became abnormally preoccupied with transition-metal catalysis.

The invention of Olin's Dexil polymers based on alternating units of $C_2B_{10}H_{12}$ with siloxanes,[11] and of similar polymers involving $C_2B_5H_7$ and siloxanes,[12] produced materials of unprecedented thermal stability and decreased viscosity–temperature dependence. These latter materials were produced by Chemical Systems, Inc., which Bob founded in 1970. At about the time that his company was acquired by the Purolator Corporation (1980), Bob became a Senior Fellow of the Loker Hydrocarbon Research Institute (1979 to the present).

Bob's poineering and continuing advances in the geometrical (and charge) systematization of polyborane and carborane cluster compounds have been a major contribution. Starting from the structures of *closo*-(polyhedral), *nido*-(one vertex B or C missing), and *nido*-(two adjacent vertices missing) species, Bob extended the coordination number patterns beyond the then established octahedral- and icosahedral-based structure relationships to the B_N *closo*-series for $5 \leq N \leq 14$, and their *nido*- and *arachno*-fragments, including both polyboranes and carboranes.[13,14] The electron-count rules for polyhedral-like species [15,16] were developed in a background that included these coordination patterns, the simplifying omission of the external outward-pointing bonds,[17] and the *styx* numbers,[18] which encompass the electron counts.[19]

In a publication of 1976, Bob greatly extended his "coordination number pattern recognition theory" of the carboranes and boranes.[20] Supplementary new rules were presented on the placement of bridge and endo-hydrogens (of BH_2 groups), the placement of carbon and other heteroelements in the framework, and the coordination of boron. He showed us enlarged patterns of relationships, predictions of new molecular geometries, and positions of heteroatoms and of bridge and endo-hydrogens. In addition, relative stabilities of isomers were judged. As exemplified in Bob's most recent surveys [3,21] this empirical approach accommodates the known structures of carboranes, and predicts fewer candidates for undiscovered or controversial structures than other theoretical proposals.

Even as Bob began his tenure at the Loker Hydrocarbon Research Institute, he was interested in the relationship between the then controversial nonclassic carbocations (e.g., norbornyl cation) and delocalized bonding in analogous polyboranes.[22] At IMEBORON-4 (Snowbird, Utah), he more than held his own against an alternative view held by a well-known organic chemist, a discussion that was ended only when the final bus had to leave!

In the case of norbornyl cation, the resolution of the classic versus the nonclassic description was eventually resolved by experiment: ^1H NMR and ^{13}C spectroscopy,[23] Raman spectroscopy,[23] X-ray photoelectron spectroscopy,[23] solid-state NMR [24] and X-ray diffraction;[25,26] and by theory.[27]

Bob continues to pursue his extended interests in the borane, carborane, and nonclassic carbonium ion analogies,[28–30] and in doing so he shows us the great value of pattern recognition, the hallmark of original science.

REFERENCES

1. Patent 3,030,289 filed by Williams and Good on September 16, 1959.
2. R. E. Williams, C. D. Good, and I. Shapiro, 140th Meeting of the American Chemical Society, Chicago, September 1961, 14N, p. 36.
3. R. E. Williams, Early Carboranes and Their Structural Legacy, *Adv. Organometal. Chem.*, **36**, 1–55, 1994.
4. W. N. Lipscomb, *Proc. Natl. Acad. Sci. USA*, **47**, 1791, 1961.
5. Serial No. 813,032 referred to in Patent 3,062,756, was issued to C. C. Clark on November 6, 1962.
6. Patent 3,030,423 to Alley and Fuchs; and Patent 3,028,432 to C. D. Ross, 1962.
7. J. J. Bobinsky, *J Chem. Ed.*, **41**, 500, 1964.
8. L. I. Zakharin, V. I. Stanko, V. A. Brattsev, Yu. A. Chapovsky, and Yu. T. Struchkov, *Bull. Acad. Sci. USSR* (Eng. Trans.), 1911, 1963.
9. $C_2B_{10}Cl_{12}$: W. N. Lipscomb, *Boron Hydries*, W. A. Benjamin, 1963, p 26.
10. $B_{10}Cl_8H_2C_2H_2$: J. A. Potenza and W. N. Lipscomb, *J. Am. Chem. Soc.*, **86**, 1874, 1964.
11. H. A. Schroeder, *Inorg. Macroml. Rev.*, **1**, 45, 1970.
12. R. E. Williams, Carborane Polymers, *Pure and Applied Chem.*, **29**, 569–583, 1972.
13. R. E. Williams, Carboranes, *Progress in Boron Chemistry*, Vol. 2, (eds, R. J. Brotherton and H. Steinberg), Pergamon Press, Oxford, 1970, pp. 37–118.
14. R. E. Williams, *Inorg. Chem.*, **10**, 210–214, 1971.
15. K. Wade, *Chem. Commun.*, 792, 1971.
16. R. W. Rudolph and W. R. Pretzer, *Inorg. Chem.*, **11**, 1974, 1972.
17. W. N. Lipscomb, *J. Chem. Phys.*, **22**, 985, 1954.
18. W. H. Eberhardt, B. L. Crawford Jr, and W. N. Lipscomb, *J. Chem. Phys.*, **22**, 989, 1954.
19. W. N. Lipscomb, *Inorg. Chem.*, **18**, 2328, 1979.
20. R. E. Williams, Coordination Number Pattern Recognition Theory of Carborane Structures, *Adv. Inorg. Chem Radiochem.*, **18**, 67—142, 1976.
21. R. E. Williams, Geometrical Systematics of *Nido*-Carboranes, -Polyboranes, and -Carbocations, *Electron Deficient Boron Carbon Clusters* (eds, G. Olah, K. Wade, and R. E. Williams), Wiley and Sons, New York, 1991, pp. 11–93.
22. R. E. Williams and L. D. Field, Non-classical Carbocations—Isoelectronic and Isostructural Carbon Copies of Polyboranes; Conservation of Chemical Shift, *Boron Chemistry*, Vol. 4 (eds, R. W. Parry and G. Kodama), Pergamon Press, New York, 1980, pp. 131–150.
23. G. A. Olah, G. K. S. Prakash, and M. Saunders, *Acc. Chem. Res.*, **16**, 440, 1983.
24. G. S. Yannoni, V. Macho, and P. C. Myhre, *J. Am. Chem. Soc.*, **104**, 907 and 7380, 1982.
25. T. Laube, *Angewandte Chemie*, **99**, 580, 1987.
26. T. Laube, *Accounts Chem. Res.*, **28**, 399–405, 1995.

27. M. Schindler, *J. Am. Chem. Soc.*, **109**, 1020, 1987.

28. G. Olah, G. K. Prakash, G. K. Surya, R. E. Williams, L. D. Field, and K. Wade, *Hypercarbon Chemistry*, John Wiley and Sons, New York, 1987 (311 pp.).

29. R. E. Williams, G. K. Prakash, G. K. Surya, L. D. Field, and G. A. Olah, The Polyborane-Carborane-Carbocation Analogy Extended; New Boron-Hydrogen-Carbon Bridge Hydrogen Containing Cations, *Mol. Struct. Energ.*, Vol. 5 (Adv. Boron, Boranes), VCH Verlagsgesell., Weinheim, (eds, J. F. Liebman, A Greenberg, and R. E. Williams), Germany, 1988, pp. 191–224.

30. R. E. Williams, *Chem. Rev.*, **92**, 117–207, 1992.

WILLIAM N. LIPSCOMB

PREFACE

In December 1995, a symposium entitled "The Borane, Carborane, Carbocation Continuum," a part of the Richard Kimbrough Research Symposium series, was held at the Loker Hydrocarbon Institute of the University of Southern California. The symposium was assembled to honor the 70th birthday of Robert E. Williams. Dr. Williams has been one of the leading contributors to the field of preparative and structural carborane chemistry over a period of four decades. The symposium title reflects one of the key research interests of the Institute—the preparation, characterization, and reactions of carbocations and the related boron analogs—the area of principal scientific contributions of Bob Williams. The choice of symposium topic also focuses attention on an emerging understanding regarding the similarity of properties and behavior of carbocations, boranes, and carboranes. These highly electron-deficient species exhibit structural features and chemical properties that sufficiently parallel each other so that a meeting of international experts in these three areas of research, to jointly discuss their fields, seemed most appropriate. The novelty of the symposium topic and the freshness of the work presented at the symposium made it natural for us to reduce the presentation to print. This volume is the result of that effort.

In the early 1950s, Lipscomb and Kaspar reported the correct structures for pentaborane and decaborane, among the many discoveries that led to the award of the Nobel Prize in 1976 to Bill Lipscomb. But, prior to the 1950s, there were no recognizable carboranes. Following his baccalaureate work at DePauw University and doctorate at the University of New Mexico, Williams accepted a job at the Pasadena laboratories of the Olin Matheson Corporation; the job involved vacuum line work and the company's interest in high-energy materials. This placed him in a good position to discover the first four carborane compounds.

Between 1956 and 1958, Williams, with the help of coworkers at Olin Matheson, determined the structures of these novel compounds using 12.8 MHz ^{11}B NMR. By the early 1960s, three classes of carboranes had been identified. The rich and novel cage structures of these compounds, resulting from their incorporation of unusual three-center two-electron bonds, which had been described and popularized by Lipscomb, gave Williams an opportunity to display a talent for which he has become widely known—that is, the ability to recognize patterns relating geometry and electron distribution in these compounds and the special insight to draw structural inferences regarding classes and families of compounds depending on molecular composition and electron count. During this early period, similar pioneering efforts of Earl Muetterties and Fred Hawthorne in carborane and metallocarborane chemistry were signal in developing an understanding of these compounds. Williams regards the boranes, carboranes, and carbocations as offering organized examples of the simplest, most elementary electron-deficient clusters in all of chemistry. Since there are many more possible electron-deficient boranes (BH) and carboranes (BCH) than there are equivalent electron-deficient carbocations (CH), the incorporation of all three groups of structures under a unified structural rubric offers widespread utility and understanding. At the same time, the close relationship of boranes and carboranes, as neutral analogs of carbocations, the study of which earned George Olah his Nobel Prize, opened up a fresh new field of cooperative studies in which Bob Williams has been able to extend his interests.

Today, integration of high-level ab initio theory and concomitant IGLO and GIAO calculations with NMR experiments is bearing fruit and has become the structure marker of these electron-deficient compounds for the 1990s. Several contributions to this volume directly reflect these developments. Williams continues to be an active participant in these currently ongoing efforts.

In the name of all participants and contributors to this volume, I wish Bob not only a happy birthday, but continuing good health. We are certain that he will continue to contribute to his beloved field for many years to come. We take special pleasure in bringing together the chapters contributed here by distinguished researchers in the field, assembled as a tribute to the outstanding efforts of Bob Williams over a period of four decades. During these four decades, Bob and I have been good friends and colleagues, and we hope that this will remain so during the next four decades.

The editor is greatful to Robert Greatrex, John D. Kennedy, and Kenneth Wade for editorial suggestions.

JOSEPH CASANOVA

CONTRIBUTORS

Joseph W. Baush, Department of Chemistry, Villanova University, Villanova, PA 19085   e-mail: bausch@rs6chem.chem.vill.edu

Vratislav Blechta, Institute of Chemical Process Engineering, Academy of Sciences of the Czech Republic, 165 02 Prague 6-Suchdol, Czech Republic

Joseph Casanova, Loker Hydrocarbon Research Institute, University of Southern California, University Park, Los Angeles, CA 90080-1661   e-mail: jcasano@calstatela.edu

Thomas P. Fehlner, Department of Chemistry University of Notre Dame, Notre Dame, IN 46556   e-mail: thomas.p.fehlner.1@nd.edu

Mark A. Fox, Chemistry Department, Durham University Science Laboratories, South Road, Durham DH1 3LE, U.K.

Jin Fusek, Institute of Inorganic Chemistry, Academy of Sciences of the Czech Republic, 250 68 Rez near Prague, Czech Republic

Paul L. Gaus, Department of Chemistry, The Ohio State University, Columbus, OH 43210-1173

Robert Greatrex, School of Chemistry, University of Leeds, Leeds LS2 9JT, U.K.   e-mail: R.Greatrex@chemistry.leeds.ac.uk

Norman N. Greenwood, Department of Chemistry, University of Leeds, Leeds LS2 9JT, U.K.

Russell N. Grimes, Department of Chemistry, University of Virginia, McCormick Road, Charlottesville, VA 22901   e-mail: rng@faraday.clas.virginia.edu

Stanislav Hermánek, Institute of Inorganic Chemistry, Academy of Sciences of the Czech Republic, 250 68 Rez near Prague, Czech Republic   e-mail: HERMANEK@UACHR.IIC.CAS.CZ

Narayan S. Hosmane, Department of Chemistry, Southern Methodist University, Dallas, TX 75275   e-mail: nhosmane@post.cis.smu.edu

Glenn T. Jordan IV, Department of Chemistry, The Ohio State University, Columbus, OH 43210-1173

John D. Kennedy, School of Chemistry, University of Leeds, Leeds LS2 9JT, U.K.   e-mail: J.Kennedy@chemistry.leeds.ac.uk

William N. Lipscomb, Department of Chemistry, Harvard University, 12 Oxford Street., Cambridge, MA 02138   e-mail: lipscomb@chemistry.harvard.edu

Fu-Chen Liu, Department of Chemistry, The Ohio State University, Columbus, OH 43210-1173

Jianping Liu, Department of Chemistry, The Ohio State University, Columbus, OH 43210-1173

Jan Machácek, Institute of Inorganic Chemistry, Academy of Sciences of the Czech Republic, 250 68 Rez near Prague, Czech Republic

John A. Maguire, Department of Chemistry, Southern Methodist University, Dallas, TX 75275

Edward A. Meyers, Department of Chemistry, The Ohio State University, Columbus, OH 43210-1173

Michael L. McKee, Department of Chemistry, Auburn University, 179 Chemistry Bldg., Auburn, AL 36849   e-mail: mckee@chem.auburn.edu

Katayoun Najafian, Computer Chemistry Center, Institut für Organische Chemie, Universität Erlangen-Nürnberg, Henkestrasse 42, D-91054 Erlangen, Germany

George A. Olah, Loker Hydrocarbon Research Institute, Department of Chemistry, University of Southern California, University Park, Los Angeles, CA 90089-1661   e-mail: olah@methyl.usc.edu

Thomas Onak, Department of Chemistry, California State University, Los Angeles, CA 90032   e-mail: tonak@calstatela.edu

Peter Paetzold, Institüt für Inorganische Chemie, Technische Hochscule Aachen, D-52056 Aachen, Germany   e-mail: Peter.Paetzold@ac.RWTH-Aachen.DE

Robert W. Parry, Department of Chemistry, University of Utah, Salt Lake City, UT 84112-1194   e-mail: parry@chemistry.utah.edu

G. K. Surya Prakash, Loker Hydrocarbon Research Institute, Department of Chemistry, University of Southern California, University Park, Los Angeles, CA 90089-1661   e-mail: prakash@methyl.usc.edu

Golam Rasul, Loker Hydrocarbon Research Institute, Department of Chemistry, University of Southern California, University Park, Los Angeles, CA 90089-1661

Paul v.R. Schleyer, Computer Chemistry Center, Institut für Organische Chemie, Universität Erlangen-Nürnberg, Henkestrasse 42, D-91054 Erlangen, Germany and Department of Chemistry, University of Georgia, Athens, GA e-mail: (Germany) paul@organik.uni-erlangen.de and (U.S.) schleyer@paul.chem.uga.edu

Sheldon G. Shore, Department of Chemistry, The Ohio State University Columbus, OH 43210-1173   e-mail: shore.1@osu.edu

A. J. Tabben, Department of Chemistry, Villanova University, Villanova, PA 19085

Kenneth Wade, Chemistry Department, Durham University Science Laboratories, South Road, Durham DH1 3LE, U.K.   e-mail: kenneth.wade@durham.ac.uk

Robert E. Williams, Loker Hydrocarbon Research Institute, Department of Chemistry, University of Southern California, University Park, Los Angeles, CA 90089-1661   e-mail: william@almaak.usc.edu

Andrei K. Yudin, Loker Hydrocarbon Research Institute, Department of Chemistry, University of Southern California, University Park, Los Angeles, CA 90089-1661

THE BORANE, CARBORANE, CARBOCATION CONTINUUM

PART I

PATTERNS OF STRUCTURE IN BORANES AND CARBORANES

1

VERTEX HOMOGENEITY: THE "HIDDEN HAND" THAT GOVERNS ELECTRON-DEFICIENT BORANE, CARBORANE, AND CARBOCATION STRUCTURES

ROBERT E. WILLIAMS

Loker Hydrocarbon Research Institute, University of Southern California,
University Park, Los Angeles, CA 90089-1661

1.1 INTRACLASS GEOMETRICAL SYSTEMATICS

1.1.1 Introduction

1.1.1.1 Vertex Connectivity, nk_C and nk_P The discovery of the *closo*-carboranes, $C_2B_nH_{n+2}$ [1] (Figure 1-1), and the related *closo*-anions, $CB_nH_{n+2}^-$ and $B_nH_n^{2-}$, took place in the early 1950s.

As this chapter concentrates upon geometrical considerations, and uses compounds as examples only, the many individual compounds will not usually be referenced unless they are recent discoveries, recently reconfirmed structures, or structures that have been disproved. For our present purposes, all heavy skeletal atoms, i.e., the borons and carbons located at the deltahedral vertices, will be identified by assigning *nk*-values to each vertex. The *n* in *nk* identifies *the number of neighboring heavy skeletal atoms* to which a given vertex, *k*, is connected. Endo-terminal, exo-terminal, and bridging hydrogens, or other substituent groups, are not included in the count. Most of the skeletal borons and carbons in Figure 1-1 are identified as occupying 5*k* and 4*k* vertex sites, while 6*k* vertices (compound **1.11**) and 3*k* vertices (compound **1.1**) are rare. Frequently, subscripts are appended to the n*k*-vertex designations, i.e., nk_C and nk_P. The subscript C, in nk_C, identifies the vertex as a cage vertex, "c," in contrast to a

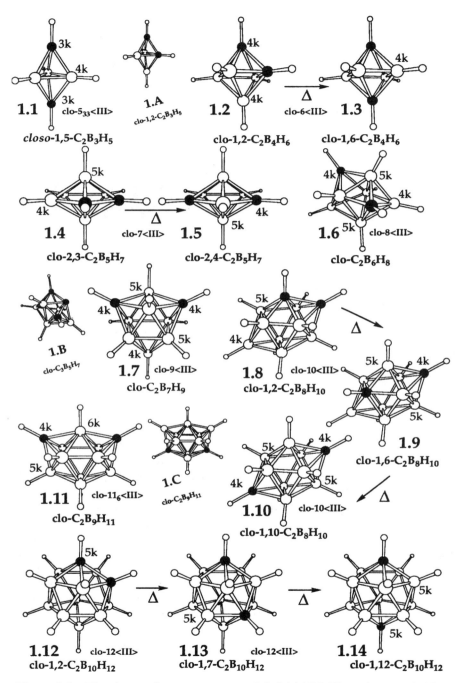

Figure 1-1   The *closo*-carboranes, structures **1.1**–**1.14** ($C_2B_nH_{n+2}$, where $n = 3$–10).

perimeter vertex, "**p**", as in nk_P. An nk_P vertex, by definition, is found around the perimeter of the larger open faces, characteristic of *arachno*- and *nido*-species. As all of the *nk* values in Figure 1-1 are cage vertices, the subscript C is unnecessary.

1.1.1.2 Closo-, Nido- *and* Arachno-*Structures are Abbreviated as clo-*n$\langle III \rangle$, *ni-*n$\langle VI, V, and IV \rangle$ *and ara-*n$\langle VI \rangle$ *Configurations*

The various carboranes and polyboranes were originally identified as members of four classes of compounds—*closo* (cage); *nido* (Gk, nest), *arachno* (Gk, web), and *hypho* (Gk, net)—as a function of their sequentially more open shapes, as well as a function of their fixed but sequentially increasing skeletal electron counts (see also Chapter 8 in this volume). In recent times, it has become apparent that the same shapes are found in different families and, thus, shapes are not definitive. The invariant skeletal electron count has thus become the dominant mode of classification (see also Chapter 2 in this volume).

The shapes of the various *closo*-carboranes in Figure 1-1 are identified by the abbreviated label, clo-*n* $\langle III \rangle$, where "clo" stands for *closo*. In this case, the *n* value reflects the total number of heavy skeletal atoms in the cluster and the $\langle III \rangle$ indicates that the largest aperture in the cluster is trigonal ("III-gonal"). Among *closo*-clusters, both $5k_c$ and $4k_c$ vertices are acceptable, but $5k_c$ vertices are favored. When troublesome $3k_c$ and $6k_c$ *cage* vertices are necessarily present, their residence is noted by adding *subscripts* following the number of heavy atoms, *n*. For example, in Figure 1-1, *closo*-$C_2B_6H_8$,compound **1.6** (with no $3k_c$ and $6k_c$ vertices), is simply labeled clo-8$\langle III \rangle$ while *closo*-$C_2B_3H_5$, compound **1.1** (with two 3*k* vertices) is labeled clo-$5_{33}\langle III \rangle$ and *closo*-$C_2B_9H_{11}$ compound **1.11** (with one 6*k* vertex) is labeled clo-$11_6\langle III \rangle$.

Certain $2k_P$ and $5k_P$ peripheral vertices may also be troublesome among *nido*-structures, and these *peripheral* vertices are noted as *superscripts* in front of the numbers of heavy atoms. The $4k_C$ *cage* vertices may also indicate less stability among *nido*-configurations. For example, the label ni-$^5 10_4 \langle V \rangle$ denotes a *nido*-10-vertex structure incorporating one troublesome $5k_p$ peripheral vertex and one troublesome $4k_c$ cage vertex. The same basic structure is characteristic of both the ni-$^{22}6\langle VI \rangle$ and ara-$6\langle VI \rangle$ configurations; however, $2k_P$ peripheral vertices are troublesome only among *nido*-compounds and not among the more electron-rich and hydrogen-rich *arachno*-carboranes.

1.1.1.3 *Early Carbon-Location and* ^{11}B *and* ^{13}C *Chemical Shift Patterns*

Several long-standing carbon and boron patterns became evident during the discovery phase of the smaller *closo*-compounds (**1.1**, **1.2**, **1.3**, and **1.5**) in the middle 1950s and had been reconfirmed by the early 1960s [1] (Figure 1-1). These patterns are:

Pattern 1.1   Carbons are usually found in the lowest coordinated sites in the molecular structures (see compounds **1.1**, **1.4–1.7**, and **1.10** in Figure 1-1).

Pattern 1.2   Carbons are usually found in nonadjacent sites in the molec-
                  ular structures (see compounds **1.1, 1.3, 1.5–1.7, 1.9–1.11,
                  1.13**, and **1.14** in Figure 1-1).

Pattern 1.3   Isomers with carbons placed in accordance with Patterns 1.1
                  and 1.2 are the most thermodynamically stable isomers.

Pattern 1.4   In the ^{11}B NMR spectrum of a given *closo*-molecule, those
                  borons that are more highly coordinated (higher *nk*-values)
                  are found to resonate at higher field, and vice versa. The ^{13}C
                  NMR spectra of skeletal carbons were later found to follow a
                  similar pattern.

Pattern 1.5   In the ^{11}B NMR spectrum of a given *closo*-molecule, boron
                  atoms that have alkyl terminal groups, BR, are found to res-
                  onate at lower field than otherwise equivalent BH groups.
                  The ^{13}C NMR spectra of skeletal carbons were later found to
                  follow a similar pattern.

The structures of Figure 1-1 have recently been cleansed of several specious structures of long standing. Structure **1.A**, which violates Patterns 1.1–1.3, and 1.5, was announced in 1966,[2] denounced in 1968–1970, [3] renounced in 1990, [4] and purged in 1996 [5] following a "lemmingesque" acceptance of almost 30 years (see Chapter 12 in this volume). On the other hand, configuration **1.C** (Figure 1-1), which violates only Pattern 1.1, was originally reported [6] as a reasonable alternative to structure **1.11**, but was later rejected in favor of the correct structure **1.11**.[7] The NMR data, reported to support the wistfully contoured compound *closo*-$C_3B_5H_7$, structure **1.B**,[8] were also challenged,[9] and recently were shown to have been misinterpreted data related to derivatives of the known *closo*-2,3-$C_2B_5H_7$, structure **1.4**.[10]

The *closo*-deltahedra, characteristic of the *closo*-carboranes, clo-$n\langle III\rangle$, may be disassembled in various ways, involving both vertex removal and connection removal, to yield the various deltahedral fragment structures characteristic of the *arachno*-carboranes (almost always with six-membered open faces), ara-$n\langle VI\rangle$ ($n = 6$–12), as well as the multifarious *nido*-carboranes with six, five, and four-membered open faces, ni-$n\langle VI\rangle$ ($n = 8, 10, 12$), ni-$n\langle V\rangle$ ($n = 6, 7, 9, 11$), and ni-$n\langle IV\rangle$. One example is:

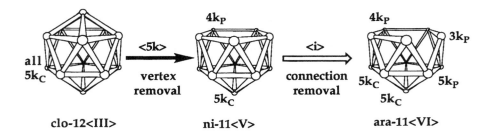

The various ways that these differing geometries are related and how the geometrical changes take place are presented in two Parts outlined in Figure 1-2.

Part 1.1 focuses on how each of the specific configurations, within each of the five families (one *closo*, three *nido*, and one *arachno*), is related to all other members of the same family of configurations. **Part 1.2** covers three different procedures for illustrating how each of the three classes of configurations is related to the others. These three procedures are labeled the original *bivertex-1* model (removal of two highest coordinate vertices), the simple and improved *vertexseco-2* model (removal of one specific vertex and one specific connection), and the heavily favored *bivertexseco-3* model (removal of two specific vertices and one specific connection).

Part 1.3 will be published later and will cover *skeletal electron bond distributions* and their effects upon *aperture contraction* among alternative *nido*-configurations because of the presence of four skeletal electron-donating groups, 4SEDs (electron-rich groups). As examples, BH and CH groups are two and three skeletal electron donors, 2SEDs and 3SEDs, respectively (the electrons donated to the clusters as skeletal electrons are identified as dots within the boxes below):

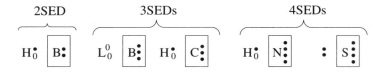

The presence of 4SEDs, such as NH and bare sulfur, does not impact clo-$n\langle\text{III}\rangle$ or ara-$n\langle\text{VI}\rangle$ configurations dramatically but frequently are associated with aperture contraction among *nido*-configurations. For example, selected ni-$n\langle\text{VI}\rangle$ and ni-$n\langle\text{V}\rangle$ configurations sometimes aperture-contract into ni-$n\langle\text{V}\rangle$ and ni-

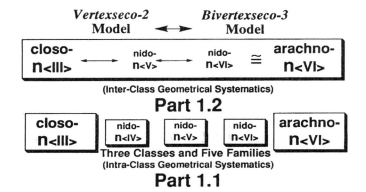

Figure 1-2   Schematic Outline of **Parts 1.1** and **1.2**.

$n\langle IV \rangle$ configurations, respectively, when one or two 4SEDs are substituted for carbon or boron atoms.

1.1.1.4 Traditional Perspective Versus "Bare Skeleton" and Skeletal Electron Pair Distribution, P-ST, Perspectives

In **Part 1.1**, we draw attention to the development of a new P-ST electron-counting formalism, even though it will not be used extensively until **Part 1.3** is published. For later references the P-ST numbers are enclosed in parentheses next to the molecular structures in the various figures throughout **Parts 1.1** and **1.2**. They simply catalog the three kinds of skeletal electron pair bonds within the specific molecules (all exo-,

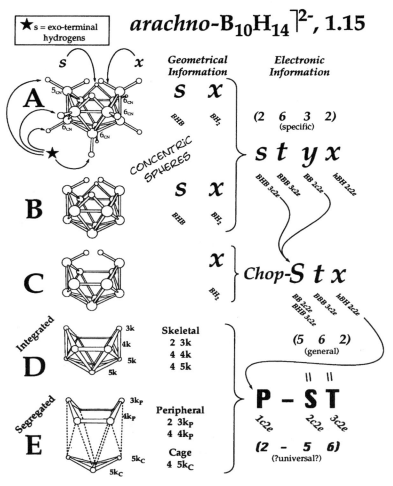

Figure 1-3   Evolution of the P-ST formalism from the *styx* formalism.

endo- and bridge hydrogens, as well as all exo-terminal electron pairs, are ignored). A review of how our current approach has evolved from past methods is illustrated in Figure 1-3.

At the top left of Figure 1-3, **A** illustrates the structure of *arachno*-$B_{10}H_{14}]^{2-}$, compound **1.15**. All exo-, endo-, and bridge hydrogens are shown and the coordination numbers of the individual borons are labeled, 6_{CN} and 5_{CN}, respectively. Lipscomb's *styx* number, 2632, yields both geometrical and electronic information: *s* and *x* identify BHB and BH_2 groups while all four numbers identify the various "kinds of" three-center two-electron (3c2e) and two-center two-electron (2c2e) bonds. Lipscomb pioneered the two concentric sphere concept wherein exo-groups occupy positions on an outer sphere while all other atoms occupy positions on an inner sphere. As all skeletal atoms (inner sphere borons) have, without exception, one exo-hydrogen (see stars, ★,) attached by one exo-2c2e bond, both the exo-hydrogens and the exo-2c2e bonds may be ignored (see situation **B** in Figure 1-3).

In more recent times, we have also chosen to ignore the presence or absence of bridge hydrogens, which results in combining the lower-case letters *s* and *y* of Lip-*styx* under a capital **S** to produce the chop-*Stx* formalism; thus, *styx* = 2632, simplifies to **Stx** = 562. In this fashion, isoelectronic species, which would differ only in the numbers and locations of bridge hydrogens, such as *arachno*-$B_{10}H_{14}]^{2-}$ and the hypothetical *arachno*-$B_{10}H_{13}]^{3-}$, and *arachno*-$C_4B_6H_{12}$ species, while having divergent *styx* values, would have the same chop-*Stx* number:

arachno-	*styx*	chop-*Stx*	P-ST
$B_{10}H_{14}]^{2-}$	2632		
$B_{10}H_{13}]^{3-}$	1642	562	2–56
$C_4B_6H_{12}$	0652		

The chop-*Stx* formalism, shown by **C** in Figure 1-3, retained mixed geometrical and electronic information, i.e., the symbol **x** identifies both BH_2 groups and one kind of 2c2e bonds.

Recently, it has become apparent that certain heteroatom groups involving bare sulfur, :S:, and nitrogen, :NH, etc. (incorporating endo lone pairs of electrons) are isoelectronic and more or less isostructural with CH_2 and $BH_2]^-$ groups that contain endo-hydrogens. Such endo lone pairs ("1c2e bonds") on nitrogen and sulfur or endo-hydrogens (attached by 2c2e bonds to carbon or boron) are structurally similar and the electron pairs involved are isolated from the other skeletal electrons that hold the cluster atoms together. We now prefer to unify these two kinds of "isolated electron pairs" under the **P** in P-ST (ignoring whether endo-hydrogens are present or absent) and, using the hyphen in **P-ST**, to segregate such electron pairs from those involved in skeletal bonding.

With the conversion to the **P-ST** formalism, the electronic and geometrical information is completely separated and the skeletal vertices may be compared as a function of vertex connectivity only (see **D** in Figure 1-3). As will become apparent below, it is even more desirable to segregate the cage vertices, $5k_C$, from the peripheral vertices, $3k_P$ and $4k_P$, within the *nido* and *arachno* deltahedral fragment configurations (see **E** in Figure 1-3).

Lipscomb's *styx* [11] numbers are useful for individual molecules, and when the *styx* numbers are totaled they exactly equal the number of skeletal electron pairs. The P-ST numbers also total the number of skeletal electron pairs but they are more general and are thus applicable to entire families of compounds. Examples of these changes are illustrated in Figure 1-4 using two examples each of the clo-9⟨III⟩, ni-9$_4$⟨V⟩, and ara-9⟨VI⟩ configurations.

In section 1.2, we will discuss the origin of the most-spherical *closo*-deltahedra with the **most-homogeneously connected vertices**, and in sections 1.3 and 1.4, the patterns applicable to the *arachno-* and *nido*-deltahedral fragments will be discussed.

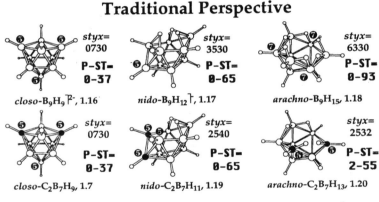

Traditional Perspective

styx= 0730
P-ST= 0-37
closo-B$_9$H$_9$²⁻, 1.16

styx= 3530
P-ST= 0-65
nido-B$_9$H$_{12}$⁻, 1.17

styx= 6330
P-ST= 0-93
arachno-B$_9$H$_{15}$, 1.18

styx= 0730
P-ST= 0-37
closo-C$_2$B$_7$H$_9$, 1.7

styx= 2540
P-ST= 0-65
nido-C$_2$B$_7$H$_{11}$, 1.19

styx= 2532
P-ST= 2-55
arachno-C$_2$B$_7$H$_{13}$, 1.20

All heavy atoms are 6-coordinate,Ⓖ, except as noted:

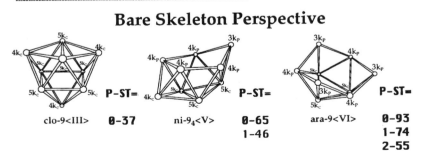

Bare Skeleton Perspective

clo-9<III>   **P-ST=**   0-37

ni-9$_4$<V>   **P-ST=**   0-65
1-46

ara-9<VI>   **P-ST=**   0-93
1-74
2-55

Figure 1-4   Two examples each of *closo*-9⟨III⟩, *nido*-9$_4$⟨V⟩, and *arachno*-9⟨VI⟩ configurations.

1.1.2 *Closo*-deltahedra with *Most-Homogeneous* Vertices and Apertures

1.1.2.1 "Arithmetically Conceivable" Closo-deltahedra and their Vertex Connectivity, k_C Temporarily ignoring the differences in the sizes of print, and the fact that some numbers are in boxes, in Figure 1-5 the seven columns within the top three rows of numbers characterize the vertices of one set of potentially ideal arithmetically conceivable deltahedra composed of various numbers of $4k_C$ and $5k_C$ vertices (matrix **A**). There are no $6k_C$ vertices within the columns of matrix **A**. There follows three subsequent three-row sets of arithmetically conceivable deltahedra (matrices **B**, **C**, and **D**) that sequentially incorporate one, two, and three $6k$ vertices, respectively. The columns within each three-row matrix are generated from the columns in the matrix above by the replacement of two $5k_C$ vertices by one $6k_C$ vertex and one $4k_C$ vertex in each case.

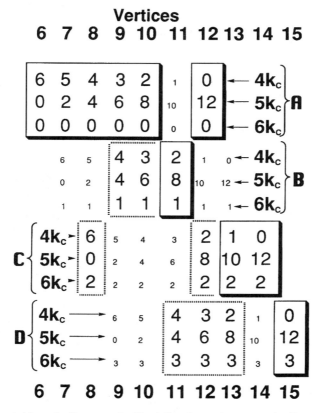

Figure 1-5 Arithmetically conceivable (all columns), geometrically possible (large print only), and most-spherical *closo*-deltahedra with most-homogeneous vertices (in solid boxes).

1.1.2.2 Geometrically Possible Deltahedra   Only 60% of the 28 combinations represented in Figure 1-5 reflect geometrically possible deltahedra (large print in boxed areas) but all are presented, including the "impossible" 40% (in small print), in order to tie together the underlying patterns that will subsequently lead to the rational selection of that specific mixture of deltahedra characteristic of the most-spherical *closo*-carboranes with ***the most-homogeneous vertices.***

Figure 1-5 lists the vertices of all arithmetically conceivable deltahedra incorporating $4k_C$, $5k_C$, and a few $6k_C$ vertices. Those that are geometrically possible are shown in large print in the boxed areas while those that are not geometrically possible are reduced in size and left as reference points only. Those that are geometrically possible as opposed to those that are geometrically unrealistic are not immediately obvious by simply gazing at the numbers in Figure 1-5.

1.1.2.3 Closo-Deltahedra with Most-Homogeneous Vertices   Among the geometrically possible *closo*-deltahedra displayed in Figure 1-5 there are at least two candidate choices for each number of vertices (between clo-8⟨III⟩ and clo-13₆₆⟨III⟩). Of the possible deltahedral choices, those with the smallest number of $6k_C$ vertices ***are always chosen*** for the actual *closo*-carborane structures (see solid boxes in Figure 1-5). Those *closo*-deltahedra with (1) the fewest $6k_C$ vertices, or, conversely, those that (2) incorporate the maximum number of $4k_C$ and $5k_C$ vertices, are simultaneously (3) the most-spherical deltahedra. However, of primary structural importance, the most-spherical deltahedra always

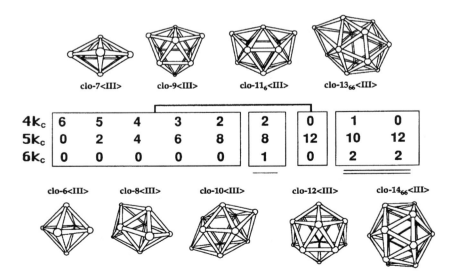

Figure 1-6   *Closo-n*⟨III⟩ patterns, where $n = 6$–14.

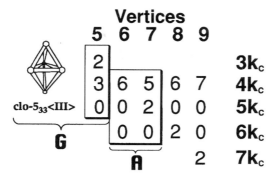

Figure 1-7   Bipyramidal *closo*-deltahedra, *closo-n*⟨III⟩.

incorporate (4) the **most-homogeneously connected cage vertices.** As will become apparent below, when *nido-* and *arachno*-configurations are considered the **maximum homogeneity of the cage vertices** (as one group) and the **maximum homogeneity of the perimeter vertices** (as a separate group) are also characteristic of the deltahedral fragment structures favored by all *arachno*-carboranes and *nido*-carboranes.

In summary then, a large number of both real and imaginary arithmetically conceivable deltahedra (the 28 columns in Figure 1-5) reduce to a lesser number of geometrically possible deltahedra (the 17 columns in large print in Figure 1-5), which reduce to the 10 most-spherical (**most-homogeneously connected**) deltahedra (solid boxed areas in Figure 1-5) that are ultimately favored for the structures of the *closo*-carboranes (illustrated in Figures 1-1 and 1-6).

To complete the picture, a matrix of deltahedral bipyramidal structures might also be considered (Figure 1-7). As far as the *closo*-carboranes are concerned, the two *closo*-configurations, from matrix **A**, with six and seven vertices are also identified as tetragonal and pentagonal bipyramids. The atypical trigonal bipyramid, clo-5_{33}⟨III⟩ (see matrix **G**), may also be added in this fashion (see compound **1.1** in Figure 1-1).

Having covered the geometrical relationships among the *closo*-carboranes (Figure 1-1) and their configurations (Figure 1-5 and 1-6), we will now address the even more homogeneous *arachno*-configurations.

1.1.3   Icosahedral-Fragment *Arachno*-configurations with *Most-Homogeneous* Vertices and Apertures

1.1.3.1   Smaller "Typical" **Arachno-3⟨III⟩** *to* **Arachno-6⟨VI⟩** *Configurations with* **Most-Homogeneous nk$_P$** *Vertices and Apertures*   Among the smaller *arachno*-deltahedral fragments, the preferred configurations, from ara-3⟨III⟩ to ara-6⟨VI⟩ (Figure 1-8), incorporate exclusively $2k_P$, $3k_P$ and $4k_P$ perimeter vertices (there are no cage vertices). The ara-4⟨IV⟩, ara-5⟨V⟩, and ara-6⟨VI⟩

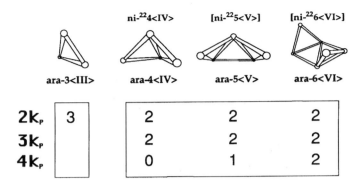

	ara-3⟨III⟩	ara-4⟨IV⟩	ara-5⟨V⟩	ara-6⟨VI⟩
2k$_P$	3	2	2	2
3k$_P$		2	2	2
4k$_P$		0	1	2

Figure 1-8 *Arachno-n*⟨III to VI⟩ patterns where *n* = 3–6 and the *nido*-^{22}4⟨IV⟩ configuration.

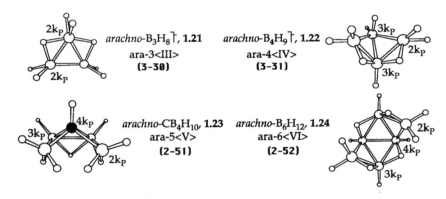

Figure 1-9 *Arachno*-species, B_3H_8⁻, structure **1.21**; B_4H_9⁻, structure **1.22**; CB_4H_{10}, **1.23**[12]; and B_6H_{12}, structure **1.24**.

structures reveal a simple pattern with an increasing number of $4k_P$ vertices with increasing size. Figure 1-9 shows examples of *arachno*-compounds that are representative of the bare skeleton models illustrated in Figure 1-8.

1.1.3.2 Larger "Typical" **Arachno-***Configurations with VI-gonal Apertures Incorporate* **5k$_C$** *Cores Surrounded by Six* **Most-Homogeneous** *Peripheral Vertices*

For some time the larger ara-6⟨VI⟩ to ara-12⟨VI⟩ configurations were simply compared with each other by noting that they were composed of 2*k*, 3*k*, 4*k*, and 5*k* vertices that (if displayed in groups representing variously sized molecules, and ignoring whether they were nk_C or nk_P vertices) gave rise to three overlapping patterns (top of Figure 1-10). We were unable to decide why these patterns might have arisen and wondered whether they might shroud some deeper meaning.

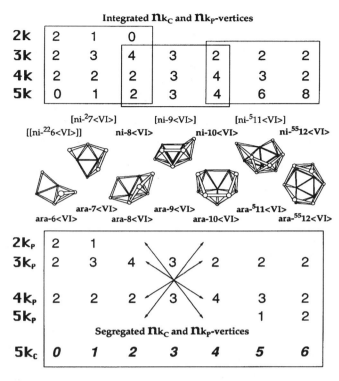

Figure 1-10 The **typical** *arachno-n*⟨VI⟩ and *nido-n*⟨VI⟩ carborane patterns where *n* = 6–12 (hypothetical configurations are shown in brackets).

Recently, we realized that the ara-6⟨VI⟩ to ara-12⟨VI⟩ species may be treated, much more informatively, in a different manner by segregating the cage vertices, nk_C, and the peripheral vertices, nk_P, from each other (bottom half of Figure 1-10). All vertices, *in excess of the six peripheral vertices, are incorporated* exclusively in **totally homogeneous $5k_C$ cage positions** (in other words, there are no $3k_C$, $4k_C$, or $6k_C$ cage vertices among any of these *arachno*-configurations). The largest configurations, ara-^{5}11⟨VI⟩ and ara-^{55}12⟨VI⟩, necessarily incorporate one and two less favorable $5k_P$ edge vertices, respectively, while the smallest configurations, ara-6⟨VI⟩ and ara-7⟨VI⟩, necessarily incorporate two and one $2k_P$ edge positions, respectively. The net result is that within the ara- and ni-*n*⟨VI⟩species, all cage vertices (when present) consist of **totally homogeneous $5k_C$ vertices** surrounded by the **most-homogeneous mixtures possible of $2k_P$, $3k_P$, $4k_P$, and $5k_P$ peripheral vertices** as illustrated in the bottom half of Figure 1-10. Hereinafter, all peripheral and cage vertices are segregated and the numbers of less favorable $5k_P$ peripheral vertices are noted by the superscripts preceding the number of skeletal atoms!

It follows that all typical arachno-deltahedral fragment structures from ara-3⟨III⟩ through ara-6⟨VI⟩, to ara-⁵⁵12⟨VI⟩ are necessarily fragments of a regular icosahedron or larger "icosahedron-like" deltahedra. A striking **centrosymmetric pattern** involving the perimeter nk_P vertices is evident in Figure 1-10. We suggest the **centrosymmetric pattern** arises from nothing more than a basic tendency toward maximizing the homogeneity of the perimeter vertices around the monotonically increasing core of totally homogeneous $5k_C$ vertices. A comparison of *closo*-configurations with *arachno*-configurations is in order.

Within the **clo-n⟨III⟩ species,** those configurations with the most homogeneously connected cage vertices, nk_C, are favored (Figures 1-1, 1-5 and 1-6); i.e., homogeneous $5k_C$ and $4k_C$ vertices are maximized in the preferred most-

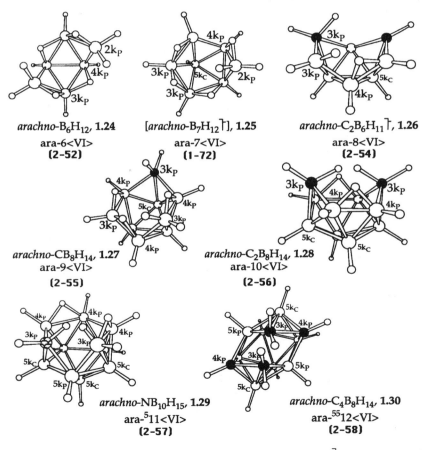

arachno-B₆H₁₂, **1.24**
ara-6⟨VI⟩
(2-52)

[*arachno*-B₇H₁₂⁻], **1.25**
ara-7⟨VI⟩
(1-72)

arachno-C₂B₆H₁₁⁻, **1.26**
ara-8⟨VI⟩
(2-54)

arachno-CB₈H₁₄, **1.27**
ara-9⟨VI⟩
(2-55)

arachno-C₂B₈H₁₄, **1.28**
ara-10⟨VI⟩
(2-56)

arachno-NB₁₀H₁₅, **1.29**
ara-⁵11⟨VI⟩
(2-57)

arachno-C₄B₈H₁₄, **1.30**
ara-⁵⁵12⟨VI⟩
(2-58)

Figure 1-11 *Arachno*-species, B₆H₁₂, structure **1.24**; B₇H₁₂⁻, structure **1.25**; C₂B₆H₁₁⁻, structure **1.26**; CB₈H₁₄, structure **1.27**; C₂B₈H₁₄, structure **1.28**; NB₁₀H₁₅, structure **1.29**; and [C₄B₈H₁₄], structure **1.30**.

spherical deltahedra. In a similar context, within the ***ara-n⟨VI⟩ configurations***, those species with the most homogeneously connected perimeter vertices, nk_P, are also favored. This maximizes the $3k_P$ and $4k_P$ vertices and results in the ***centrosymmetric pattern*** observed in Figure 1-10. Figure 1-11 shows representative examples of *arachno*-compounds having the bare skeleton structures displayed in Figure 1-10.

1.1.3.3 Larger "Abnormal" **Arachno-***Configurations with VII-gonal Apertures Incorporate* $5k_C$ *Cores Surrounded by Seven* **Most-Homogeneous** *Peripheral Vertices*

It is noteworthy to observe that a second ***centrosymmetric*** pattern is observed (bottom of Figure 1-12) when the "**abnormal**," proposed and unknown *arachno*-structures with VII-gonal open faces, ara-7⟨VII⟩ to ara-12⟨VII⟩, are compared and the cage vertices, nk_C, and peripheral vertices, nk_P, are segregated. The ara-7⟨VII⟩ and ara-8⟨VII⟩ structures have been proposed in the past and, along with the known structure of the "**abnormal**"-*arachno*-B_9H_{15} isomer, compound **1.31**, ni-9⟨VII⟩, all fit neatly into Figure 1-12.[13]

The hypothetical structures assigned to the ara-^{5}10⟨VII⟩, ara-^{55}11⟨VII⟩, and ara-^{55}12⟨VII⟩ formulae were generated by selecting an appropriately increasing core of $5k_C$ vertices and subsequently minimizing the numerical spread of the seven surrounding nk_P perimeter vertices. Only one acceptable unique configuration per ara-^{5}10⟨VII⟩, ara-^{55}11⟨VII⟩, and ara-^{55}12⟨VII⟩ formula was found and, again, a ***centrosymmetric*** pattern resulted. While the **typical** structures with VI-gonal open faces (Figure 1-10) are the dominant *arachno*-configurations, it is

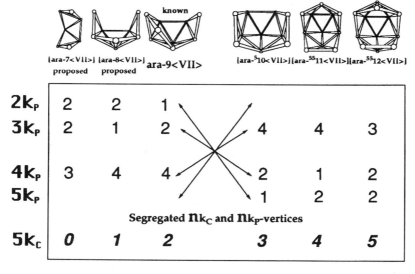

Figure 1-12   The "**abnormal**" *arachno-n⟨VII⟩* carborane pattern where $n = 7$–12 (only two examples of ara-9⟨VII⟩ are known).

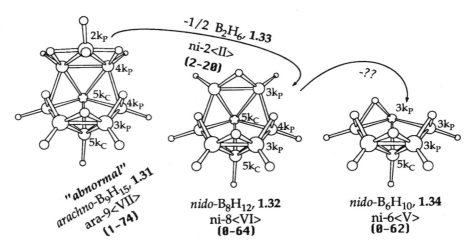

Figure 1-13   The relationship of "**abnormal**" *arachno*-B_9H_{15}, structure **1.31**, to *nido*-B_8H_{12}, structure **1.32**, and *nido*-B_2H_6, structure **1.33**. *nido*-B_6H_{10}, structure **1.34**.

probable that additional *arachno*-structures, with VII-gonal apertures (Figure 1-12) will be discovered in the future.

The "**abnormal**" isomer, *arachno*-B_9H_{15}, structure **1.31**, might better be viewed as an unstable adduct of *nido*-B_8H_{12}, structure **1.32**, and $[BH_3]$. In 1966, [14] we reported that "**abnormal**" *arachno*-B_9H_{15}, struture **1.31**, which decomposes in days at room temperature, was stabilized for weeks in the presence of B_2H_6, structure **1.33**, at >25 atm (Figure 1-13).[15]

Having concluded the discussion of the *arachno*-configurations with most-homogeneous cage and most-homogeneous peripheral vertices (Figures 1-10 and 1-12) as well as the *closo*-configurations with the most-homogeneous vertices possible (Figures 1-1, 1-5, and 1-6), the multifarious *nido*-configurations with three different sizes of larger apertures, but with the most-homogeneous vertices possible, will now be confronted.

1.1.4   *Heterogeneous Nido*-configurations Involve VI-gonal, V-gonal, and IV-gonal Sized Apertures

In spite of the variable aperture sizes among the various *nido*-configurations, when taken as a whole, each family with uniform apertures—i.e., ni-$n\langle$VI$\rangle$, ni-$n\langle$V$\rangle$, and ni-$n\langle$IV$\rangle$—favors configurations incorporating vertices that are as homogeneous as possible.

1.1.4.1   *Ni-n$\langle$V$\rangle$ and Ni-n$\langle$VI$\rangle$ Configurations, with the Most-Homogeneous-Possible* nk_P *and* nk_C *Vertices*   The bare skeleton ni-$n\langle$VI$\rangle$ carborane configurations (n = 8, 10, 12) with maximally homogeneous vertices are illustrated

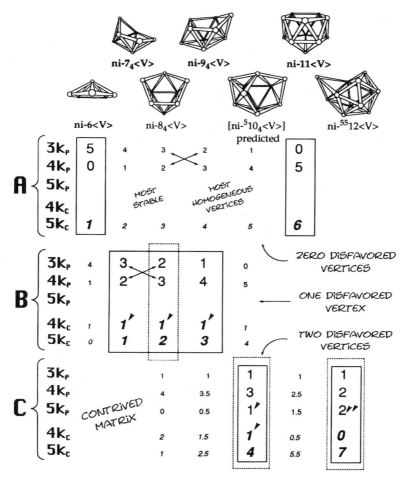

Figure 1-14 *Nido-n*⟨V⟩ carborane configurations, where $n = 6$–12 (*nido-8₄*⟨V⟩ and *nido-⁵⁵12*⟨V⟩ are known only when one 4SED is present; *nido-⁵i0₄*⟨V⟩ is presumed to be involved in certain rearrangements.

along with the ara-*n*⟨VI⟩ configurations in Figure 1-10. The more complex ni-*n*⟨V⟩ configurations are addressed separately in Figure 1-14.

Figure 1-14 lists three five-row matrices of numbers (matrices **A, B,** and **C**) containing candidate columns of numbers that monotonically change from left to right and to which the structures of known compounds may be related. Only one ni-*n*⟨V⟩ structure is available per value of *n*. The top matrix, **A,** is **ideal** in as much as **zero 4k_C vertices are tolerated** and **both perimeter and cage vertices are totally homogeneous** in the known (boxed) ni-6⟨V⟩ and ni-11⟨V⟩ configurations. Digressing, remember that only 5k_C cage vertices are favored and that $4k_C$ cage vertices are totally avoided by ni-$n\langle$VI$\rangle$, ara-$n\langle$VI$\rangle$, and ara-$n\langle$VII$\rangle$ icosahedral fragment structures (Figures 1-10 and 1-12). The columns of numbers related to the most-stable ni-6$\langle$V$\rangle$ and ni-11$\langle$V$\rangle$ configurations *(with planar five-membered open faces derived from the icosahedron)* are incorporated at the extreme ends (boxed columns) of the top six-column matrix **A** with no $4k_C$ vertices. The four intermediate columns in matrix **A** do not give rise to geometrically realistic configurations and are therefore printed in small sized numbers to fill out the matrix from which the two ideal, realistic, totally homogeneous, icosahedral fragment structures, ni-6$\langle$V$\rangle$ and ni-11$\langle$V$\rangle$, are derived. Had the other four configurations in matrix **A** of Figure 1-14 been realistic, then a **centrosymmetric** pattern (see crossed arrows) would have resulted, as was seen in Figures 1-10 and 1-12 and will be seen later in Figure 1-20.

The columns containing the *nk* values related to the intermediate, ni-$7_4\langle$V$\rangle$, ni-$8_4\langle$V$\rangle$, and ni-$9_4\langle$V$\rangle$, configurations are derived from the middle five-column matrix, **B**. **One less-favored $4k_C$ cage vertex is necessarily tolerated in each case**, as acknowledged by the subscript 4 following the number of atoms in each

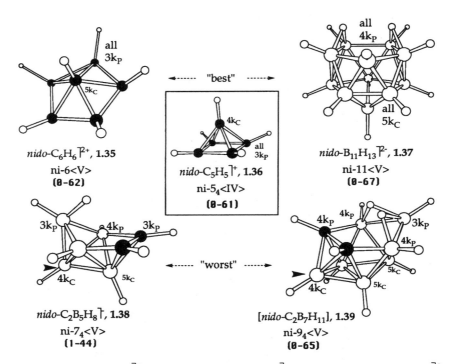

Figure 1-15   *Nido*-C$_6$H$_6$]$^{2+}$, structure **1.35**; *nido*-C$_5$H$_5$]$^+$, structure **1.36**; *nido*-B$_{11}$H$_{13}$]$^{2-}$, structure **1.37**; *nido*-C$_2$B$_5$H$_8$]$^-$, structure **1.38**, and *nido*-C$_2$B$_7$H$_{11}$, structure **1.39**.

deltahedral fragment cluster. Within this middle matrix, the extreme smallest and largest columns do not represent geometrically realistic configurations and are therefore printed in small-sized numbers. Had the left four columns of matrix **B** all represented realistic configurations, then another *centrosymmetric* pattern (see crossed arrows) would have resulted.

The bottom five-column matrix, **C**, includes two absurd columns, as well as one unrealistic column. All three are in small print and could have been left out, but the construction of contrived matrix **C** does allow the inclusion of two columns that give rise to the configurations characteristic of the ni-^{5}10$_4\langle$V$\rangle$ and ni-^{55}12$\langle$V$\rangle$ species, which, in turn, are *derived from the two unique deltahedra that necessarily incorporate 6k$_C$ vertices,* clo-11$_6\langle$III$\rangle$ and clo-13$_{66}\langle$III$\rangle$ (Figure 1-6). In the ni-^{5}10$_4\langle$V$\rangle$ and ni-^{55}12$\langle$V$\rangle$ configurations, **two less-favored vertices are tolerated** in each case. One 4k$_C$ vertex plus one 5k$_P$ vertex are incorporated into the former, while two less-favored 5k$_P$ vertices are tolerated in the latter (Figure 1-14). The disfavored 5k$_P$ vertices are noted as superscripts preceding the numbers of atoms in the deltahedral fragment clusters. The ni-8$_4\langle$V$\rangle$ and ni-^{55}12$\langle$V$\rangle$ structures have been observed, but only when one heteroatom, which donates four skeletal electrons, 4SED, is present.

Typical molecular examples of ideal ni-6$\langle$V$\rangle$ and ni-11$\langle$V$\rangle$ configurations (matrix **A** in Figure 1-14) are illustrated at the top of Figure 1-15, while examples of the less-than-ideal ni-7$_4\langle$V$\rangle$ and ni-9$_4\langle$V$\rangle$ configurations (matrix **B** in Figure 1-14) are illustrated at the bottom. No representative examples of pure carborane or polyborane compounds with ni-8$_4\langle$V$\rangle$, ni-^{5}10$_4\langle$V$\rangle$, or ni-^{55}12$\langle$V$\rangle$ configurations are known (identified by the dashed-line boxes in matrices **B** and **C** in Figure 1-14). Examples of typical *nido*-carboranes with ni-8$\langle$VI$\rangle$, ni-10$\langle$VI$\rangle$, and ni-^{55}12$\langle$VI$\rangle$ configurations (Figure 1-10) are illustrated in Figure 1-16.

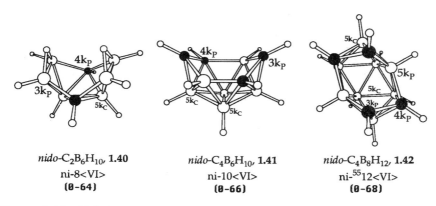

nido-C$_2$B$_6$H$_{10}$, **1.40**
ni-8<VI>
(0-64)

nido-C$_4$B$_6$H$_{10}$, **1.41**
ni-10<VI>
(0-66)

nido-C$_4$B$_8$H$_{12}$, **1.42**
ni-^{55}12<VI>
(0-68)

Figure 1-16 *Nido*-C$_2$B$_6$H$_{10}$, structure **1.40**; *nido*-C$_4$B$_6$H$_{10}$, structure **1.41**; and *nido*-C$_4$B$_8$H$_{12}$, structure **1.42**.

1.1.4.2   Rationalization of* Nido-n⟨V⟩ *and* Nido-n⟨VI⟩ *Structural Choices
When only borons, carbons and hydrogens are present, *nido*-carboranes and
nido-polyboranes with seven or more borons and carbons might seem to have
two *nido*-structural options, i.e., to assume either ni-n⟨VI⟩ configurations (Figure 1-10) or ni-n⟨V⟩ configurations (Figure 1-14) depending upon whether n is
odd or even. This alternating choice could either be due to some sophisticated
"white-collar" theoretical reason or may be due to a set of "blue-collar" coincidences *(caveat emptor)*. In Figure 1-17, the **fourteen** possible candidate ni-n⟨V⟩
and ni-n⟨VI⟩ configurations (n = 6–12) are compared. The odd–even relationship is highlighted by using large print to identify the known configurations and
small print for the unknown carborane and polyborane structures. With the
"defection" of the ni-6⟨V⟩ configuration from the even configurations the
"odds" outnumber the "evens" by 4:3. **Ten** of the **fourteen** configurations in
Figure 1-17 are icosahedral fragments and are so identified by triangles, ◀. Two
of the favored structures, ni-7_4⟨V⟩ and ni-9_4⟨V⟩, are not icosahedral fragments,

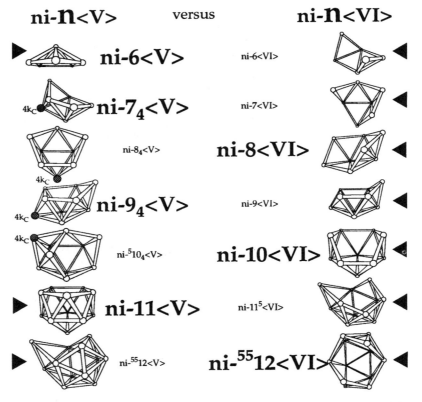

Figure 1-17   *Nido-n_{odd}⟨V⟩ versus Nido-n_{even}⟨VI⟩ structural patterns (nido-6⟨V⟩ is an
exception).*

while three structures, $n = 6$, 11, and 12, have a choice between two candidate icosahedral fragment configurations.

We suggest that the "odd–even" relationship, illustrated in Figure 1-17, is probably coincidental. As a point of departure, there may possibly be a *preference* for icosahedral or "icosahedral-like" fragment structures with both pentagonal and hexagonal apertures surrounding a core of $5k_C$ cage vertices and an absence of $4k_c$ cage vertices.

It is interesting (Figure 1-17) that perimeter vertex homogeneity cannot be greater than in ni-11$\langle V \rangle$ (all perimeter vertices are $4k_P$) and in ni-6$\langle V \rangle$ (all perimeter vertices are $3k_P$) (matrix **A** in Figure 1-14), which, in turn, accounts for the structural dominance of the ni-11$\langle V \rangle$ and ni-6$\langle V \rangle$ configurations among the *nido*-compounds. The two icosahedral-like configurations ni-^{55}12$\langle V \rangle$ and ni-^{55}12$\langle VI \rangle$ have quite similar peripheral vertex homogeneity (Figure 1-17), but the latter configuration is somewhat more symmetrical and is possibly favored by *nido*-carboranes. In an abridged format, the apertures of the icosahedral fragment structures displayed in Figure 1-17 are reproduced in the left-hand column, **A**, of the four-column narrative sequence [**A**, **B**, **C**, **D**] illustrated in Figure 1-18. In three cases in column **A** ($n = 6$, 11, 12), two competitive icosahedral fragment structures are illustrated, and the two question marks identify the two preferred non-icosahedral configurations ($n = 7$, 9), to be explained below.

Proceeding to the second column, **B**, in Figure 1-18, the less-favored ni-^{22}6$\langle VI \rangle$ and ni-^{5}11$\langle VI \rangle$ configurations are excluded due to their lack of perimeter vertex, nk_P, homogeneity when compared with the totally homogeneous perimeter vertices and much greater symmetry of the favored ni-6$\langle V \rangle$ and ni-11$\langle V \rangle$ configurations. The ni-^{55}12$\langle VI \rangle$ configuration may be slightly favored because of its greater symmetry (Figure 1-19). Both question marks of column **A** remain in column **B** of Figure 1-18.

Proceeding to the third column, **C**, in Figure 1-18, the impact of the presence or absence of BH_2 groups associated with $2k_P$ vertices is emphasized. The BH_2 groups are located at $2k_P$ vertices, as would be present in both the ni-^{22}6$\langle VI \rangle$ and the ni-^{2}7$\langle VI \rangle$ configurations (see column **A** in Figure 1-18 and Figure 1-17), and they always incorporate one, and frequently two, neighboring bridge hydrogens. If we glance ahead to Figure 1-33, it is seen that $2k_p$ vertices are found in the series ni-^{22}4$\langle IV \rangle$, ni-^{22}5$\langle V \rangle$, ni-^{22}6$\langle VI \rangle$, and ni-^{2}7$\langle VI \rangle$, of which only the first, ni-^{22}4$\langle IV \rangle$, is associated with known *nido*-compounds. Unobserved configurations are illustrated by wire-frame renderings. The *nido*-borane molecular formulae, B_nH_{n+4}, plus the presence of $2k_p$ vertices provide the explanation. Only known analogs of *nido*-B_4H_8, ni-^{22}4$\langle IV \rangle$ (illustrated as a cylindrical bond model in Figure 1-33), would have a high enough hydrogen/boron ratio to accommodate a BH_2 group and associated bridge hydrogens. In Figure 1-20 the hydrogen/boron ratios of the various *nido*-boranes are plotted as a function of the number of borons. Those with ratios greater than **1.9** incorporate BH_2 groups at $2k_P$ vertices, while those with ratios less than **1.9** do not have BH_2 groups at $2k_P$ vertices.

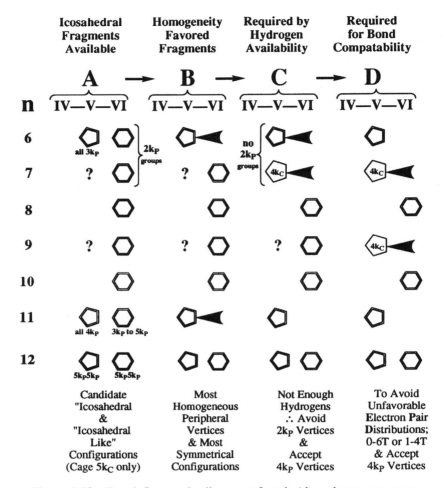

Figure 1-18   Four influences leading to preferred *nido*-carborane structures.

For these reasons, the ni-^{22}6⟨VI⟩ and ni-^{2}7⟨VI⟩ configurations are rejected in favor of the ni-6⟨V⟩ and ni-7$_4$⟨V⟩ configurations (column **C** in Figure 1-18 and Figure 1-17) that do not incorporate any $2k_P$ vertices and hence do not require BH$_2$ groups at $2k_P$ vertices. The non-icosahedral ni-7$_4$⟨V⟩ configuration thus becomes the preferred configuration in spite of the presumed negative impact of incorporating one $4k_c$ vertex into the ni-7$_4$⟨V⟩ structure. *Nido*-compounds larger than ni-^{2}7⟨VI⟩ , e.g., ni-8⟨VI⟩ and ni-9⟨VI⟩, etc., have sufficient hydrogens as they have no $2k_P$ vertices. One question mark remains in column **C** of Figure 1-18.

Finally, the right-hand column, **D**, of Figure 1-18 illustrates our conclusion that the icosahedral configurations for both ni-^{2}7⟨VI⟩ and ni-9⟨VI⟩ fragments are

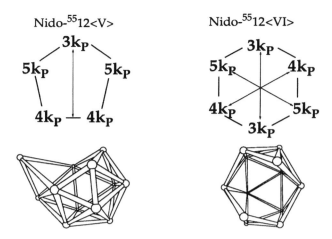

Figure 1-19 The *nido-*[55]12⟨VI⟩ configuration has greater symmetry than the *nido-*[55]12⟨V⟩ configuration.

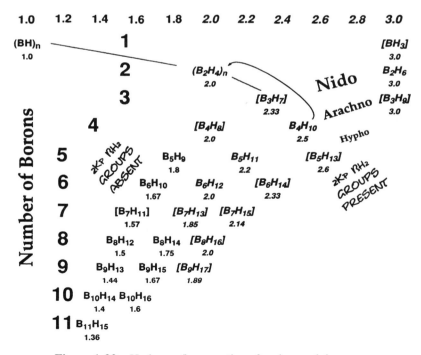

Figure 1-20 Hydrogen/boron ratios of various polyboranes.

seemingly "inhospitable" to the requisite kinds of skeletal electron pair bonds necessarily present. For our current purposes, we point out that the nido-n-compounds, where $n = 7$ and 9, must involve $2n + 4$ skeletal electrons ($n + 2$ skeletal electron pairs), which generate 9 and 11 skeletal electron pair bonds, respectively. For the ni-7$\langle$V or VI$\rangle$ and ni-9$\langle$V or VI$\rangle$ configurations, the 9 and 11 bonds may, in principal, be distributed in three different ways in each case:

	"lone pair" 1c2e bonds	B—B 2c2e bonds	B—B / B 3c2e bonds	
	0	6	3	
ni-7$\langle$VI or V$\rangle$	1	4	4	$\Sigma = 9$
	2	2	5	
	0	6	5	
ni-9$\langle$VI or V$\rangle$	1	4	6	$\Sigma = 11$
	2	2	7	

No exo-terminal hydrogens or their connecting electron pairs are counted. All endo-hydrogens and bridging hydrogens are also ignored as if they were not present.

The resulting three sets of 1c2e, 2c2e, and 3c2e electron pair bonds cannot satisfactorily be distributed about the ni-^{2}7$\langle$VI$\rangle$ and ni-9$\langle$VI$\rangle$ configurations with larger apertures without resorting to open 3c2e BBB bonds (once accepted, but rejected in recent times). In other words, just as **geometrical considerations** had rejected as impossible a hypothetical clo-11$\langle$III$\rangle$ structure (with no 6k vertices, Figures 1-5 and 1-6) and, as a consequence, had to adopt a clo-11$_6\langle$III$\rangle$ configuration with one 6k vertex, we now submit that **electronic considerations** reject both of the icosahedral ni-^{2}7$\langle$VI$\rangle$ and ni-9$\langle$VI$\rangle$ structures with larger apertures and favor the non-icosahedral ni-7$_4\langle$V$\rangle$ and ni-9$_4\langle$V$\rangle$ configurations, despite the concomitant incorporation of one less-favorable 4k_C vertex in each case.

The right-hand column, **D**, in Figure 1-18 also illustrates that all ni-n-structures larger than $n = 6$ have the observed alternating structural pattern (Figure 1-17) and no question marks remain. In summary, there appear to be several "blue-collar" reasons for the alternating structural pattern other than some esoteric "white-collar" reason tied to the odd and even numbers of skeletal atoms.

In a complementary rationale, it is possible to justify notionally the final *nido* odd–even pattern (column **D** in Figure 1-18) by assuming that the following five presumptions are correct:

1. All *closo*-species ($n = 6$–14) have clo-$n\langle\text{III}\rangle$ structures (Figures 1-5 and 1-6) with the most-homogeneous nk_C vertices possible. (The $5k_C$ vertices are ideal, $4k_C$ vertices are acceptable, while $6k_C$ and $3k_C$ vertices are tolerated only when more-acceptable alternatives do not exist.)

2. All *arachno*-species ($n = 6$–12) have ara-$n\langle\text{VI}\rangle$ icosahedral structures (Figure 1-10) with $5k_C$ cage vertices only (no $4k_C$ vertices) and the most-homogeneous nk_P perimeter vertices possible.

3. If all *nido*-carborane structures were hypothetically subject to gross geometrical considerations only, then perhaps all of the ni-$n\langle\text{VI}\rangle$ structures (characteristic of the ara-$n\langle\text{VI}\rangle$ species in Figure 1-10) might also be the chosen icosahedral *nido*-configurations (column **A** in Figure 1-18) **unless** either (a) certain other alternative configurations were **overwhelmingly attractive** or (b) selected icosahedral ni-$n\langle\text{VI}\rangle$ structures were **electronically forbidden**.

4. The ni-$6\langle\text{V}\rangle$ and ni-$11\langle\text{V}\rangle$ structures (matrix **A** in Figure 1-14) are **overwhelmingly attractive** as both their nk_C vertices and their nk_P vertices are totally homogeneous (see column **B** in Figure 1-18).

5. The icosahedral ni-$^2 7\langle\text{VI}\rangle$ and ni-$9\langle\text{VI}\rangle$ structures (Figure 1-10) are apparently **electronically forbidden** and thus give way to ni-$7_4\langle\text{V}\rangle$ and ni-$9_4\langle\text{V}\rangle$ structures (matrix **B** in Figure 1-14), even though the latter configurations incorporate one less-favored $4k_P$ vertex in each case (column **D** in Figure 1-18). These conclusions are driven by vertex homogeneity and account for the alternating pattern.

1.1.4.3 Nido-$n\langle IV\rangle$ *configurations with* **Most-Homogeneous-*Possible*** nk$_P$ *and* nk$_C$ *Vertices*

Boron and carbon are two skeletal electron donors, 2SEDs, and three skeletal electron donors, 3SEDs, respectively. When two 4SEDs (four skeletal electron donors), such as nitrogen or sulfur, are present, ni-$n\langle\text{IV}\rangle$ configurations have been observed when $n = 6$ and 9. Similar structures have been seen when transition elements are present ($n = 7, 10, 11$). Figure 1-21 displays such fragments and their associated nk-values. Much more on the effects of 4SED heteroatoms will be covered in **Part 1.3** (to be published later).

In Figure 1-21, another *centrosymmetric* pattern emerges, from which the largest and smallest structures are excluded. The nk_P vertices in the *centrosymmetric* pattern are as homogeneous as possible and, again, there is an axis of symmetry.

Two heteroatom 4SEDs were unable to convert the ni-$11\langle\text{V}\rangle$ configuration into the ni-$^{55}11_4\langle\text{IV}\rangle$ configuration (Figure 1-21), but then the ni-$^{55}11_4\langle\text{IV}\rangle$ configuration (which is observed when certain transition elements are present) is not derived from a regular icosahedron.

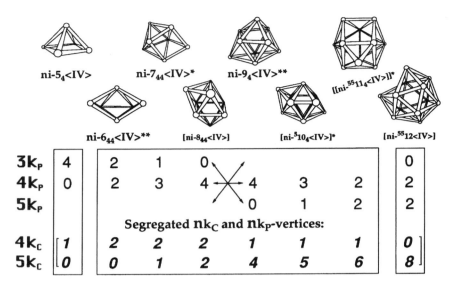

Figure 1-21 *Nido-n*⟨IV⟩ patterns wherein $n = 5, 6, 7, 9$; hypothetical configurations, wherein $n = 8, 12$, are in brackets; structures marked* are observed when transition element atoms are present; structures marked** are observed when two 4SED heteroatoms are present.

1.2 *INTERCLASS* GEOMETRICAL SYSTEMATICS

1.2.1 Introduction

The structures of the first boron cluster compounds (boron hydrides) were determined in the early 1950s.[11] Since that time, there have been many schemes to classify them as a function of their molecular formulae (Stock's hydrogen-poor, B_nH_{n+4}, and hydrogen-rich, B_nH_{n+6}, compounds [16]), their electron counts, and by their geometries. At the time, it was recognized that all *known* boron hydrides had skeletal structures that could be considered as fragments of a regular icosahedron, e.g., $B_{10}H_{14}$, configuration **2.1**,[11] with the exception of B_5H_9, configuration **2.2** (Figure 1-22) and the related carborane, configuration **2.3,** and carbocation, configuration **2.4**, which are fragments of a regular octahedron. The geometrical explanations for the labels $4k_C$, $3k_P$, and ni-5_4⟨IV⟩, surrounding B_5H_9, structure **2.2**, were explained in **Part 1.1**.

   All four classes of polyboranes, carboranes, and carbocations (*closo*, *nido*, *arachno*, and *hypho*) may be identified as members of several general molecular formulae. For example, all of the *closo*-carborane compounds **1.1** through **1.14** in Figure 1-1, **Part 1.1**, subscribe to the *closo*-$C_{0\,to\,2}B_nH_{n+2}$ formula (see below). It was noted that each class differed from its neighbors by either a dihydrogen (H_2)[16] a borane group (BH_3),[17] a hydride (H^-), a Lewis base (L), or an electron pair (e^2), and that adding or removing any number of protons (H^+), to produce anions or cations did not change the class.

$$closo\text{-}C_{0\ to\ 2}\ B_nH_{n+2}$$

$+ H_2$ or $+ BH_3$ or $\Big|\ + H^-$ or $+\ \substack{\text{electron}\\\text{pair}}$

$$nido\text{-}C_{0\ to\ 4}\ B_nH_{n+4}$$

$+ H_2$ or $+ BH_3$ or $\Big|\ + H^-$ or $+\ \substack{\text{electron}\\\text{pair}}$

$$arachno\text{-}C_{0\ to\ 6}\ B_nH_{n+6}$$

$+ H_2$ or $+ BH_3$ or $\Big|\ + H^-$ or $+\ \substack{\text{electron}\\\text{pair}}$

$$hypho\text{-}C_{0\ to\ 8}\ B_nH_{n+8}$$

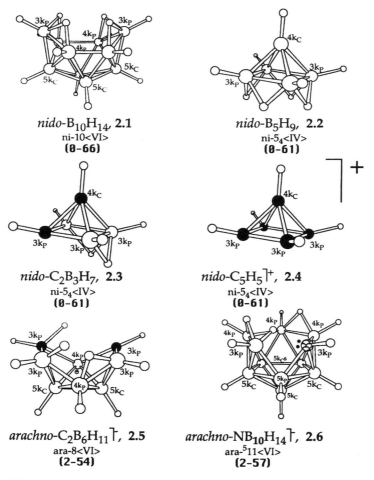

nido-$B_{10}H_{14}$, **2.1**
ni-10<VI>
(0-66)

nido-B_5H_9, **2.2**
ni-5_4<IV>
(0-61)

nido-$C_2B_3H_7$, **2.3**
ni-5_4<IV>
(0-61)

nido-$C_5H_5{}^{\rceil+}$, **2.4**
ni-5_4<IV>
(0-61)

arachno-$C_2B_6H_{11}{}^{\rceil}$, **2.5**
ara-8<VI>
(2-54)

arachno-$NB_{10}H_{14}{}^{\rceil}$, **2.6**
ara-^{5}11<VI>
(2-57)

Figure 1-22 Structures of *nido*-$B_{10}H_{14}$, **2.1**; *nido*-B_5H_9, **2.2**; *nido*-$C_2B_3H_7$, **2.3**; *nido*-$C_5H_5{}^{\rceil+}$, **2.4**; *closo*-2,3-$C_2B_5H_7$, **2.5**; and *closo*-1,7-$C_2B_6H_8$, **2.6**.

Shortly after the structures of the closed cage carboranes, $C_2B_nH_{n+2}$ (Figure 1-1) and their isostructural polyborane dianions, $B_nH_n]^{2-}$, had become established, we reported that the sequential removal of *two highest coordinated adjacent vertices,* labeled the 1971 ***bivertex-1*** procedure, (a), Figure 1.23,[18] could interrelate the sequentially more-open classes of compounds, *closo, nido,* and *arachno.* The ***bivertex*** portion of the label refers to the two highest coordinated vertices that must be removed, while the ***-1*** portion indicates that all *nido-* and

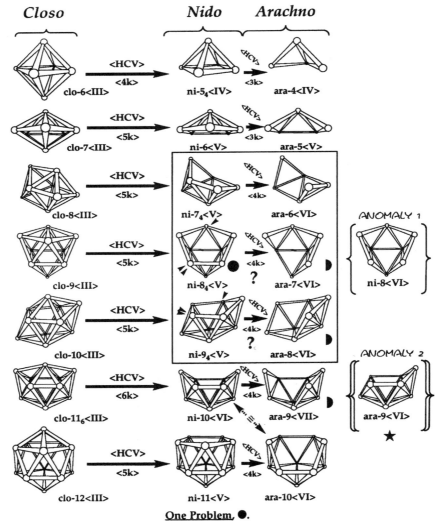

Figure 1-23 Geometrical systematics: the 1971 *bivertex-1* model.

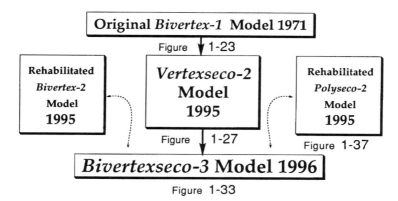

Figure 1-24   Evolution of interclass and intraclass geometric patterns.

arachno-fragments may be produced via *one unique pathway* (in Chapters 2 and 4 in this volume, **bivertex-1** is replaced by "*debor*"). A second, somewhat less rigorous sequence, (b), the multiple connection removal (the **polyseco-1** procedure), espoused in 1985 and reviewed by Hermánek in 1992, [19] could also be used to relate the *closo-*, *nido-*, and *arachno*-configurations.

   Both the *bivertex-1* (Figure 1-23) (a) and *polyseco-1* (b) procedures could forecast the correct molecular shapes *in almost all cases* (provided one selected and sequentially removed the two highest connected *adjacent vertices*, or one happened to remove both the proper kind and the proper number of adjacent connections). The *bivertex-1* procedure projected minimally distorted spatial configurations and predicted that the *nido-*5, 7, 8, and 9-vertex carboranes would not have icosahedral fragment structures (c) (boxed area in Figure 1-23) since the *closo*-deltahedral precursors were not regular icosahedra. For these reasons, the *shibboleth* that "*all polyborane structures were icosahedral fragments*" (c) and the *polyseco-1* procedure (b) were abandoned and the more informative *bivertex-1* procedure (a) was emphasized in 1971.[18]

   Recently, a much more rigorous *vertexseco-2* model and an all-encompassing *bivertexseco-3* model have been developed. Following the elucidation of the *vertexseco-2* pattern, the insights gained during its development have made possible the development of a somewhat improved *polyseco-2* model (Figure 1.24).

1.2.2   The 1971 *Bivertex-1* Model

Our objective, in the late 1960s, had been to deduce a set of geometrical patterns, now termed the *bivertex-1* model (Figure 1-23), that (1) **without exception** would account for *all* of the then-known skeletal structures among the many diverse classes of polyboranes and carboranes; (2) would illustrate how the various structures were geometrically related to each other as a function of their *closo-*, *nido-*, and *arachno*-classes; and (3) as a consequence would allow *accurate prediction* of the missing *nido-* and *arachno*-configurations (see boxed area of Figure 1.23), which we felt confidant would be discovered in the future.

By 1968, we thought that we had deduced such a ubiquitous pattern.[18] In 1971, a one-pathway two-step geometrical procedure was reported (Figure 1-23), which illustrated (1) how the deltahedral structures, characteristic of the *known closo*-carboranes (Figure 1-1), upon **removal of one highest connected vertex,** ⟨HCV⟩, produced the deltahedral fragment structures favored by the *known nido*-carboranes, *nido*-polyboranes, and selected *nido*-carbocations (Figure 1-23), and subsequently showed (2) that the **removal of a second highest connected vertex,** ⟨HCV⟩ *(necessarily adjacent to the open face)*, produced the structures favored by the known *arachno*-carboranes, *arachno*-polyboranes, and *arachno*-carbocations.

1.2.2.1 Vertex Connectivity, **nk, nk$_C$** *and* **nk$_P$**

The skeletal connectivities, *nk*, of the vertices removed are given below the arrows in Figure 1-23, e.g., the labels ⟨4k⟩, ⟨5k⟩, and ⟨6k⟩, etc., indicate that 4-connected, 5-connected, and 6-connected, vertices, respectively, are being removed. When desired, the origin of a given vertex is further defined, with a subscript (see Figure 1-22), as being either a cage vertex, *nk$_C$*, or a peripheral vertex, *nk$_P$*, about the open face (where *n* equals the number of other skeletal heavy atoms to which the given vertex under consideration is connected). All exo-terminal, endo-terminal, and bridging hydrogens or alkyl groups are ignored in determining the connectivity, *n*, of each vertex.

1.2.2.2 **Closo-, Nido-** *and* **Arachno-n⟨III, IV, V, VI, VII⟩** *Identification Scheme*

All of the *closo*-configurations in the left-hand column of Figure 1-23 are labeled clo-*n*⟨III⟩, reflecting the fact that their largest aperture is a triangle (III-gon). All vertices in the *closo*-structures are the preferred 4k$_C$ or 5k$_C$ vertices, except in the *closo*-11-vertex case where one 6k$_C$ vertex is necessarily incorporated (Figures 1.5–1.7). In this latter case, the presence of the atypical, destabilizing 6k vertex is noted by a subscript (see clo-11$_6$⟨III⟩).

In later sections the variously structured *nido*-species will be identified as members of the ni-*n*⟨IV⟩ (Figure 1-21), ni-*n*⟨V⟩ (Figure 1-14), and ni-*n*⟨VI⟩ (Figure 1-10) families of configurations, reflecting the fact that their largest open faces are tetragonal, pentagonal, and hexagonal, respectively. Most *arachno*-configurations are labeled as ara-*n*⟨VI⟩ configurations since almost all have hexagonal open faces (Figure 1-10).

1.2.2.3 **Bivertex-1** *Model;* **Predicted Nido-** *and* **Arachno-structures**

The boxed area in Figure 1-23 shows the structures predicted [18] in 1971 for the unknown *nido*-7-, 8-, and 9-vertex configurations, as well as the *arachno*-6-, 7-, and 8-vertex configurations. In the subsequent 18 years, examples of all six predicted structures were reported, in spite of the fact that our preferred ni-8$_4$⟨V⟩ configuration for carboranes (**Anomaly 1** in Figure 1.23) was incorrect.

1.2.2.4   Anomaly 1, the Known Structure of Nido-B₈H₁₂, compound 2.7 (ni-8₄⟨V⟩ versus ni-8⟨VI⟩)

1.2.2.4   Anomaly 1, the Known Structure of **Nido-B₈H₁₂,** *compound 2.7 (ni-8₄⟨V⟩ versus ni-8⟨VI⟩)*   The first exception to the *bivertex-1* model, labeled **Anomaly 1** in Figure 1-23, involved the structure of *nido*-B_8H_{12}, compound **2.7**, *ni*-8⟨VI⟩. This structure was known, prior to 1971, to have a hexagonal (VI-gonal) open face (Figure 1-25) rather than the pentagonal, *ni*-8₄⟨V⟩, open face we had predicted (see *ni*-8₄⟨V⟩ accompanied by a solid circle, ●, in the boxed area in Figure 1-23). In 1971, we rationalized this incongruity by assuming that our *bivertex-1* patterns were fundamentally correct, barring extenuating circumstances, and that potentially serious congestion, caused by the presence of four bridge hydrogens about the *"favorably convex"* pentagonal open face of *nido*-B_8H_{12}, had somehow forced the skeleton to expand from our preferred *ni*-8₄⟨V⟩ configuration to the *ni*-8⟨VI⟩ configuration (see structure **2.7** in Figure 1-25) in order to accommodate the four bridge hydrogens. As if in support of this contention, the parent compounds, *nido*-B_9H_{13} (*ni*-9₄⟨V⟩) and *nido*-B_7H_{11} (*ni*-7₄⟨V⟩), had never been reported, presumably, we think, because of the congestion caused by the four bridge hydrogens that would "necessarily" be occasioned by the *"unfavorably concave,"* pentagonal open faces (see Figure 1-23).

Our contention was speciously confirmed when Zimmerman and Sneddon [20] subsequently reported the isoelectronic compound *nido*-$S(CpCo)_2B_5H_7$,

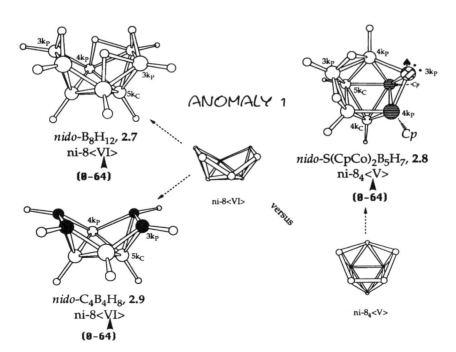

Figure 1-25   The *nido*-8-⟨VI⟩ compounds, B_8H_{12}, structure **2.7**, and $C_4B_4H_8$, structure **2.9**, versus the *nido*-8₄⟨V⟩ compound $S(CpCo)_2B_5H_7$, structure **2.8**.

structure **2.8**, (Figure 1-25), which had exactly our predicted ni-8_4⟨V⟩ structure. We were elated to note that compound **2.8** had only two bridge hydrogens, which we eagerly rationalized as insufficient to cause aperture expansion due to bridge hydrogen congestion. Our hubris was short lived. Fehlner [21] soon reported an alkyl derivative of *nido*-$C_4B_4H_8$, compound **2.9** (Figure 1-25), which was isoelectronic and isostructural with *nido*-B_8H_{12}, compound **2.7**, *but had no bridge hydrogens at all.* Even with no bridge hydrogens, *nido*-$C_4B_4H_8$, compound **2.9**, had the aperture-expanded ni-8⟨VI⟩ configuration. Bausch (see Chapter 9 in this volume) finds, *via ab initio calculations,* that at least one carborane anion with a ni-8_4⟨V⟩ structure is only one kcal or so less stable than its ni-8⟨VI⟩ analog. It is quite possible that the ni-8⟨VI⟩ structure is only slightly favored over the ni-8_4⟨V⟩ structure and that ni-8⟨VI⟩ configurations are promoted by both (1) hydrogen congestion, when many hydrogens are present, and (2) carbons "looking for" $3k$ vertices, when many carbons are present.

1.2.2.5 Anomaly 2, the "Abnormal" Structure of "Normal" **Arachno-B_9H_{15},** *2.10 (Arachno-9⟨VII⟩ versus Arachno-9⟨VI⟩)*

The second exception in the *bivertex-1* model, labeled **Anomaly 2** in Figure 1-23, involves the "abnormal," misnamed "normal," isomer of *arachno*-B_9H_{15}, structure **2.10** (it was misnamed "normal" simply because it was discovered first). The compound, "abnormal" *arachno*-B_9H_{15}, structure **2.10**, ara-9⟨VII⟩ in Figure 1-26, had the predicted VII-gonal open face or aperture (based upon our proscribed high-coordinated $4k_P$ vertex removal from ni-10⟨VI⟩). On the other hand, an alternative configuration, "typical" ara-9⟨VI⟩, with the smaller VI-gonal open face (see star, ★, outside of the boxed area in Figure 1-23), resulting from the removal of a *low-connected* $3k_P$ vertex from the ni-10⟨VI⟩ configuration, is known to be characteristic of *all other typical arachno*-9-vertex structures, including the also misnamed "iso-isomer" of *arachno*-B_9H_{15}, compound **2.11**, ara-9⟨VI⟩, as well as compounds **2.12** and **2.13** in Figure 1-26.

For the purposes of this chapter, we will henceforth refer to the misnamed "normal isomer" of *arachno*-B_9H_{15}, as the **"abnormal"** isomer of *arachno*-B_9H_{15}, compound **2.10**, ara-9⟨VII⟩, with the seven-membered open face (see Figure 1-12), and will simply include the misnamed "iso-isomer" of *arachno*-B_9H_{15}, compound **2.11**, ara-9⟨VI⟩ (with the smaller open face) as the **"typical"** isomer or the expected *arachno*-structure (see Figure 1-10). Additional support for these designations will be furnished in later sections. Except for this one compound, i.e., **abnormal** *arachno*-B_9H_{15}, compound **2.10**, ara-9⟨VII⟩, and one carborane analog, all other **typical** *arachno*-species, *if they are large enough,* have six-membered open faces, ara-n⟨VI⟩ (Figures 1-10 and 1-26). In a later section, we will show how **Anomaly 1** loses its anomalous character and comfortably conforms with the new patterns without complications.

1.2.2.6 Choice of Removal Between Two Equivalent **$4k_P$** *Vertices*

Two series in the horizontal rows of Figure 1-23 involve the ni-8_4⟨V⟩ → ara-7⟨VI⟩ and ni-9_4⟨V⟩ → ara-8⟨VI⟩ interconversions. In both cases, the second step

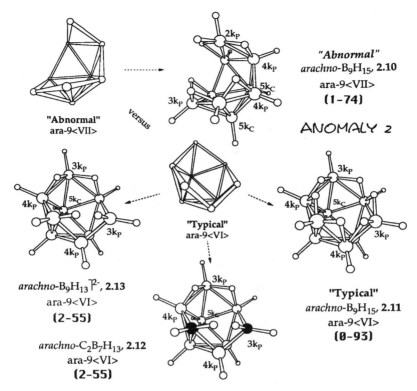

Figure 1-26 The "**abnormal**" (formerly "normal") *arachno*-9-⟨VII⟩ compound, B₉H₁₅, **2.10**; and "**Typical**" (formerly "iso-") *arachno*-9⟨VI⟩ compounds, B₉H₁₅, structure **2.11**; C₂B₇H₁₃, structure **2.12**; and B₉H₁₃⌉²⁻, structure **2.13**.

involves the removal of *one* $4k_P$ vertex, but in each case there are *two* candidate $4k_P$ vertices eligible for removal. These $4k_P$ vertices are identified respectively, by single arrowheads, ➤, and double arrowheads, ➤➤, on the structures ni-8₄⟨V⟩ and ni-9₄⟨V⟩. In each case, the $4k_P$ vertex labeled with a double arrowhead, ➤➤, *must be selected for removal* to yield the correct ara-7⟨VI⟩ and ara-8⟨VI⟩ configurations (labeled with half-circles, ◗, in Figure 1-23). In this instance, as well as for the improved *vertexseco-2* and *bivertexseco-3* patterns, the removal of the correct $4k_P$ vertex is critically important. *Restated, when two different kinds of otherwise equivalent $4k_P$ vertices are candidates for removal, that specific $4k_P$ vertex which, upon being removed, converts any residual $4k_C$ vertices present into $4k_P$ vertices must be selected for removal.* This is in keeping with the observation that all resulting *arachno*-configurations (Figure 1-10) are fragments of regular icosahedra, wherein all cage vertices are necessarily $5k_C$ vertices, and $4k_C$ vertices are not tolerated.

1.2.2.7   Coincidence 1, **nido-*10⟨VI⟩* ≅ arachno-*10⟨VI⟩***   It was noted that the ni-10⟨VI⟩ and ara-10⟨VI⟩ shapes were quite similar despite their difference in class (see diagonal two-headed dotted arrow in Figure 1-23). *This first coincidence raised the suspicion that a shape-based classification system (closo = "cage," nido = "nest," arachno = "web," hypho = "net") might not be sufficiently distinctive, and later caused us to rank "isoelectronicity" rather than shape as of paramount importance in classification.*

1.2.3   Disenchantment with the *Bivertex-1* Model

By 1993, the *bivertex-1* pattern [18] (Figure 1-23) of **sequentially removing two adjacent highest connected vertices** had accumulated an embarrassing, *and growing,* number of exceptions. The original *bivertex-1* procedure (Figure 1-23) had been embraced and expanded by Wade.[22] In Chapter 2 in this volume, Wade illustrates a somewhat different view of these structural relationships. Shore popularized the *bivertex-1* pattern (see Chapter 15 in this volume) and later Rudolph endorsed its usefulness.[23] Rudolph summarily rotated Williams' pattern (Figure 1-23) by 45° and added a *nido*-tetrahedral structure that Williams had rejected. Wade [22] treated Williams' simple geometrical removal of two highest connected adjacent vertices in a more sophisticated molecular orbital sense. He preferred to remove (first) $BH]^{2+}$ groups (vertices) from the various *closo*-$B_nH_n]^{2-}$ deltahedra to produce *nido*-$B_nH_n]^{4-}$ deltahedral fragments, following which he removed (second) additional adjacent $BH]^{2+}$ groups to produce the *arachno*-$B_nH_n]^{6-}$ deltahedral fragment structures. Wade expanded his molecular orbital-electron accountability to include transition elements and found that many inorganic and organometallic clusters, beyond the boundaries of carborane chemistry, seemingly followed the same patterns.

Eventually, the combination of both geometrical and electron-counting patterns, as applied to all electron-deficient clusters of many elements, became known as Wade's Rules. Over the next 20 years, the *bivertex-1* requirements for geometric rigor, originally espoused by Williams, were relaxed and any fairly "spherical" *closo*-deltahedron minus any vertex, *high connected or low connected,* became an acceptable *nido*-configuration, and any fairly spherical deltahedron minus any two adjacent vertices became an acceptable *arachno*-configuration.

Of more concern, specific deltahedral fragment shapes gradually became indelibly identified as either *nido*- or *arachno*-configurations (depending on the ideology of the investigator), and published comments referred to a *nido*-cluster (based on electron count) as having an *arachno*-structure (based on geometry) or an *arachno*-cluster (electronically) having a *nido*-structure (geometrically). These conflicts generally involved *nido*-species. *Such conflicts would be hailed as exceptions to Wade's Rules!* The discovery of the new *nido* and *arachno* structural patterns, which emphasized structures with most-homogeneous vertices (Figures 1-10–1-21), suggested that the original *bivertex-1* model (Figure 1-23) could be improved.

1.2.4 The 1995 *Vertexseco-2* Model

We are now able to relate, in two simple steps, the ***conforming*** *closo*-structures (Figure 1-7) with the ***conforming*** *arachno*-structures (Figure 1-10) in a surprisingly rigorous manner (see Figure 1-27) that indicates two alternative **nonconforming** *nido*-structures (we call this approach the *vertexseco-2* model; one path, two optional *nido*-configurations). *The first step involves the removal of one specific vertex, followed by the second step that involves the removal of one specific connection.* The preferred structures (for electron-deficient boron, carbon, and hydrogen compounds) are displayed as cylindrical bond models, while those that are rejected are shown as wire-frame models (see also Figures 1-17–1-18).

1.2.4.1 The Vertexseco-2 Model The *vertexseco-2 rules* are:

1.

    (a) If $6k_C$ vertices are present, then one must ***always remove*** a $5k_C$ vertex next to *all* $6k_C$ vertices (this applies to the clo-11_6 and $13_{66}\langle III \rangle$ deltahedra).

    (b) If there are no $6k_C$ vertices present, one must ***always remove*** any $5k_C$ vertex, $\langle 5k_C \rangle$ (this applies to clo-7, 8, 9, 10, and $12\langle III \rangle$ deltahedra).

    (c) If no $5k_C$ vertices are present, then one must ***always remove*** any $4k_C$ vertex (this applies to the clo-$6\langle III \rangle$ deltahedron).

[With the exception of the last case, this usually results in a ni-$n\langle V \rangle$ configuration that may or may not be the "correct" *nido*-structure (see Figure 1-27).

2.

    (d) One must ***always remove*** that one specific edge connection, $\langle i \rangle$, that results in the conversion of any residual $4k_C$ cage vertices present in the intermediate or "candidate" *nido*-structures into $4k_P$ edge vertices.

    (e) If no $4k_C$ cage vertices are present, then one must ***always remove*** the highest connected edge connection, $\langle i \rangle$, around the pentagonal open face.

[Following this very specific one-pathway, two-step, preliminary *vertexseco-2* procedure, two sets of candidate ni-$n\langle V \rangle$ and ni-$n\langle VI \rangle$ structures are produced. The second set, ni-$n\langle VI \rangle$, has the same structures as the known-to-be-correct ara-$n\langle VI \rangle$ configurations. In this fashion, the **correct** *arachno*-configurations are ***always produced in every case*** from the **correct** *closo*-configurations, ***with one partial exception*** (**Anomaly 3**, which will be addressed below).]

The *vertexseco-2* procedure relates the invariant, most-homogeneous *closo*- and *arachno*-configurations, shown in the first and fourth columns of Figure 1-27. The two optional columns of candidate *nido*-configurations, ni-$n\langle V \rangle$ and ni-$n\langle VI \rangle$, are illustrated in the second and third columns.

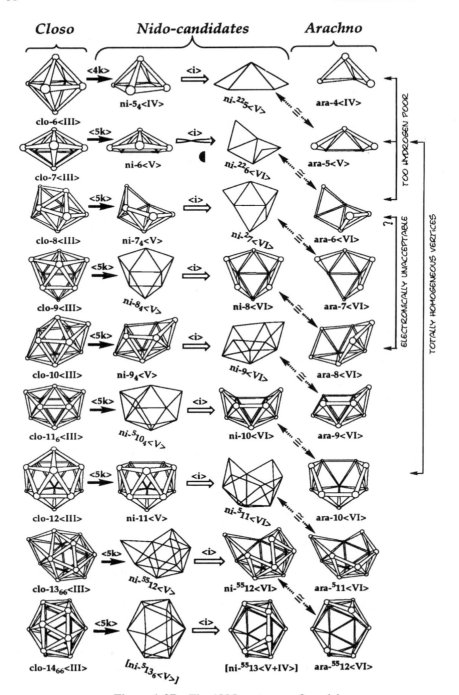

Figure 1-27   The 1995 *vertexseco-2* model.

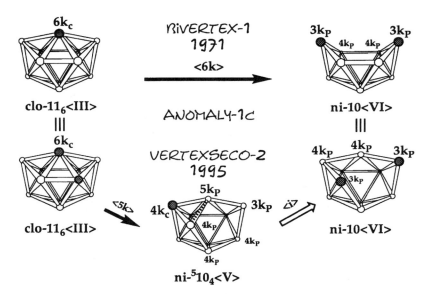

Figure 1-28 *Bivertex-1* and *vertexseco-2* paths both yield the same *nido*-10⟨VI⟩ Configurations.

1.2.4.2 Anomalies 1a, 1b, and 1c (ni-8⟨VI⟩, ni-10⟨VI⟩, ni-⁵⁵12⟨VI⟩) are Not Anomalous in the* Vertexseco-2 *Model The removal of one of the four 5*k* vertices, next to the 6*k* vertex of clo-11₆⟨III⟩ in Figure 1-27, yields an "inelegant" ni-10⟨V⟩ fragment that uniquely possesses no symmetry whatsoever (see Figure 1-28). The asymmetric ni-⁵10₄⟨V⟩ structure is certainly irregular as *all other configurations* displayed in Figure 1-27 have at least one element of symmetry. In view of the total lack of symmetry in the ni-⁵10₄⟨V⟩ configuration, we are unsure whether $N_2B_8H_{10}$ would have structure **2.15** or structure **2.15′** (see Figure 1-29), but currently favor the latter.

1.2.4.3 Anomaly 2 is Reassigned as "Abnormal" **Anomaly 2** in Figure 1-23 becomes less important as the **typical** isomer, ara-9⟨VI⟩, compound **2.11** in Figure 1-26, becomes the predicted isomer in the *vertexseco-2* procedure (Figure 1-27) and the **abnormal** isomer, ara-9⟨VII⟩, compound **2.10** in Figure 1-26, is firmly excluded as the only configuration that cannot be incorporated into Figure 1-27 as it alone incorporates an **"abnormal"** VII-gonal open face!

1.2.4.4 Anomaly 3, the Structures of arachno-B_6H_{12} (compound 2.21 versus compound 2.22) An incorrect space isomer of *arachno*-B_6H_{12}, i.e., compound **2.21** in Figure 1-30, is projected by the *vertexseco-2* procedure if one connection is simply removed to produce the ni-²²6⟨VI⟩ and ara-6⟨VI⟩ configurations from the ni-6⟨V⟩ configuration (see hollow arrow at top left in Figure 1-30). To

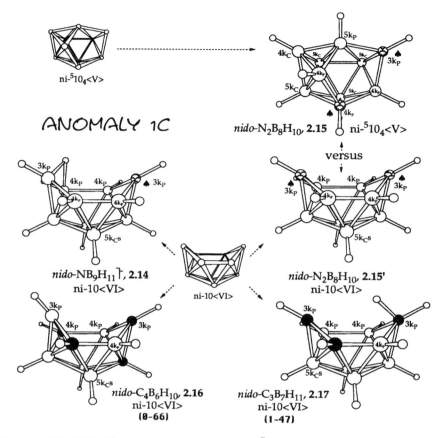

Figure 1-29 *Nido*-10-vertex compounds, $NB_9H_{11}]^-$, struture **2.14**; $N_2B_8H_{10}$, structure **2.15**; $C_4B_6H_{10}$, structure **2.16**; and $C_3B_7H_{11}$, structure **2.17** (VI-gonal versus V-gonal apertures).

rectify this problem, and to produce the correct configuration, an exception must be made and the top half of the configuration is converted into its mirror image (see twisted hollow arrow at top right in Figure 1-30 and Figure 2.6). We identify this mismatch as **Anomaly 3**, which we rationalize by comparison with the correct isomer, structure **2.22**, in Figure 1-30.

The projected but incorrect isomer of *arachno*-B_6H_{12}, compound **2.21**, ara-$6\langle VI \rangle$ is very similar to the correct isomer of *arachno*-B_6H_{12}, compound **2.22**, ara-$6\langle VI \rangle$ in Figure 1-30; both isomers incorporate almost identical architectural features. The same kinds and numbers of exo- and endo-hydrogens, as well as two 66- and two 5'6-bridge hydrogens, are present in both isomers *but the incorrect isomer, compound 2.21, incorporates one unfavorable $5k_P$ vertex that makes it a misfit (see Figure 1-10) in the pattern relating k-values within the arachno-*

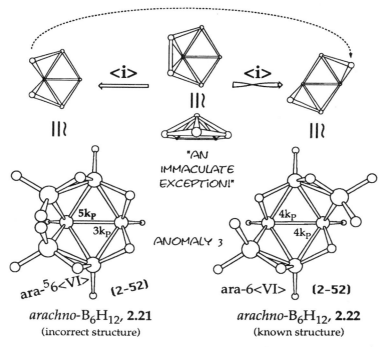

Figure 1-30 Relationship between projected isomer, compound **2.21**, and correct isomer, compound **2.22**, of *arachno*-B_6H_{12}.

compounds. It is apparent that merely *swing-shifting* the top half of the incorrect isomer, compound **2.21** (which might relieve any steric crowding between the two BH_2-groups), yields the correct isomer, compound **2.22**. The *k*-values of the correct isomer ($2k_P$, $3k_P$, and $4k_P$ vertices) are more homogeneous than the *k*-values within the incorrect isomer ($2k_P$, $3k_P$, and $5k_P$ vertices). In this case, the exception does indeed prove the rule: *"vertex homogeneity is truly of overriding importance."*

In Figure 1-27, the *vertexseco-2* procedure for relating *closo*-configurations rigorously with *arachno*-configurations is shown and two alternative *candidate nido*-structures are illustrated in each case, i.e. (1) the simple $\langle 5k_C \rangle$ removal produces a set of ni-$n\langle V \rangle$ structures and (2) the $\langle 5k_C \rangle$ removal plus $\langle i \rangle$ removal produces a set of ni-$n\langle VI \rangle$ structures. These available alternative *nido*-configurations are also listed in Figure 1-17, and the reasons that one is preferred and the other rejected are explained in the text associated with Figure 1-18.

For reasons illustrated in Figure 1-18, the eight correct *nido*-configurations (and the predicted ni-[55]$13\langle V + IV \rangle$ configuration) in Figure 1-27 are shown in columns 2 and 3 as cylindrical-bond illustrations above horizontal labels. The four **less-preferred** ni-$n\langle V \rangle$ configurations are indicated in column 2 by wire-frame renditions. These rejected ni-$n\langle$V$\rangle$ configurations are frequently observed, when 4SEDs are present. The four **rejected** ni-$n\langle$VI$\rangle$ configurations and the ni-^{22}5$\langle$V$\rangle$ configuration, with larger apertures, have not been observed, but are illustrated as wire-frame configurations in column 3. It is fairly simple to remember the preliminary one-pathway *vertexseco-2* model, illustrated in Figure 1-27, following which, one *nido*-candidate must be eliminated in each case. The oscillating "odd–even" ni-7$_4\langle$V$\rangle$ to ni-^{55}12$\langle$VI$\rangle$ pattern in Figure 1-27 is probably coincidental, as explained in the text accompanying Figure 1-18. One criticism of the *vertexseco-2* pattern was the opinion that by not removing a second vertex, Wade's procedure of sequentially removing two adjacent BH $\rceil^{2+}$ groups to produce, in order, *nido*-B$_n$H$_n\rceil^{4-}$ and *arachno*-B$_n$H$_n\rceil^{6-}$ deltahedral fragment structures, from the various *closo*-B$_n$H$_n\rceil^{2-}$ deltahedra, lost much of its illustrative power (see Chapter 2 in this volume). For this reason, we initiated a search for ways to "resurrect" the sequential removal of two vertices.

In Figure 1-31, an example of the *bivertex-1* sequence (Figure 1-23) is shown in row **A.** Two highest connected vertices are sequentially removed in accor-

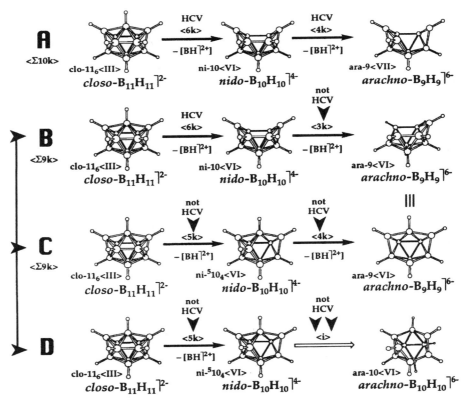

Figure 1-31   Geometrical and electron accountability alternatives.

dance with Wade's $BH]^{2+}$ preferences. That $6k_C$ plus $4k_P$ yield the "sum-total-value" of $\Sigma 10k$ is noted beneath the letter **A**. In row **B**, the removal process is modified to get the preferred, typical ara-9$\langle$VI$\rangle$ configuration and the sum-total-value becomes $\Sigma 9k$. In row **C**, both of the vertices that are removed are not the highest connected, yet the preferred ara-9$\langle$VI$\rangle$ configuration is generated and the sum-total-value is again $\Sigma 9k$. Both $6k_C + 3k_P$ (row **B**) and $5k_C + 4k_P$ (row **C**) equal $\Sigma 9k$ when the desired ara-9$\langle$VI$\rangle$ configuration is generated. In row **D**, an example of the *vertexseco-2* model is illustrated. The simplicity and accuracy of the final *vertexseco-2* pattern (Figure 1-27) suggested that the *bivertex-1* (Figure 1-23) and *vertexseco-2* models (rows **B**, **C**, and **D** in Figure 1-31) might be merged and improved.

1.2.5 Merging the *Bivertex-1* and *Vertexseco-2* Models into the *Bivertexseco-3* Model

In retrospect, by 1993 the original 1971 *bivertex-1* geometrical systematics (Figure 1-23) had been expanded to include deltahedral fragments derived from the smaller clo-$5_{33}\langle$III$\rangle$ deltahedron, as well as from the larger clo-$13_{66}\langle$III$\rangle$ and clo-$14_{66}\langle$III$\rangle$ deltahedra. The subscripts in clo-$5_{33}\langle$III$\rangle$ indicate that, of the five vertices, two are unfavorable $3k$ vertices and, of the 13 and 14 vertices, two are disfavored $6k$ vertices in each case. Unfortunately, as Williams' *bivertex-1* geometrical systematics expanded in scope from 1971 (Figure 1-23) to 1995, the number of *exceptions* to the simple sequential removal of *the two highest coordinated vertices,* which produce the *nido-* and *arachno*-fragments via the *bivertex-1* procedure, had also expanded.

1.2.5.1 Vertex and Connection Removal Produce Similar Aperture Sizes

We were faced with upgrading the original *bivertex-1* scheme (Figure 1-23) and merging it with the preliminary *vertexseco-2* scheme (Figure 1-27). In the latter, one highly specific $5k$ vertex is removed (first), followed by the removal of one highly specific connection (second). There is a way that these two patterns can be "hybridized" in such a fashion that they mutually reinforce each other. Figure 1-32 are illustrates the various geometric manipulations that can give rise to the variously sized apertures, from triangular (III-gonal) to hexagonal (VI-gonal). In the top four rows of Figure 1-32, it is seen that the VI-, V-, and IV-gonal apertures may be produced (see horizontal arrows) from a surface composed of triangles (III-gons), either by the removal of $6k_C$, $5k_C$ and $4k_C$ vertices ($\langle 6k \rangle$, $\langle 5k \rangle$, $\langle 4k \rangle$) or by the removal of 3, 2, or 1 connections ($\langle$iii$\rangle$, $\langle$ii$\rangle$, $\langle$i$\rangle$). The removal of a $3k_C$ vertex, $\langle 3k \rangle$, does not change the size of the largest aperture as a III-gon is produced. Second, any aperture on a large enough surface may be expanded to the next larger aperture size by the removal of one connection, $\langle$i$\rangle$ (see vertical arrows).

At the bottom of Figure 1-32 are two examples that illustrate how the subsequent removal of a $3k_P$ vertex from a VI-gonal open face (produced by the prior removal of a $6k$ vertex) changes the VI-gonal open face into a second VI-gonal

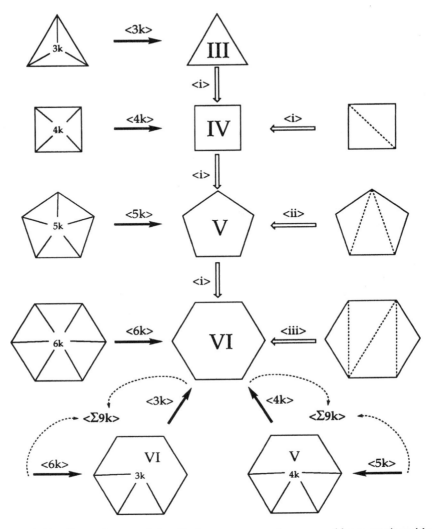

Figure 1-32 Four-, five-, and six-sided apertures may be prepared by removing either connections, $\langle i \rangle$, $\langle ii \rangle$, or $\langle iii \rangle$, or vertices, $\langle 4k \rangle$, $\langle 5k \rangle$, or $\langle 6k \rangle$.

open face. In a contrasting sequence, the subsequent removal of a $4k_P$ vertex from a V-gonal open face (produced by the prior removal of a $5k$ vertex) also results in a VI-gonal open face. The point here is that the *closo*-structures, clo-$n\langle \text{III} \rangle$, in Figure 1-23 may be converted into ara-$n\langle \text{VI} \rangle$ structures either by sequential removal of $5k_C$ and $4k_P$ vertices or by sequential removal of $6k_C$ vertices and $3k_P$ vertices, so long as the sum-total-value of the two vertices equals $\Sigma 9k$, i.e., $5k + 4k = \text{``}\Sigma 9k\text{''} = 6k + 3k$ (see also Figure 1-31).

Had we recognized, prior to 1971, that two vertices totaling $\Sigma 9k$ should sequentially be removed rather than the two "highest connected" vertices, $\langle HCV \rangle$ (Figure 1-23), then we would have sequentially removed $6k_C$ and $3k_P$ vertices from clo-$11_6 \langle III \rangle$ and gotten the preferred or typical ara-$9 \langle VI \rangle$ structure rather than the "abnormal" ara-$9 \langle VII \rangle$ structure, and **Anomaly 2** (Figure 1-23) would never have arisen. Second, had we been able to remove $6k_C$ and $3k_P$ vertices from the most homogeneous *closo*-9-vertex structure in Figure 1-23, to produce the ni-$8 \langle VI \rangle$ configuration, then **Anomaly 1** would never have materialized. Of course, there is a problem: the most-homogeneous clo-$9 \langle III \rangle$ structure has no $6k$ vertex available for removal. There is, however, a *"second-choice"* nine-vertex structure, clo-$9_6 \langle III \rangle$, which does incorporate a $6k$ vertex for removal. Configuration clo-$9_6 \langle III \rangle$ is arithmetically conceivable and geometrically possible, see column 4:4:1 in matrix **B** of Figure 1-5 (see also wire-frame illustration in Figure 1-33). Utilizing both the optimal and the less-acceptable *closo*-9-vertex structures should cause little confusion. The optimal clo-$9 \langle III \rangle$ configuration (illustrated with cylindrical bonds) incorporates the most-homogeneous vertices available for actual nine-vertex *closo*-carboranes and, via $5k_C$ vertex removal, yields the most-homogeneous ni-$8_4 \langle V \rangle$ configuration. In contrast, the less-acceptable *"second-choice"* clo-$9_6 \langle III \rangle$ configuration (illustrated as a wire-frame model), while not as homogeneous, and thus not the preferred structure for *closo*-9-vertex carboranes, does serve as the precursor for the most-homogeneous ni-$8 \langle VI \rangle$ deltahedral fragment configuration. The conclusion here is that structures with the most-homogeneous vertices possible are characteristic of all classes and families of carboranes thus far-discovered. It does not follow, however, that all most-homogeneous *nido*-structures can be derived from the most-homogeneous *closo*-structures, just as they cannot be derived from the *"formerly hallowed"* icosahedron.

Figure 1-33 represents a blending of an upgraded *bivertex-1* pattern (Figure 1-23) and *vertexseco-2* (Figure 1-27) pattern. In Figure 1-33, we see an almost three-dimensional triangulation of the various cylindrical-bond structures, which highlights their individual uniqueness; no alternative shapes will fit! Figure 1-33 is composed of five columns of structures: two *closo*, two *nido*, and one *arachno*. The left-hand column is labeled "virtual precursors" at the top and "second choices" further down. With the exception of the two "dual-purpose" clo-$11_6 \langle III \rangle$ and clo-$5_{33} \langle III \rangle$ configurations, the left-hand column illustrates unacceptable *"second-choice"* *closo*-structures (i.e., clo-$9_6 \langle III \rangle$, clo-$10_6 \langle III \rangle$, clo-$12_{66} \langle III \rangle$, and clo-$13_{666} \langle III \rangle$; see Figure 1-5), which incorporate $6k$ vertices, and which may be removed to produce the most-homogeneous ni-$n \langle VI \rangle$ and ara-$n \langle VI \rangle$ configurations (Figure 1-33). The three contrived "virtual-precursor" structures, clo-$6_{33} \langle III \rangle$, clo-$7_{633} \langle III \rangle$, and clo-$8_{63} \langle III \rangle$, are simply "retroconstructed" by capping the open faces of the ni-$^{22}5 \langle V \rangle$, ni-$^{22}6 \langle VI \rangle$, and ni-$^{22}7 \langle VI \rangle$ configurations with $6k_C$ vertices. Column 2 in Figure 1-33 lists the most-homogeneous *closo*-structures characteristic of actual *closo*-carboranes (Figures 1-1 and 1-6).

Column 3 is occupied by the most-homogeneous-possible ni-$n \langle V \rangle$ configurations (Figure 1-14). All of the ni-$n \langle V \rangle$ structures are rigorously associated to

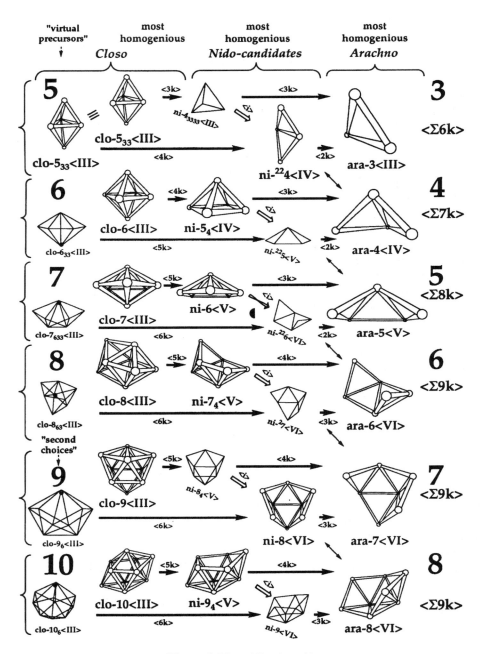

Figure 1-33a *(Continued.)*

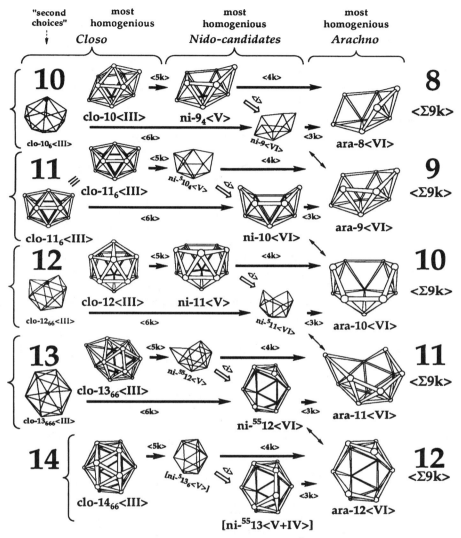

Figure 1-33b   The 1996 *bivertexseco-3* model.

three neighboring configurations: (1) related to the most-homogeneous clo-$n\langle III\rangle$ configurations by the removal of one highly specific $5k$ vertex, (2) related to the same-sized ni-$n\langle VI\rangle$ and ara-$n\langle VI\rangle$ configurations by the removal of a highly specific connection, $\langle i\rangle$, and (3) related to the one-vertex-smaller ara-$n\langle VI\rangle$ structure by the removal of one highly specific $4k$ vertex.

Column 4 illustrates the most-homogeneous-possible ni-$n\langle VI\rangle$ structures. They may be (1) derived from the same-sized ni-$n\langle V\rangle$ configurations by the removal of one highly specific connection, (2) related to the one-vertex-smaller ara-$n\langle VI \rangle$ structure by the removal of a highly specific $3k$ vertex, and, of course, (3) derived from the less-acceptable "one-$6k$-vertex-larger, second most-homogeneous" *closo*-configurations. All told, the two invariant candidate structural choices, i.e. the ni-$n\langle V \rangle$ and ni-$n\langle VI \rangle$ alternatives, are "trapped" between the *fixed, most-homogeneous* ara-$n\langle VI \rangle$ configurations and the *fixed, most-homogeneous* clo-$n\langle III \rangle$ structures by no less than three geometrical relationships with surrounding structures. The nido-structural choices emphasized in the *vertexseco-2* scheme (Figure 1-27), are reemphasized in Figure 1-33 and explained in the text surrounding Figure 1-18.

In Figure 1-33 a right-hand column of labels, $\langle \Sigma 9k \rangle$, may be seen, emphasizing that two vertices must sequentially be removed in either of two ways. For example, either a $5k_C$ vertex plus a $4k_P$ vertex (which totals $\Sigma 9k$) or a $6k_C$ vertex plus a $3k_P$ vertex (which also totals $\Sigma 9k$) must be removed (Figure 1-33). In Figure 1-33, note that the $\Sigma 9k$ value decreases monotonically from $\Sigma 9k$ to $\Sigma 6k$ as the *arachno*-open-faces decrease in size from ara-$6\langle VI \rangle$ to ara-$3\langle III \rangle$. **Anomaly 1** and **Anomaly 2** are no longer worthy of concern. **Anomaly 1** has become moot, and **Anomaly 2** is relegated to one abnormal configuration in both the *vertexseco-2* (Figure 1-27) and *bivertexseco-3* schemes (Figure 1-33). The only surviving anomaly, **Anomaly 3**, constituted a minor problem in Figure 1-27 but is *greatly diluted* in Figure 1-33. The configuration for ni-$^{22}6\langle VI \rangle$ (shown in Figure 1-33) "fits" with three of the four neighboring structures and requires only minor alteration to "fit" with the fourth. **Anomaly 3** is equivalent to a "single defect" in an otherwise "unflawed crystal."

The *vertexseco-2* scheme, displayed in Figure 1-27, might fancifully be considered as analogous to a game of two-dimensional *tic-tac-toe (or noughts and crosses)* as all three shapes must "line up" in a coherent manner. In a similar fashion, the *bivertexseco-3* scheme may be considered to resemble "three-dimensional" tic-tac-toe wherein the various geometric shapes have to line up multidirectionally in a unified manner. For alternative patterns, see Chapter 3 in this volume.

In reviewing Figure 1-33, it is seen that in each set of two rows, the *arachno*-shapes usually have VI-gonal open faces and the n in the $\langle \Sigma nk \rangle$ column is usually 9. *This "connectivity-sum-value", $\langle \Sigma 9k \rangle$, equals the size of the arachno-open-face, VI-gon, plus 3.* When the *arachno*-open-face diminishes in size to V-, IV-, and III-gons, then, as expected, the connectivity-sum-value also decreases to $\langle \Sigma 8k \rangle$, $\langle \Sigma 7k \rangle$, and $\langle \Sigma 6k \rangle$, respectively. In this connection, the one abnormal *arachno*-configuration with a VII-gonal open face, ara-$9\langle VII \rangle$, illustrated in Figures 1-12, 1-23, and 1-26, and heretofore identified as **Anomaly 2**, comes to mind. As illustrated in Figure 1-34, such a structure with a VII-gonal open face should lead to the larger connectivity-sum-value of $\langle \Sigma 10k \rangle$ (see stars, $\star$). In the future, perhaps additional "abnormal" *arachno*-shapes with VII-gonal open faces will be discovered (see Figure 1-13) and perhaps a pattern-relating connectivity-sum-values of $\Sigma 10k$ will result.

Figure 1-33, which applies to electron-deficient species, is not amenable to the inclusion of a number of electron-precise hydrocarbons as was the original

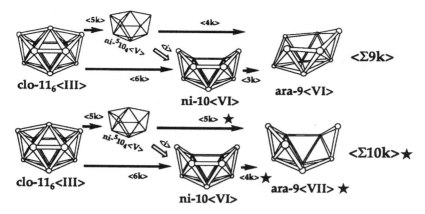

Figure 1-34 *Closo*-11_6 to normal *arachno*-9⟨VI⟩ conversions yield sum-total-values of Σ9k while closo-11_6 to abnormal *arachno*-9⟨VII⟩ conversions yield sum-total-values of Σ10k.

systematics (Figure 1-23). The writer is not disappointed with this eventuality; it is anticipated that related patterns that "bridge" electron-deficient clusters and electron-precise structures will be developed (see Chapter 2 in this volume).

1.2.5.2 Addition of the clo-5₃₃⟨III⟩, ni-²²4⟨IV⟩, and ara-3⟨III⟩ Configurations In 1972, Olah et al. (see Chapter 5 in this volume) reported the structure of the first ni-^{22}4⟨IV⟩ compound, *nido*-C_4H_5]$^+$, structure **2.23**. It could be derived from the deltahedron, typified by *closo*-1,5-$C_2B_3H_5$, structure **2.24**, clo-5_{33}⟨III⟩ (Figure 1-35), following the originally proscribed removal of one highest connected $4k_C$ vertex. Compound *nido*-C_4H_5]$^+$, structure **2.23**, is known to organic

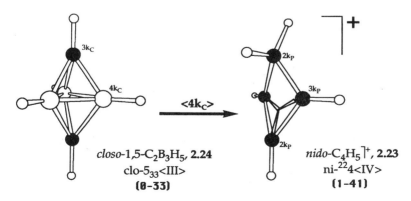

Figure 1-35 Conversion of the *closo*-1,5-$C_2B_3H_5$, **2.24**, Structure, *closo*-5⟨III⟩, into the *nido*-C_4H_5]$^+$, **2.23**, Configuration, *nido*-4⟨IV⟩.

chemists as "homocyclopropenium" cation and is related to Paetzold's derivatives of *nido*-NB$_4$H$_6$ (see Chapter 17 in this volume).

1.2.5.3   Former Anomaly 1 now becomes Anomalies 1a, 1b, and 1c

In three cases, the known-to-be-correct *nido*-configurations, ni-8⟨VI⟩, ni-10⟨VI⟩, and ni-^{55}12⟨VI⟩ for *nido*-carboranes prevail as opposed to the ni-*n*⟨V⟩ alternatives. We have discussed the structure ni-8⟨VI⟩, identified as **Anomaly 1** in Figures 1-23 and 1-25. We now identify Anomaly 1 as Anomaly 1a. A second example of this same kind of exception, **Anomaly 1b**, involves the structure ni-^{55}12⟨VI⟩. Representative molecular examples of **Anomaly 1b** are illustrated in Figure 1-36.

A few years ago, following the development of the preliminary *vertexseco-2* model (Figure 1-27), it was found that the clo-11$_6$⟨III⟩ to ni-10⟨VI⟩ conversion, by simple 6k_C vertex removal (as illustrated in Figure 1-23), could also be achieved by 5k_C vertex removal plus connection, ⟨i⟩, removal. Thus, a third example of the same kind of exception was added and transiently identified as **Anomaly 1c** (Figures 1-28 and 1-29).

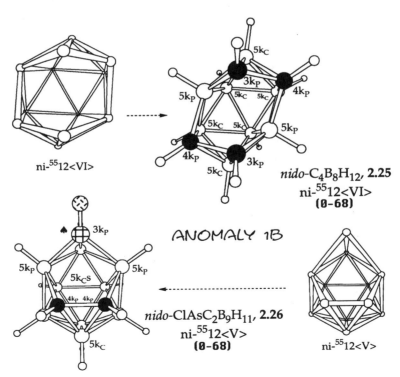

Figure 1-36   *Nido*-C$_4$B$_8$H$_{12}$, **2.25**, ni-^{55}12⟨VI⟩, and *nido*-ClAsC$_2$B$_9$H$_{11}$, **2.26**, ni-^{55}12⟨V⟩.

1.2.5.4   Coincidental Shape Redundancy   A growing number of cases where molecules of the same size within neighboring classes have the same connectivities and almost identical shapes became apparent in Figures 1-27 and 1-33. The *nido-* to *arachno-*examples of these are illustrated with diagonal arrows in Figure 1-33. Because these "coincidences" are rife, exclusively shape-based classification has become suspect. ***Shape-based classes are no longer sufficiently distinctive.*** *Shifting definitive classification from* **indistinct** *shapes to* **less variable** *numbers of skeletal electrons seems desirable.*

1.2.6   The 1995 *Polyseco-2* Model

To complete a comparative evaluation of the various vertex-removal and connection-removal schemes (Figure 1-24), we considered the possibility that the *polyseco* model [19] might also be upgraded in light of the 1995 *vertexseco-2* model (Figure 1-27) and the 1996 *bivertexseco-3* model (Figure 1-33). (See Chapter 4 in this volume, where Hermánek illustrates additional advantages of his seco-approach.)

From the *vertexseco-2* model, we know that the most rigorous correlation between *closo-*deltahedra and *arachno-*deltahedral fragments is the removal (first) of a very specific $5k_C$ vertex, $\langle 5k \rangle$ (producing a pentagonal aperture) and (second) the removal of a very specific connection, $\langle i \rangle$ (which produces a hexagonal aperture). Alternatively, we know that pentagonal apertures can also be produced from deltahedra (if they are large enough) by the removal of two adjacent surface connections, $\langle ii \rangle$, just as well as by the removal of a $5k$ vertex, $\langle 5k \rangle$ (Figure 1-32).

In causing the *polyseco* model to emulate the *vertexseco-2* sequence of Figure 1-27, one highest connected vertex in each of the clo-6$\langle$III$\rangle$ to clo-12$\langle$III$\rangle$ deltahedra must be identified (see left-hand column of Figure 1-37) and two adjacent highest "connected connections", $\langle ii \rangle$, radiating from that selected highest connected vertex must be removed. This is somewhat comparable to the first step in the *vertexseco-2* procedure (Figure 1-27), in that a pentagonal aperture is produced. This first step in the *polyseco-2* procedure (Figure 1-37) yields a similar set of deltahedral fragments (albeit severely distorted and not reduced in number of atoms) as does the first step of the *vertexseco-2* procedure (Figure 1-27). A totally abnormal ni-^{2}6$_4\langle$V$\rangle$ configuration is produced, however (see brackets in Figure 1-37). This configuration resembles a proposed structure for *nido-*(μ-BH$_3$)B$_5$H$_8$⊤ (see Chapter 15 in this volume).

The second step of the *polyseco-2* procedure, the further removal of one more additional connection, $\langle i \rangle$, is then carried out in exactly the same fashion as the second step in the *vertexseco-2* procedure (Figure 1-27). The correct, albeit distorted, *arachno-*configurations are produced as expected. A number of configurations are illustrated in Figure 1-37 that are different from those illustrated in Hermanek's 1992 review [19a] (see also Chapter 4 in this volume). They could not have been foreseen prior to guidance derived from the *vertexseco-2* model (Figure 1-27). All of the coincidentally redundant skeletal structures, which

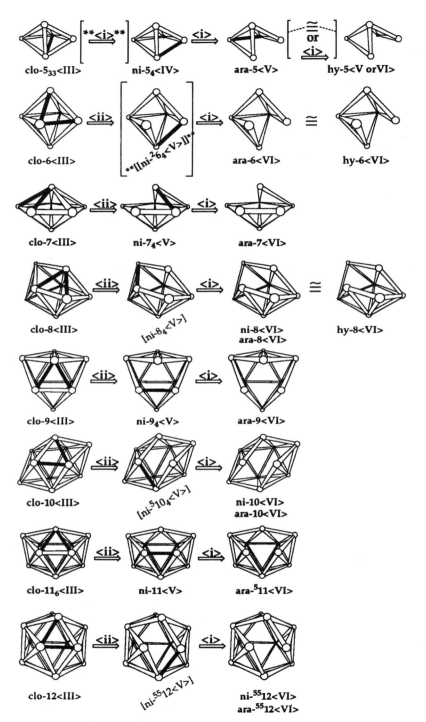

Figure 1-37 Revised 1995 *polyseco-2* model.

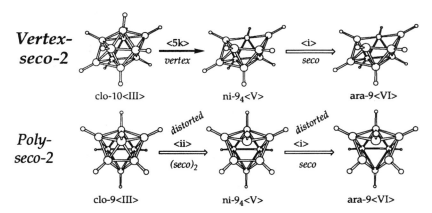

Figure 1-38 A comparison of the *vertexseco-2* and *polyseco-2* procedures for producing the arachno-9⟨VI⟩ configuration from the *closo*-10 and *closo*-9 deltahedra, respectively.

were noted in the *vertexseco-2* model (Figure 1-27) and the *bivertexseco-3* model (Figure 1-33), are also reproduced using the *polyseco-2* model (Figure 1-37), i.e., the *nido-* and *arachno*-configurations, ni-n⟨VI⟩ and ara-n⟨VI⟩, are identical when the numbers of vertices, n, are 8, 10, and 12.

There are annoying differences, however. The **polyseco-2** scheme does not work well with the smaller, $n = 5$ and $n = 6$, deltahedra and their deltahedral fragment structures. For example, in the first step of the **vertexseco-2** procedure (Figure 1-27), a $5k$ vertex is removed and the "correct structure" for the ni-6⟨V⟩ species is predicted. However, in the following step (⟨i⟩ removal), a partially incorrect structure for the ara-6⟨VI⟩ species is predicted (**Anomaly 3** in Figure 1-30).

In exactly the reverse order, in the **polyseco-2** procedure (Figure 1-37), an incorrect ni-^{2}6$_4$⟨V⟩ structure is predicted following the removal of two adjacent connections,⟨ii⟩, in the first step, but following the second step, i.e., the removal of one more connection, ⟨i⟩, from the incorrect ni-6⟨V⟩ configuration, the correct configuration for the ara-6⟨VI⟩ species is produced. *It is as if the two patterns, vertexseco-2 and polyseco-2, "collide" at the ni-6<V> to ara-6<VI> interface; each pattern projects one correct six-vertex configuration and one wrong six-vertex configuration, but they are the reverse of each other!* A limited comparison between the 1995 *vertexseco-2* (Figure 1-27) and the 1995 *polyseco-2* (Figure 1-37) models result in the "same" ara-9⟨VI⟩ configuration, as is illustrated in Figure 1-38.

1.2.7 Conclusions

- The *closo*-carboranes, **clo-n⟨III⟩**, adopt a *unique* series of closed configurations in which all of the vertices are as homogeneously connected as possible (Figure 1-6). *There are no random components to the closo-structural choices.*

- All *arachno*-carboranes, from **ara-6⟨VI⟩** to **ara-12⟨VI⟩**, adopt a *unique* series of icosahedral fragment configurations in which a monotonically increasing number of $5k_C$ cage vertices are surrounded with six nk_P peripheral vertices that are as homogeneously connected as possible. This results in a unique *centrosymmetric* distribution of nk_P peripheral vertices (Figure 1-10). *There are no random components to the arachno-structural choices.*

- The *nido*-carboranes (boron and carbon skeletal atoms only) fall into three patterns:
  - *most favored* **ni-n⟨V⟩** structures, where n = 6, 11 (Figure 1-14; matrix **A**);
  - *intermediately favored* **ni-n⟨VI⟩** structures, where n = 8, 10, 12 (Figure 1-10);
  - *least favored* **ni-n_4⟨V⟩** structures, where n = 7, 9 (Figure 1-14; matrix **B**).

There are *"positive"* reasons that the **nido-6** and **nido-11** compounds (Figure 1-14) assume the *most favored* **ni-6⟨V⟩** and **ni-11⟨V⟩** configurations. They are icosahedral fragment structures and the nk_C and nk_P vertices are totally homogeneous in contrast to their less homogeneous ni-^{22}6⟨VI⟩ and ni-^{5}11⟨VI⟩ conformers.

In contrast, there are *"negative"* reasons that the **nido-7** and **nido-9** compounds assume the *less favored* **ni-7$_4$⟨V⟩** and **ni-9$_4$⟨V⟩** configurations. The alternative ni-^{2}7⟨VI⟩ and ni-9⟨VI⟩ structures do not appear to be electronically compatible with the electron pair bonds available to these species (Figure 1-18) and therefore alternative ni-n⟨V⟩ structures are necessarily chosen (Figure 1-14), even though one less-favored $4k_C$ vertex must be accommodated in each case. Another reason that **nido-6** and **nido-7** compounds "choose" **ni-6⟨V⟩** and **ni-7$_4$⟨V⟩** configurations is that *nido*-compounds incorporate too few skeletal electrons and hydrogens to be compatible with the ni-^{22}6⟨VI⟩ and ni-^{2}7⟨VI⟩ structures. *There are apparently no random components to the nido-structural choices when ni-n<V> and ni-n<VI> configurations are involved.*

- The alternating choices between the ni-n_{odd}⟨V⟩ and ni-n_{even}⟨VI⟩ configurations are almost certainly due to "blue-collar" coincidences rather than some esoteric "white-collar" structural pattern.

- A quarter century of accumulated exceptions have caused some disenchantment with the 1971 *bivertex-1* pattern (Figure 1-23), which involved the sequential removal of two adjacent highest connected vertices from the *closo*-deltahedra to produce the *nido*- and *arachno*-deltahedral fragment configurations.

- A simple, accurate, 1995 *vertexseco-2* model (sequential removal of one specific $5k_C$ vertex followed by the removal of one connection) has been presented (Figure 1-27).

- Hybridization of the 1995 *vertexseco-2* model and a rehabilitated *"bivertex-2"* pattern "cross-links" them into the even more accurate and rigorous "three-dimensional" 1996 *bivertexseco-3* pattern (removal of two specific vertices and one specific connection) (Figure 1-33).

- The *vertexseco-2* and *bivertexseco-3* models yield less-distorted structures and seem to be simpler and more rigorous than the *polyseco-2* procedure, but all should be considered in future evaluations (see also Chapters 2–4 in this volume for additional insights and viewpoints).

ACKNOWLEDGMENTS

Physical support from the Loker Hydrocarbon Research Institute and moral support from Professors George A. Olah and G. K. S. Prakash is acknowledged. I am also indebted to Professor Joe Casanova for volunteering to edit this volume and to all of the individual authors who contributed chapters. I am especially grateful to Professor Robert Bau for vastly improving the text and to Professors Bau, Robert W. Parry, Peter Paetzold, and Sheldon G. Shore for debating and refining selected arguments.

REFERENCES

1. Williams, R. E., Early Carboranes and Their Structural Legacy, *Adv. Organometal. Chem.*, 1994, **36**, 1–55.

2. (a) Grimes, R. N., *J. Am. Chem. Soc.*, 1966, **88**, 1070–1071; (b) Grimes, R. N., *J. Am. Chem. Soc.*, 1966, **88**, 1895–1899; (c) Bramlett, C. L. and Grimes, R. N., *J. Am. Chem. Soc.*, 1966, **88**, 4269; (d) Grimes, R. N., Bramlett, C. L., and Vance, R. L., *Inorg. Chem.*, 1969, **8**, 55–58; (e) Grimes, R. N., *Carboranes*, Academic Press, London, 1970, Chap. 4, p. 34–36.

3. Williams, R. E., *Progress in Boron Chemistry* (Brotherton, R. J. and Steinberg, H., eds), Pergamon Press, Oxford, 1970, Vol. 2, Chap. 2, p. 57.

4. Bausch, J. W., Prakash, G. K. S., and Williams, R. E., BUSA-II Meeting, Research Triangle, NC, June 1990.

5. Hofmann, M., Fox, M. A., Greatrex, R., Schleyer, P. v. R., Bausch, J. W., and Williams, R. E., *Inorg. Chem.*, 1996, **35**, 6170–6178, and references therein.

6. Berry, T. E., Tebbe, F. N., and Hawthorne, M. F., *Tetrahedron Letters*, 1965, **12**, 715 (see footnote 3).

7. Tsai, C. and Strieb, W. E., *J. Am. Chem. Soc.*, 1966, **88**, 4513.

8. (a) Thompson, M. L. and Grimes, R. N., *J. Am. Chem. Soc.*, 1971, **93**, 6677; (b) Thompson, M. L. and Grimes, R. N., *Inorg. Chem.* 1972, **11**, 1925.

9. Bausch, J. W., Prakash, G. K. S., and Williams, R. E., *Inorg. Chem.*, 1992, **31**, 3763.

10. Fox, M. A. and Greatrex, R., *J. Chem. Soc., Dalton Trans.*, 1994, 3197.

11. Lipscomb, W. N., *Boron Hydrides*, W. A. Benjamin Co., New York, 1963.

12. Fox, M. A., Greatrex, R., Hofmann, M., and Schleyer, P. v. R., *Angew. Chem.*, 1994, **106**, 2384; *Angew. Chem.* (int. ed.), 1994, **33**, 2298.

13. The **"abnormal"**-B_9H_{15} isomer, compound **1.31**, ara-9⟨VII⟩ has some redeeming features; its edge vertex connectivity is not as homogeneous as is that of the **typical** *arachno*-B_9H_{15} isomer in Figure 1-10, ara-9⟨VI⟩, compound **1.18**, but its unique **"abnormal"** structure may actually be a response to bridge hydrogen congestion as the structurally **"abnormal"**-*arachno*-B_9H_{15} isomer, compound **1.31**, is more thermally stable than the **typical** *arachno*-B_9H_{15} isomer, compound **1.18**.

14. Williams, R. E., Onak, T. P., Dunks, D. B., Spielman, J. R., and Gerhart, F. J., *Inorg. Chem.*, 1966, **5**, 118.

15. When the B_2H_6, compound **1.33**, was removed, the **"abnormal"** *arachno*-B_9H_{15}, compound **1.31**, decayed into *nido*-B_8H_{12}, compound **1.32**, which further decayed into *nido*-B_6H_{10}, compound **1.34**.

16. Stock A., *Hydrides of Boron and Silicon*, Cornell University Press, Ithaca, NY, 1933.

17. Parry, R. W. and Edwards, L. J., *J. Am. Chem. Soc.*, 1959, **81**, 3554–3560.

18. (a) Williams, R. E., *Inorg. Chem.*, 1971, **10**, 210–214; (b) Williams, R. E., *Adv. Inorg. Chem. Radiochem.*, 1976, **18**, 67; (c) Williams, R. E., *Chem. Rev.*, 1992, **92**, 177.

19. (a) Hermánek, S., *Chem. Revs*, 1992, **92**, 325–363; (b) Hermánek, S., "Seco versus Debor Concept," May 23, 1985, The University of Munich, BRD.

20. Zimmerman, G. J. and Sneddon, L. G., *J. Am. Chem. Soc.*, 1981, **103**, 1102.

21. Fehlner, T. P., *J. Am. Chem. Soc.*, 1977, **99**, 8355; 1980, **102**, 3424.

22. (a) Wade, K., *J. Chem. Soc., Chem. Commun.*, 1971, 792; (b) Wade, K., *Adv. Inorg. Chem. Radiochem.*, 1976, **18**, 1; (c) Wade, K., *New Scientist*, 1974, **62**, 615.

23. (a) Rudolph, R. W. and Pretzer. W. R., *Inorg. Chem.*, 1972, **11**, 1974–1978; (b) Rudolph, R. W., *Accounts Chem. Res.*, 1976, **9**, 446–452.

2

THE BORANE–CARBORANE STRUCTURAL PATTERN: SOME CORRELATIONS AND IMPLICATIONS

Mark A. Fox and Kenneth Wade

Chemistry Department, Durham University Science Laboratories, South Road, Durham DH1 3LE, U.K.

2.1 INTRODUCTION

This book is based on material presented at a symposium held in December 1995 to honor and celebrate the 70th birthday of Dr. Robert E. Williams, a senior scientist at the Loker Hydrocarbon Research Institute at USC, who throughout his distinguished career has contributed greatly to our understanding of boron cluster chemistry and carbocation chemistry. He did so originally as a young experimentalist showing how carboranes, hitherto unknown mixed hydrides of boron and carbon in which both elements were present in the poly-hedral molecular skeleton, could be made from boranes and unsaturated hydro-carbons.[1] He correctly deduced their unprecedented polyhedral shapes from their ^1H and ^{11}B NMR spectra in pioneering applications of the technique, and went on to draw attention to systematic patterns in those shapes in a seminal paper published in 1971.[2] The systematics he then outlined have been refined and extended over the years,[3,4] and his chapter (Chapter 1) in this book shows how comprehensive his treatment of borane, carborane, and carbocation struc-tures has become. Other chapters focus on specific structural, bonding, synthet-ic, or other aspects of borane, carborane, and carbocation chemistry that illustrate its rich variety. The aims of this particular chapter are to look at some familiar features of borane clusters in different ways, to comment on specific aspects of the borane structural and bonding pattern that appear to be either significant or controversial, to outline approaches taken up in more detail in later chapters, and also to place boron cluster chemistry in context within cluster chemistry in general.

2.2 THE BORANE–CARBORANE STRUCTURAL/BONDING PATTERN: MAIN FEATURES

Key features of the structural chemistry of cluster compounds that were first recognized for boron clusters and later seen to hold for cluster compounds in general were:

1. The structures of most clusters can be seen to be based on a well-defined series of triangular-faced polyhedra (deltahedra) or fragments thereof.
2. Whether the deltahedron was complete, as in *closo*-clusters, or a fragment with one or more nontriangular faces, as in *nido*- (nest-like) or *arachno*- (cobweb) clusters, depended on the number of electrons available to the system, which number can usually be deduced readily from the molecular formula.[5,6]

Thus, borane anions $B_nH_n^{2-}$ and isoelectronic carboranes $CB_{n-1}H_n^-$ or $C_2B_{n-2}H_n$ have the familiar *closo*- deltahedral skeletons (Figure 2-1), in which their n skeletal boron or carbon atoms are held together by $(n + 1)$ skeletal electron pairs (a pair from each BH unit and a pair from the anionic charge of $B_nH_n^{2-}$). More-open (polyhedron fragment) structures (Figure 2-1) are adopted by *nido*- and *arachno*-systems B_nH_{n+4} and B_nH_{n+6} for which there are higher numbers of electrons available to hold the skeletal atoms together: $(n + 2)$ skeletal pairs for the n skeletal atoms of *nido*-species, $(n + 3)$ skeletal pairs for *arachno*- species. In all cases, the deltahedron on which the structure is based has a number of vertices that is one fewer than the number of skeletal electron pairs available.

One can understand these relationships in that the parent *closo*-polyhedra represent the most effective ways of making good bonding use of all of the skeletal electron pairs available, bearing in mind that each BH or CH unit can contribute three atomic orbitals (one – an *sp* hybrid – pointing toward the cluster center, and two *2p* orbitals perpendicular to this, tangential to the pseudospherical surface of the cluster) (Figure 2-2).[6] The radially oriented orbitals combine in phase to generate one bonding molecular orbital (MO) of *A* symmetry. The remaining skeletal bonding MOs arise primarily from suitable combinations of the tangential atomic orbitals (AOs), stabilized in some cases by suitable (out-of-phase) combinations of appropriate radial AOs. Figure 2-3 illustrates the MOs of octahedral $B_6H_6^{2-}$ as an example.[7]

The deltahedra in Figure 2-1— the skeletal shapes of *closo*-anions $B_nH_n^{2-}$ or $CB_{n-1}H_n^-$ or neutral carboranes $C_2B_{n-2}H_n$ ($n = 5$–12)— thus represent the shapes that make the most effective use of the $3n$ AOs and $(n + 1)$ skeletal bond pairs available. Established as the thermodynamically preferred shapes for *closo*-

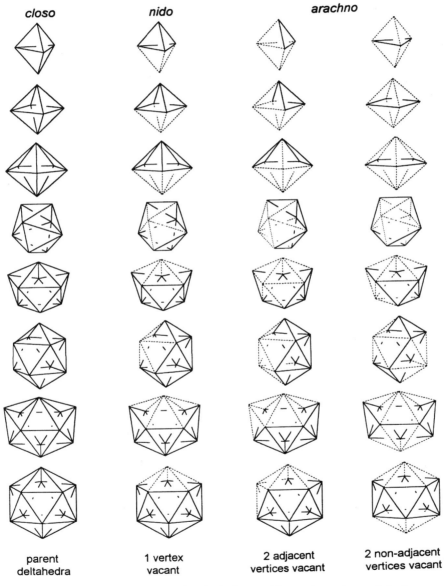

closo	nido	arachno	

| parent deltahedra | 1 vertex vacant | 2 adjacent vertices vacant | 2 non-adjacent vertices vacant |

Figure 2-1 Skeletal shapes of typical *closo*-, *nido*-, and *arachno*-polyhedra.

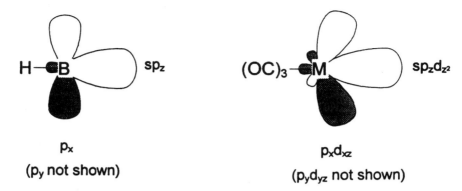

Figure 2-2 Atomic orbitals used by cluster-forming units BH and isolobal species $M(CO)_3$. Only one of the two tangential orbitals is shown in each case.

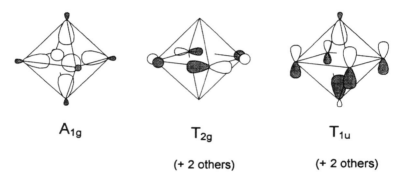

Figure 2-3 Skeletal bonding molecular orbitals of $B_6H_6^{2-}$.

borane anions and carboranes,[8–11] they set the pattern of preferred structures for all other clusters composed of units that, like the BH or CH units of boranes and carboranes, made use of three AOs (one radial, two tangential) and had $(n + 1)$ electron pairs to hold the n skeletal atoms together. Since it is the *lobal* characteristics of the atomic orbitals that influence the way they mix to form MOs, the cluster-forming units that form clusters with *closo*-shapes do not have to be isoelectronic with BH or CH units, but *isolobal* with them (Figure 2-2),[6,12–14] as effectively demonstrated by M. F. Hawthorne in his classic studies on metallacarboranes,[15,16] in which metal-carbonyl or -cyclopentadienyl residues, such as $Fe(CO)_3$ or $Co(C_5H_5)$, were shown to be able to replace BH units of carboranes. There is insufficient space here—and it would anyway be inappropriate—to survey all the structural and bonding ramifications that have been explored by studies on metallaboranes and -carboranes,[17–22] though other chapters in this book deal with selected such systems. Suffice it to say that, although isolobal replacement of BH and CH units by metal residues has afforded large numbers of *closo*-metallaboranes and -carboranes with deltahedral structures like those in Figure 2-1, different structures become possible, and have been found, where the metal residues are not strictly isolobal with the BH units they formally replace, and also in the case of *n*-atom *n*-skeletal pair (*hypercloso*) systems, where a range of new deltahedra have emerged, e.g., from the studies of Hawthorne and colleagues,[23] Greenwood,[18] Kennedy,[19] Mingos and colleagues,[24] and others (see, e.g., Chapter 3). Such systems provide a reminder that the division of the atomic orbitals on cluster-forming units into those that contribute to skeletal bonding and those that do not, though justifiable for BH or CH units, may prove an oversimplification in certain cases where transition metal units are involved.[18–25]

In the discussion below, attention is focused initially on aspects of the borane–carborane–carbocation structural and bonding pattern that are sometimes overlooked or misunderstood, as discussion of such aspects may provide reminders of the salient features of that pattern.

2.3 THE *CLOSO*-DELTAHEDRA: FRONTIER ORBITAL CONSIDERATIONS

A useful starting point is to focus on the characteristics of the deltahedra (Figure 2-1) on which borane–carborane clusters, and so many other clusters, are based. Questions commonly raised include: What is special about these polyhedra? Why don't *closo*-systems include the tetrahedron? Why do neutral boron halides B_nX_n, with *n* skeletal atoms and *n* skeletal bond pairs, have such structures?

The *closo*-deltahedra are special, first, in being deltahedra. Their triangular faces maximize the skeletal connectivities, k, of the skeletal atoms, ensuring that they are in bonding contact with as many neighbours as possible, a structural condition that minimizes the energy (maximizes the stability) of such systems.[26] Second, their skeletal connectivities are either the same for all skeletal atoms (as in the case of $B_6H_6^{2-}$ for which $k = 4$, and $B_{12}H_{12}^{2-}$, $k = 5$) or as nearly the same as possible (Table 2-1). This ensures as even a distribution of charge as

TABLE 2-1 Number of Vertices with Skeletal Connectivities k as a Function of n for *Closo*-clusters $B_nH_n^{2-}$ or $C_2B_{n-2}H_n$

	n							
k	5	6	7	8	9	10	11	12
3	2	0	0	0	0	0	0	0
4	3	6	5	4	3	2	2	0
5	0	0	2	4	6	8	8	12
6	0	0	0	0	0	0	1	0

possible. Atoms with lower k values are more negatively charged than those with higher k values (the latter can be thought of as being involved in more three center two electron, 3c2e, bonds which necessarily spread charge away from that atom more than two-center two-electron, 2c2e, bonds do).[27]

Reference to Table 2-1, which lists the number of atoms of each skeletal connectivity k for the *closo* deltahedra with 5–12 vertices, shows a nearly perfect progression from the octahedron of $B_6H_6^{2-}$ to the icosahedron of $B_{12}H_{12}^{2-}$ as 5-connected vertices progressively replace 4-connected vertices. The only anomalous structure is the octadecahedron of $B_{11}H_{11}^{2-}$ or $C_2B_9H_{11}$, which might have been expected to have 10 vertices of connectivity 5 and one of connectivity 4 instead of eight with $k = 5$, two with $k = 4$, and one with $k = 6$. The reason for this anomaly is geometrical—it is not possible to construct an 11-vertex deltahedron with only one vertex with $k = 4$, and all of the remainder with $k = 5$. One result is that $B_{11}H_{11}^{2-}$ is significantly less stable than its neighbours in the *closo*-series, $B_{10}H_{10}^{2-}$ and $B_{12}H_{12}^{2-}$.[28,29] Discussions of patterns in skeletal connectivities, particularly in *nido*- and *arachno*-species, feature prominently in Williams' chapter in this book (Chapter 1).

That the tetrahedron, the smallest possible deltahedron and, moreover ·(like the octahedron), a platonic solid of great importance in chemistry generally, is not the smallest *closo*-polyhedron in the borane anion/carborane series $B_nH_n^{2-}/C_2B_{n-2}H_n$ is worth noting. A tetrahedral anion $B_4H_4^{2-}$ is neither known nor expected. Uniquely among deltahedra, all of its vertices are of connectivity $k = 3$, the same as the number of AOs each skeletal atom contributes for skeletal bonding. This causes the symmetry-dictated interactions between the three AOs that each BH or CH unit at its vertices could supply, as in the hypothetical species B_4H_4 or C_4H_4, to generate in total six, not five, skeletal bonding MOs, of symmetries A, T, and E, and so nondegenerate, triply degenerate and doubly degenerate, respectively. Closed-shell electronic configurations therefore correspond to four or six skeletal electron pairs [rather than the five skeletal pairs that would conform with an $(n + 1)$ rule]. In localized bond terms (because the skeletal connectivity k equals the number of AOs each atom provides for skeletal bonding), these can be thought of as corresponding to four 3c2e bonds in the four faces of the tetrahedron, or six 2c2e bonds along the six edges (an "electron-precise" structure). Otherwise, the tetrahedron can be viewed as a *nido*-fragment of a trigonal bipyramid lacking a low k (axial) atom (Figure 2-1).

The tetrahedron is, of course, represented in boron cluster chemistry by the tetrahalides B_4X_4 [30,31] (and in carbon chemistry by derivatives of tetrahedrane [32]). Higher halides B_nX_n are also known that have similar *closo*-structures to those of analogous hydride anions $B_nH_n^{2-}$, and the question arises as to why these halides are exceptional in requiring only n skeletal pairs. The most thermally stable higher halides are the compounds with eight or nine boron atoms, B_8X_8 or B_9X_9, and this appears to be significant.[31,33] One can understand why such *closo*-halide clusters can exist by looking at the HOMOs of the anions $B_nH_n^{2-}$. Where these are degenerate (as in most cases), neutral species

B_nH_n or B_nX_n would have unpaired electrons in their HOMOs, which would be a source of reactivity and instability. A neutral octahedral chloride B_6Cl_6, for example, would have two unpaired electrons in the triply degenerate HOMO (Figure 2-3), and so would be expected to undergo a Jahn-Teller distortion to relieve the degeneracy, or otherwise participate in reactions that would lead to closed-shell products like tetrahedral B_4Cl_4, D_{2d} dodecahedral B_8Cl_8, or D_{3h} tricapped trigonal prismatic B_9Cl_9. That these latter two have closed-shell electronic configurations with no unpaired electrons reflects the nondegeneracies of the HOMOs of $B_8H_8^{2-}$ and $B_9H_9^{2-}$. The D_{2d} skeleton of $B_8H_8^{2-}$ can tolerate 8, 9, or (in principle) 10 skeletal pairs because the HOMO and LUMO of $B_8H_8^{2-}$ are both nondegenerate. Similarly, the D_{3h} skeleton of $B_9H_9^{2-}$ can tolerate 9, 10, or 11 skeletal electron pairs without cleavage of any polyhedron edges. This is not to say that the skeletons do not change at all with the number of skeletal electron pairs. The skeleton of B_8Cl_8 retains the $(3n - 6)$, i.e., 18 edges of $B_8H_8^{2-}$, though its overall shape is larger and less nearly spherical than that of the latter. Since $(n + 1)$ skeletal pairs are still the optimum number to hold together these n-atom clusters, the species $B_8H_8^{2-}$ and $B_9H_9^{2-}$ have more-compact, near-spherical skeletons than their neutral halide counterparts. The edge-bonding characteristics of the HOMO of $B_8H_8^{2-}$ (the orbital that is empty in B_8Cl_8) are neatly reflected by the edge lengths in the less-spherical skeleton of B_8Cl_8. Edges that are strongly bonded by the HOMO electrons of $B_8H_8^{2-}$ are longer in B_8Cl_8; those in $B_8H_8^{2-}$ for which the HOMO electrons are antibonding are shorter in B_8Cl_8. Similar comparisons can be made between $B_9H_9^{2-}$ and B_9Cl_9 (see Figure 2-4).

These are not the only structural features on which frontier orbitals shed light. There is a neat complementarity about the edge-bonding or -antibonding characteristics of the frontier orbitals — where the HOMO of $B_8H_8^{2-}$ is bonding, the LUMO is antibonding, and vice versa (the HOMO is transformed into the LUMO by rotating each p AO by 90° in the same direction, clockwise in Figure 2-4, whence edge-bonding characteristics change as follows: $\sigma \rightarrow \pi^*$, $\pi \rightarrow \sigma^*$, $\sigma^* \rightarrow \pi$, and $\pi^* \rightarrow \sigma$). Because of this, the distortion to a less-spherical structure that occurs going from 9 to 8 skeletal pairs would also be possible going to 10 skeletal pairs ($C_4B_4H_8$ or isoelectronic species) though other shapes are, of course, possible for these more-electron-rich systems. In the nine-atom *closo*-series, a bismuth cluster, Bi_9^{5+}, which formally contains 11 skeletal pairs and for which a *nido*-structure would have been expected, actually shows the same type of distortion from the near-spherical shape of $B_9H_9^{2-}$ as is shown by the nine-atom nine-skeletal pair cluster B_9Cl_9.[34] Reference to other nine-atom clusters held together by unusual numbers of skeletal electrons is made in a later section of this chapter, concerned with Zintl anions (p. 77).

The versatility in the number of skeletal electron pairs tolerated that is shown by borane clusters with eight or nine skeletal atoms is, in principle, possible also for 11-atom clusters based on the octadecahedron of $B_{11}H_{11}^{2-}$ (Figure 2-1), the frontier orbitals of which are expected to be nondegenerate (unless accidental degeneracies occur) because of the C_{2v} symmetry of this polyhedron. Neutral halides $B_{11}X_{11}$ of this shape appear possible. However, we have already seen that

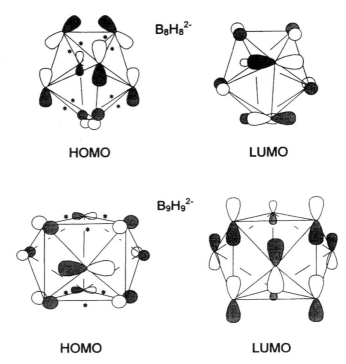

Figure 2-4　Polyhedron edge lengths in $B_8H_8^{2-}$ (B_8Cl_8) and $B_9H_9^{2-}$ (B_9Cl_9) reflect frontier orbital bonding / antibonding character. Bonds with asterisks are shorter and other bonds are longer in the halides B_nCl_n than in the hydrides $B_nH_n^{2-}$.

this shape makes less effective use of the skeletal bonding electrons than neighboring polyhedra, so neutral halides appear unlikely to become very important members of the B_nX_n family of boron subhalides.

One further feature of the chemistry of *closo*-borane anions $B_nH_n^{2-}$ that is worth noting is their low proton affinity. They are not readily protonated to anions $B_nH_{n+1}^{-}$, still less so to neutral species B_nH_{n+2}.[35] The delocalization of their skeletal electrons leaves no significant local regions of negative charge to facilitate protonation, and, indeed, their electrons are already strongly bound despite the overall anionic charge they bear. In localized bond terms, three skeletal pairs are in 2c2e bonds and $(n - 2)$ pairs are in 3c2e bonds in *closo*-clusters, though the many resonance canonical forms ensure that the skeletal electrons are evenly distributed, giving slightly more negative charge to the low-connectivity atoms that form proportionately fewer 3c2e bonds. *Nido*- and *arachno*-boranes, by contrast, offer regions of greater concentration of electronic charge around their open faces,[8,26,36,37] a point that will be taken up later.

2.4   NIDO- AND ARACHNO-CLUSTERS

The pattern by which the structures of n-atom clusters with $(n + 2)$ or $(n + 3)$ skeletal pairs are generally understood is that first popularized by Williams in 1971 [2] and subsequently elaborated,[3,4] though substantially modified by him in this volume. This is that the shapes of *nido*-systems [n atoms, $(n + 2)$ pairs] and *arachno*-systems [n atoms, $(n + 3)$ pairs] can be regarded as fragments of the *closo*-parent appropriate for their skeletal electron count, with one or two sites vacant (deboronated relative to the *closo*-parent). This deboronated polyhedral fragment (**debor**) approach derives from the earlier icosahedral fragment approach to boron cluster chemistry, whereby the structures of the earliest characterized boranes (apart from B_5H_9) could be seen as placing their boron atoms on fragments of an icosahedron, the shape that was then rare in chemistry, though known in the structures of elemental boron, boron carbide, and some metal borides.[38] (The apparently exceptional B_5H_9, whose square pyramidal skeleton could be viewed as being based on an octahedron with one vacant vertex, appeared less exceptional in that B_6 octahedra were also common in metal boride structures.)

An alternative approach to cluster shapes, the **seco** principle, focuses on the 2-center links between individual skeletal atoms.[39] Instead of progressing from an n-atom *closo*-parent structure by successive removal of atoms to generate *nido*- and *arachno*-residues, it starts with a *closo*-parent with the same number of atoms as are present in the fragment, reaching the open *nido*- or *arachno*-structure by successive edge-breaking steps (effectively progressing by diagonal rather than horizontal steps between skeletons in Figure 2-1). In both the **debor** and **seco** approaches, the number and distribution of the 2-center links (not necessarily 2c2e *bonds*) between skeletal atoms in *nido*-, *arachno*- or yet more open clusters are seen to be important. In the **debor** approach, the emphasis is on the retention of a recognizable three-dimensional fragment of the original polyhedron of the parent *closo*-species $B_nH_n^{2-}$ from which BH^{2+} units have been removed (to be replaced by suitably dispersed H^+ ligands around the open face). In the **seco** approach, as links in the parent species are severed, the fragment left is free to relax to a different three-dimensional shape provided that specified 2-center links are retained.

These two approaches can be rationalized by recourse to molecular orbital or localized (3c2e and 2c2e) bond treatments. The **debor** approach is better understood using MO arguments. The **seco** approach, focusing on localized links, is better understood in localized bond terms. A brief summary of the rationalizations that these approaches permit may be helpful here.

2.5   THE DEBOR APPROACH; MO IMPLICATIONS

If one considers the removal of a BH^{2+} unit from the octahedral *closo*-species $B_6H_6^{2-}$ to generate a *nido*-$B_5H_5^{4-}$ residue as an example of the **debor** approach to a *nido*-species, one can see from Figure 2-3 that loss of a BH^{2+} unit will affect all three types of skeletal bonding MOs. The A_{1g} combination of radial orbitals will lose one of the six contributing radial AOs. Two of the T_{2g} and two of the T_{1u} MOs will lose one of the four contributing tangential AOs. In all cases, the remaining interactions are, of course, still bonding. There is no change in the total number of bonding orbitals, though their energies and symmetry labels will change. The unchanged number of skeletal bonding electron pairs is therefore intelligible. Moreover, since the four protons formally added to the *nido*-$B_5H_5^{4-}$ fragment to convert it into the neutral borane B_5H_9 are arranged symmetrically about the open face, over the centers of the B—B edges, and each H^+ contributes a $1s$ AO, then three of the four symmetry-dictated combinations of these four hydrogen AOs mimic the lobal characteristics of the BH^{2+} unit lost, and so compensate for the missing BH^{2+} unit. In a sense, the BH^{2+} unit is replaced by a square isolobal $(H_4)^{4+}$ unit in going from $B_6H_6^{2-}$ to B_5H_9 (Figure 2-5).[6,40]

Similar arguments can be used to rationalize the relationship between other *closo*-$B_nH_n^{2-}$ species and daughter *nido*-$B_{n-1}H_{n+3}$ fragments, in which the extra four hydrogen atoms replacing the missing BH unit are disposed in a suitable manner over the four-, five-, or even six-membered open face to provide replacements for the missing BH radial and tangential AOs.

Since, in considering the *closo*-polyhedra, it was noted that these are deltahedra in order to maximize the number of bonding contacts, it may seem strange that the preferred *nido*-structure is generally that in which a BH^{2+} unit has been removed from the *highest connectivity* vertex. If maximizing the number of two-center links between skeletal atoms is so important for *closo*-species, why should the maximum possible number of links be broken in forming the *nido*-residue? This question can be answered in two ways. First, since an uneven electron distribution is expected in a *closo*-cluster containing more than one type of skeletal atom (i.e., for all anions $B_nH_n^{2-}$ except $B_6H_6^{2-}$ and $B_{12}H_{12}^{2-}$)—BH units of higher connectivities are in regions of less negative charge than BH units of lower connectivities—then one might expect removal of a BH^{2+} unit to be easier from a higher connectivity site. Second, removal of a BH unit as BH^{2+} formally leaves to the cluster the skeletal electron pair that the neutral BH unit brought to the cluster in the first place. That electron pair, accommodated in the fragment on the open face, will be attracted to that open face the more skeletal nuclei there are adjacent to that face. Expressed another way, the frontier orbitals of a *nido*-fragment $B_{n-1}H_{n-1}^{4-}$, which concentrate electronic charge around the open face, will be stabilized—have lower energies—the more skeletal atom nuclei there are surrounding that open face.[26,36,37,40–42]

Extension of this type of argument from a *nido*- to an *arachno*-fragment leads to an apparent problem. If it pays to sever as many links as possible going from a *closo*-parent anion $B_nH_n^{2-}$ to a *nido*-fragment $B_{n-1}H_{n-1}^{4-}$, why is the next BH^{2+} unit lost, going on to an *arachno*-fragment $B_{n-2}H_{n-2}^{6-}$, removed from a vertex *adjacent* to the first vacant vertex, when loss of a BH^{2+} unit from a *nonadjacent* vertex would sever more links? (Illustrations of possible *arachno*-skeletons leaving *nonadjacent* high k vertices vacant are shown in the last column of Figure 2-1.)

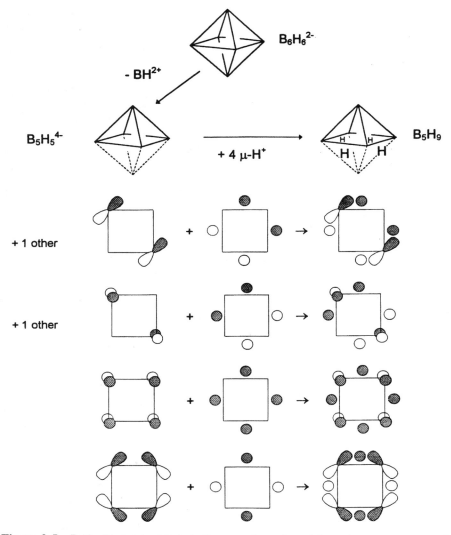

Figure 2-5 B_5H_9: Skeletal orbitals in the open face viewed from the vacant vertex of $B_5H_5^{4-}$ on removing BH^{2+} from $B_6H_6^{2-}$ (see Figure 2-3) and in B_5H_9 showing how $(\mu^2\text{-H})_4$ atomic orbitals can match the lobal characteristics of $B_5H_5^{4-}$ molecular orbitals.

The answer to this question lies with the extra bridging or terminal (*endo*) hydrogen atoms in *arachno*-boranes B_nH_{n+6}. As already pointed out for *nido*-boranes B_nH_{n+4}, the four *endo*-hydrogen atoms—the μ^2 hydrogen atoms in B_5H_9, for example—stabilize the system by providing a set of symmetry-adapted orbitals that replace the radial and tangential orbitals of the lost BH^{2+} unit. The six *endo*-hydrogen atoms in an *arachno* boranes, such as B_4H_{10} or B_5H_{11}, act more effectively, in this respect, to stabilize the polyhedral $B_{n-2}H_{n-2}^{6-}$ fragment if there is but one open face (as results from the loss of two adjacent BH^{2+} units from the *closo*-$B_nH_n^{2-}$ parent) than if there are two open faces, as would result from the loss of a *trans*-related pair of BH^{2+} units from $B_6H_6^{2-}$, or by loss of the axial pair of BH^{2+} units with $k = 5$ from the pentagonal bipyramidal $B_7H_7^{2-}$ (Figure 2-1). Molecular orbital calculations have been carried out to assess the relative stabilities of the various possible *arachno*-fragments $B_{n-2}H_{n-2}^{6-}$ that can result from removal of either adjacent or nonadjacent BH^{2+} units from parent *closo*-borane anions $B_nH_n^{2-}$ ($n = 5$–12).[36] These calculations suggest that the most stable skeletons for $B_{n-2}H_{n-2}^{6-}$ are generally those that result from loss of two *high-connectivity nonadjacent* BH^{2+} units. However, when the extra six H^+ ions are added to convert these hypothetical species $B_{n-2}H_{n-2}^{6-}$ into neutral molecules $B_{n-2}H_{n+4}$ of the type actually known as *arachno*-boranes, like B_4H_{10} or B_5H_{11}, then the stabilizing effect of these six *endo*-hydrogen atoms is greater if they are grouped around one open face than if they are grouped around two.[36,37] The importance of this effect is seen by comparing B_5H_{11} (tent-shaped arrangement of boron atoms) with the formally analogous cyclopenta-dienyl anion $C_5H_5^-$ (pentagonal arrangement of carbon atoms), or B_6H_{12} with benzene (the latter is formally derived from an alternative eight-vertex polyhe-dron to the D_{2d} dodecahedron familiar to boron chemists—a hexagonal bipyra-mid of D_{6h} symmetry—by loss of the two nonadjacent high-connectivity sites, see Figure 2-6 below). The difference is not simply that carbon and boron atoms have different site preferences (in carboranes, carbon atoms occupy lower connectivity sites, where the electron density is higher, whilst the more elec-tropositive boron atoms occupy the higher connectivity sites, where the electron density is lower), but reflects the influence that hydrogen atoms can have on skeletal structure. The more *endo*-hydrogen atoms there are to accommodate, the greater is the need for the cluster skeleton to become asymmetrical, building up charge on the open face of the $B_{n-2}H_{n-2}^{6-}$ residue to accommodate the extra six H^+ ions needed to produce the neutral molecule $B_{n-2}H_{n+4}$. There are many other examples elsewhere in cluster chemistry of the influence that hydrogen or other ligands can have on the skeletal shape of particular clusters; see, e.g., Refs 43–47. Within the borane–carborane–carbocation family, it is important to remember the effect of *endo*-hydrogen ligands. Otherwise, general rules drawn from the known behavior of boranes and carboranes with a low carbon content will be found to break down in carbon-rich systems.

Comparing B_5H_{11} with $C_5H_5^-$, one other point is worth making. This is that although all eight skeletal electron pairs in the cyclopentadienyl anion $C_5H_5^-$ (five pairs in 2c2e C—C bonds, three pairs in the aromatic π-system) undoubt-edly play a skeletal *bonding* role, this is not the case in B_5H_{11}, in which two or three of the skeletal pairs (depending whether one includes the pair bonding the unique *endo*-hydrogen atom on the apical boron atom, which can be regarded as either *endo* terminal or semibridging) bond terminal *endo*-hydrogen atoms; in the tent-shaped anion $B_5H_5^{6-}$, they would be lone pairs on atoms on the open

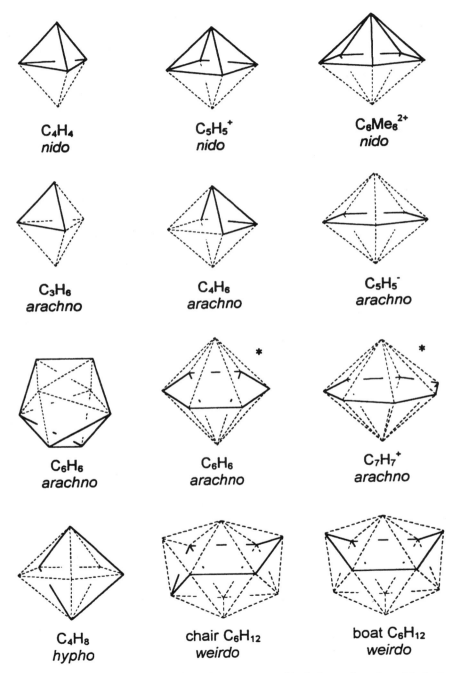

Figure 2-6 Neutral and ionic hydrocarbon systems with skeletons based on deltahedra including bipyramids (with asterisks) not normally seen in borane clusters.

face. This has led some workers [48] to misunderstand the conventional skeletal electron count, which is concerned with *all* of the atoms on the pseudo-spherical surface on which the boron atoms of boranes $B_{n-1}H_{n+3}$ (*nido*), $B_{n-2}H_{n+4}$ (*arachno*), $B_{n-3}H_{n+5}$ (*hypho*), etc., lie, in the same way that Lipscomb's *styx* rules [8] are concerned with *all* of these atoms. Since the terminal *endo*-hydrogen atoms are located on the pseudo-spherical surface of $B_nH_n^{2-}$ near the site from which the BH^{2+} units have been lost, and the skeletal electron count is concerned with the bonding between *all* of the atoms on that surface, then it is logical and appropriate to include in the formal skeletal electron count the electrons that bind these *endo*-terminal hydrogen atoms, just as one also includes the electrons that bind the *(endo)* bridging hydrogen atoms, even though, in the case of the *endo*-terminal BH bond pairs, the electrons are nonbonding as far as boron–boron interactions are concerned.

Echoes of the borane–carborane–carbocation structural pattern in relatively electron-rich systems are common, where compounds have skeletons that are recognizable fragments of the *closo*-borane polyhedra, as in the case of cyclohexane C_6H_{12}, the boat and chair forms of which can be seen by leaving five sites vacant of the 11-vertex *closo*-polyhedron of $C_2B_9H_{11}$, whilst cyclobutane C_4H_8, treated as a *hypho*-species based on a pentagonal bipyramid with three vacant equatorial sites, is correctly predicted to have a folded ring structure deviating by 36° from planarity.[49] These and a few other hydrocarbon systems that reflect the borane–carborane structural pattern are shown in Figure 2-6.

To summarize this discussion, MO treatments can be used to help one understand how successive removal of BH^{2+} units from *closo*-parents $B_nH_n^{2-}$ leave a residue that can accommodate the same total number of skeletal electron pairs, even though, in the case of *arachno*-boranes $B_{n-2}H_{n+4}$, some of these skeletal electron pairs end up bonding *endo*-hydrogen atoms terminally to selected boron atoms. The **debor** approach to *nido-, arachno-,* and ultimately more-open borane structures has considerable merit and in our opinion should not be discarded lightly.

2.6   THE SECO APPROACH: LOCALIZED BONDING IMPLICATIONS

In contrast to the **debor** approach, in which the emphasis is on the three-dimensional shapes of *nido-* or *arachno*-species as fragments of *closo*-parent deltahedra, in the **seco** approach the emphasis is on the number of bonding contacts (polyhedron edges) that can be accounted for by the number of skeletal bond pairs available. This, in turn, is explored most easily by describing the bonding in terms of 2c2e and 3c2e localized bonds rather than in terms of molecular orbitals. However, there is a problem caused by the large number of ways in which the localized bonds can be distributed around the polyhedra,[27] so it is better to look at the total number of bonds than at their location, at least in the case of *closo*-species as exemplified by anions $B_nH_n^{2-}$.

In descriptions of the bonding in such systems in terms of 2c2e BB bonds and 3c2e BBB bonds, if each boron atom is involved in three skeletal bonds (using all three AOs available for skeletal bonding), then Lipscomb's *styx* rules (*s* BHB, *t* BBB, *y* BB, and *x* BH bonds) [8] lead to the conclusion that the total of $(n + 1)$ skeletal bond pairs must be used in three 2c2e BB bonds ($y = 3$) and $(n - 2)$ 3c2e BBB bonds ($s = x = 0$ since no *endo*-hydrogen atoms need to be accommodated in BHB or *endo*-BH bonds).[27] These skeletal bonds can account for a total of $3 + 3(n - 2) = 3(n - 1)$ polyhedron edges, since a 2c2e bond can account for one edge and a 3c2e bond can account for three edges. Delta-hedra with n vertices have $3(n - 2)$ edges, so all edges can be accounted for.

Adding a pair of electrons to generate a *nido*-species $B_nH_n^{4-}$, and retaining the condition that each boron atom is involved in three skeletal bonds, leads to the conclusion that there must be six 2c2e BB bonds and $(n - 4)$ 3c2e bonds. The new total of $(n + 2)$ skeletal bond pairs can now account for a maximum number of polyhedron edges of $6 + 3(n - 4)$, i.e., $3(n - 2)$ edges, a reduction of three in the number of edges that can be accounted for, though still just sufficient, in principle, for the $3(n - 2)$ edges of an n-vertex deltahedron. In practice, how-ever, satisfactory distributions of these bonds may not always be possible.[27] For *arachno*-species $B_nH_n^{6-}$, for which the *styx* rules require nine 2c2e BB bonds and $(n - 6)$ 3c2e BBB bonds to be used, the maximum number of polyhedron edges that can be accounted for is $9 + 3(n - 6)$, i.e., $3(n - 3)$ edges, certainly too few for a *closo*-deltahedral geometry to be retained.

There is thus a reduction by three in the total number of bonding contacts (polyhedron edges) that can be accounted for as an electron pair is added, though it is not immediately obvious that *any* edges need to be cut in going from $(n + 1)$ to $(n + 2)$ skeletal pairs, and fewer than three edges may need to be cut, in practice, in going from a *closo*- to a *nido*- structure, or from a *nido*- to an *arach-no*-structure. For example, on going from a *nido*-system $B_nH_n^{4-}$ to an *arachno*-system $B_nH_n^{6-}$, it becomes realistic to consider converting a 2c2e BB bond on one polyhedron edge into two lone pairs (one on each of the atoms originally connected) when the extra skeletal electron pair is added (the lone pairs become terminal *endo*-BH links in the neutral molecule B_nH_{n+6}). This uncertainty about the number of polyhedron edges that need to be cut on adding a pair of electrons, and the need to allow the resulting network of atoms to rearrange to a different three-dimensional structure from that of the *closo*- or *nido*- parent, in our opinion reduces the value of the **seco** approach compared with the **debor** approach to borane-type clusters, though both Williams and Hermánek illustrate other respects in which it is of value in other chapters of this volume (pp. 3 and 117).

2.7 METAL CARBONYL CLUSTERS RELATED TO BORANE CLUSTERS

The very wide variety of other cluster-forming atoms or groups (generally metallic, though some nonmetallic) that can replace the BH or CH units of boranes and carboranes is evident from books and reviews of metallaborane and related chemistry.[10,17–22] Hawthorne showed that transition metal units ML_n could replace selected BH units of carboranes, that the pentagonal C_2B_3 open face of *nido*-carborane anions $C_2B_9H_{11}^{2-}$ could coordinate to metal ions in a similar manner to cyclopentadienide anions $C_5H_5^-$, and that, like the latter, they could form sandwich complexes with suitable naked metal cations in dicarbollide complexes $[M(C_2B_9H_{11})_2]^{n-}$.[16,50–52] These studies led to the recognition of the isolobal relationship between BH units and trigonal pyramidal $M(CO)_3$ units (M = Fe, Ru, or Os), or between CH units and related pyramidal $M(CO)_3$ units of the cobalt subgroup metals (Co, Rh, Ir).[5,6,13,14] This, in turn, allowed metal-carbonyl clusters themselves, $M_x(CO)_y^{n-}$, to be recognized as members of the same cluster family as boranes and carboranes,[5] and it has become customary to classify them [and related hydrides $M_x(CO)_yH_z$, anions $M_x(CO)_yH_z^{n-}$ etc.] as *closo-*, *nido-*, *arachno-*, or other cluster types according to the numbers of skeletal electrons they contain, although the range of polyhedra is less evenly represented.[6,13,43–47] Triangular M_3, tetrahedral M_4, and octahedral M_6 systems tend to predominate, formally containing six, six, and seven skeletal bond pairs, respectively. Although there is not room to discuss their structures and bonding in detail here, it is worth noting the ways in which metal carbonyl clusters resemble or differ from borane-type clusters.

Since transition metals normally make as full use as possible of all nine of their valence shell AOs, then if they are to reserve three of these for cluster bonding, the remaining six AOs are available to bond *exo*-ligands or accommodate nonbonding electrons. The electron count used to deduce the number of skeletal electrons for a unit ML_n is thus $(v + x - 12)$, where v is the number of valence shell electrons and x is the number supplied by the n *exo*-ligands L. This may be contrasted with the skeletal electron count of $(v + x - 2)$ for main group elements like boron or carbon, which are assumed to use four valence shell AOs. Units $M(CO)_3$ (M = Fe, Ru, or Os) or $M'(C_5H_5)$ (M' = Co, Rh, or Ir) are thus sources of two skeletal electrons like BH units, whilst units $M'(CO)_3$ or $M''(C_5H_5)$ (M'' = Ni, Pd, or Pt), like CH units, provide three skeletal electrons.

One simple respect in which metal carbonyl clusters differ from borane clusters is that they are bigger—big enough, even in the case of octahedral clusters, to accommodate a core first- row atom, such as carbon, at the center of the polyhedron. Second-row or even heavier atoms can be accommodated in larger clusters. Some examples are shown in Figure 2-7.[53–63] Such a core atom can make all of its valence shell electrons available for skeletal bonding and so supplement those made available by the surrounding ML_n units. Metal carbonyl carbide clusters $M_xC(CO)_y$ are relatively common—they are accessible by pyrolysis of lower nuclearity metal carbonyls, and their core carbon atoms result from a disproportionation reaction of carbonyl ligands, $2CO \rightarrow C + CO_2$. There is no room for similar core atoms to be accommodated in the smaller borane clusters, though, in principle, there should be enough room in icosahedral or larger borane clusters. (However, in such larger clusters, a first-row element like boron or carbon, with only four valence shell AOs, has less scope for stabilizing the skeletal bonding MOs than a heavier element with d AOs would have.)

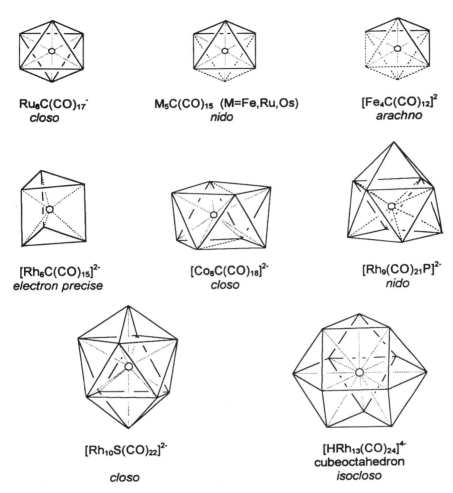

Figure 2-7 Examples of metal-carbonyl clusters containing core atoms.

Another respect in which metal-carbonyl clusters differ from boranes and carboranes is in their capacity to accommodate atoms capping triangular faces of the fundamental polyhedron, as in the case of the hexanuclear osmium-carbonyl $Os_6(CO)_{18}$, which has a capped trigonal bipyramidal structure (bicapped tetrahedral structure) [64], and the heptanuclear $Os_7(CO)_{21}$, which has a capped octahedral structure [65] (Figure 2-8). Their capping $Os(CO)_3$ units make available the two electrons apiece that they can offer for cluster bonding by using their three cluster orbitals to lock onto the three metal atoms in the capped face, exploiting, for the purpose, electrons that would otherwise not be involved in cluster metal–metal bonding. (Each of the capped atoms effectively makes a fourth AO available to the capping atom for cluster bonding.) Such

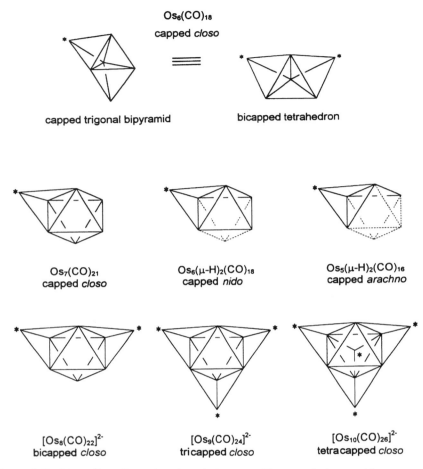

Figure 2-8 Examples of metal-carbonyl clusters with capped *closo*-, *nido*-, or *arachno*-skeletons. Capping atoms are labeled by asterisks.

n-atom *n*-skeletal pair clusters with capped structures may be contrasted with the neutral boron halides B_8Cl_8 and B_9Cl_9 discussed above, where a distorted *closo* ("*hypercloso*," highly electron deficient) geometry is adopted rather than a capped structure based on a smaller deltahedron. Capping of B_3 triangular faces of *closo*-borane clusters is unlikely, in *n*-atom *n*-skeletal pair boron clusters B_nX_n ,etc., because there are no spare electrons on the boron atoms to use for capping, though metal units can more distantly cap such faces by forming B—H—M 3c2e bonds.[20,21,66]

This capacity on the part of metals to form capped structures allows them to form a wide range of clusters that, though based on the polyhedron expected from a skeletal electron count (most commonly, an octahedron or a fragment thereof), can alternatively be viewed as a fragment of the bulk metal. Some illustrative examples are shown in Figure 2-8.[67]

A further respect in which metal carbonyl clusters differ from borane clusters is in the frequency with which they adopt structures that can be accounted for by assuming that each metal–metal link corresponds to a 2c2e M—M bond, and assuming that each metal atom uses all nine valence shell AOs [i.e., assuming the 18 electron or effective atomic number (EAN) rule].[67,68]

Despite these respects in which metal-carbonyl clusters differ from borane clusters, polyhedron skeletal electron pair theory (PSEPT),[6,13] i.e., electron counting as in borane clusters, remains the approach by which one can most readily either account for known structures or predict structures from formulae, though in the latter exercise it is vital to consider the range of isomeric structures possible (e.g., capped *nido*, or even bicapped *arachno*, instead of *closo*).

The metal for which the widest range of metal-carbonyl clusters has been structurally characterized is osmium, thanks, in particular, to the attention that osmium-carbonyl cluster chemistry has received from B.F.G. Johnson, J. Lewis, M. McPartlin, and P. Raithby.[43,67] A recent study [68] of their structures and the effectiveness with which they use the skeletal electrons available for metal–metal bonding has shown this to vary smoothly with the number of additional electrons per metal atom provided by the ligands (CO, H, or other ligands), core atoms, and anionic charges for the whole series of compounds structurally characterized. This smooth relationship is expected to prove of value in future discussions of the relative stabilities of such metal cluster systems. Electron numbers clearly remain important in this area of cluster chemistry.

It may be worth adding here that late transition metal-carbonyl clusters, in which each metal atom uses three AOs for cluster bonding, may be contrasted with octahedral early transition metal *halide* or related clusters ($Nb_6Cl_{12}^{2+}$, $Mo_6Cl_8^{4+}$, etc.) in which each metal atom uses *four* AOs for metal–metal bonding.[6,47,69–71] These latter octahedral metal halide clusters, in which each metal atom has a skeletal connectivity of four (equal to the number of skeletal AOs), are typically held together by eight 3c2e MMM bonds in the octahedral faces or 12 2c2e MM bonds along the octahedral edges. Clusters like $Nb_6Cl_{12}^{2+}$ and $Mo_6Cl_8^{4+}$ thus play roles in octahedral cluster chemistry analogous to those played by B_4Cl_4 and P_4 or tetrahedrane in tetrahedral cluster chemistry, with skeletal electron pair numbers that match the numbers of polyhedron edges or faces.

2.8 ORGANOMETALLIC CLUSTER SYSTEMS

Another area of cluster chemistry containing many echoes of the borane structural and bonding pattern is that of mixed metal–carbon clusters, i.e., that branch of organometallic chemistry in which metal ions bind to the (π)-bonding electrons of unsaturated organic ligands.[72] Figure 2-9 shows some representa-

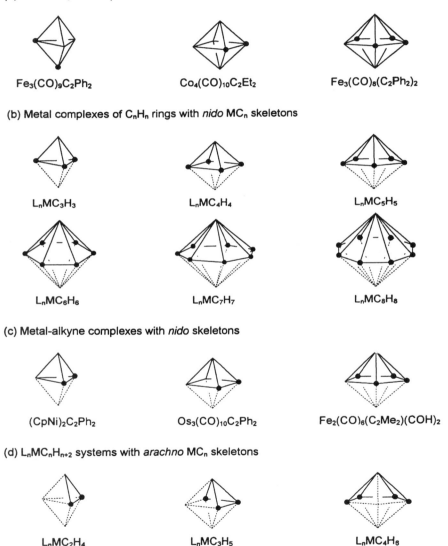

(a) Metal-alkyne complexes with *closo* skeletons

$Fe_3(CO)_9C_2Ph_2$ $Co_4(CO)_{10}C_2Et_2$ $Fe_3(CO)_8(C_2Ph_2)_2$

(b) Metal complexes of C_nH_n rings with *nido* MC_n skeletons

$L_nMC_3H_3$ $L_nMC_4H_4$ $L_nMC_5H_5$

$L_nMC_6H_6$ $L_nMC_7H_7$ $L_nMC_8H_8$

(c) Metal-alkyne complexes with *nido* skeletons

$(CpNi)_2C_2Ph_2$ $Os_3(CO)_{10}C_2Ph_2$ $Fe_2(CO)_6(C_2Me_2)(COH)_2$

(d) $L_nMC_nH_{n+2}$ systems with *arachno* MC_n skeletons

$L_nMC_2H_4$ $L_nMC_3H_5$ $L_nMC_4H_8$

Figure 2-9   Metal–hydrocarbon π-complexes with *closo-*, *nido-* or *arachno*-skeletons, showing positions of carbon atoms by black circles and vacant vertices by broken lines (metal atoms occupy other vertices).

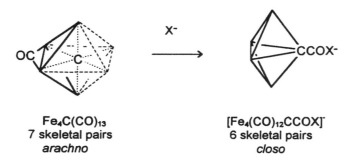

Fe$_4$C(CO)$_{13}$
7 skeletal pairs
arachno

[Fe$_4$(CO)$_{12}$CCOX]$^-$
6 skeletal pairs
closo

Figure 2-10 Conversion of a core carbon atom into a vertex skeletal atom in an Fe$_4$C cluster system.

tive examples. The commonest examples are penta-hapto (η^5)-cyclopentadienyl–metal complexes, the pentagonal pyramidal C$_5$M structures of which resemble the *nido*-structure of B$_6$H$_{10}$, whilst η^4-butadiene complexes have structures (and skeletal electron counts) like that of *arachno*-B$_5$H$_{11}$. Although these and other organometallic π-complexes are usually treated from a metal-oriented viewpoint, by considering how many (π) electrons the organic ligands contribute toward the 18 normally associated with the metal's valence shell, it is often helpful to look at the overall polyhedral shape of the M$_x$C$_y$ polyhedron. One can, for example, deduce the most likely orientation of alkyne ligands with respect to metal clusters [73] by treating the complexes as composite M$_x$C$_y$ clusters, as in the case of species such as (PhCCPh)Fe$_3$(CO)$_9$ (which has a *closo*-trigonal bipyramidal Fe$_3$C$_2$ skeleton, with one axial and one equatorial carbon atom) [74] or PhCCPhOs$_3$H$_2$(CO)$_9$ (*nido*-Os$_3$C$_2$ skeleton, with basal carbon atoms).[75,76] A family of Fe$_4$C clusters is known in which the carbon can occupy either a core or vertex site, and containing a butterfly-shaped (*arachno*) Fe$_4$ unit (Figure 2-10).[55,77–79] The core carbon atom of the carbonyl carbide complex Fe$_4$C(CO)$_{13}$, which contains four Fe(CO)$_3$ units and one μ^2-carbonyl ligand, moves out to occupy a vertex site upon reaction with a variety of reagents (ROH, R$_2$NH, Et$_3$BH$^-$, or MeI) which convert that two electron, metal-bridging carbonyl ligand into a one-electron ligand—COX (X = OR, NR$_2$, H, or Me, respectively)—attached to that carbon atom (Figure 2-10).

2.9 ZINTL ANIONS AND RELATED NAKED MAIN GROUP CLUSTERS

A further expanding area of cluster chemistry in which borane structural patterns have been found to be both relevant and helpful is in the chemistry of "naked" (ligand-free) anionic or cationic clusters formed by elements below or to the right of boron in the Periodic Table.[80–85] Such cluster anions as Sn$_9^{4-}$,

Pb_9^{4-}, Bi_3^{3-}, and Bi_7^{3-} tend to be referred to as Zintl anions, and the alloy phases in which they have been structurally characterized are referred to as Zintl phases, in recognition of the pioneering work carried out on these and related systems by Zintl,[86,87] who prepared their solutions in liquid ammonia by dissolving the appropriate metals in reducing solutions of alkali metals in liquid ammonia. The range of such systems now known is very extensive and rapidly growing. Refs [82–85] include recent surveys of their structures and bonding.

Such systems have lone pairs of *exo*-electrons where borane clusters have *exo*-B—H bonds, and they generally have the characteristic electron counts expected by analogy with related boranes. For example, for anionic clusters M_n^{x-} formed by the heavier members, M, of Group 13 (Al, Ga, In, or Tl), the anionic charge for a *closo*-structure is $-(n + 2)$; *nido*-and *arachno*-structures are expected for even higher negative charges, $-(n + 4)$ and $-(n + 6)$, respectively. Such high negative charges can only be tolerated if such clusters are surrounded by charge-balancing cations, as in alloys of these metals with alkali metals. For Group 14 elements (Si, Ge, Sn, Pb), anionic charges of -2, -4, and -6 correspond to *closo*-, *nido*-, and *arachno*-structures, respectively, whilst for Group 15 elements (P, As, Sb, Bi) the charges expected are $(n - 2)$ (*closo*), $(n - 4)$ (*nido*), and $(n - 6)$ (*arachno*). Unlike the Group 13 and 14 systems, these last Group 15 systems may be anionic or cationic, depending on the number of atoms involved; most known examples are cationic. A few cationic Group 16 systems are also known.

As in other areas of cluster chemistry, tetrahedral tetranuclear systems held together by six skeletal bond pairs are particularly important. These are formally *nido*-systems but are actually electron-precise clusters best described by a 2c2e bond along each edge, and exemplified by neutral molecules such as P_4 or anions M_4^{4-} (M = Si, Ge, Sn, or Pb) [88], or M_4^{8-} (when M is from Group 13), [89] or mixed metal anionic systems such as $[TlSn_3]^{5-}$.[90] Tetranuclear clusters held together by seven skeletal bond pairs, as in Se_4^{2+}, Te_4^{2+}, [91,92] Sb_4^{2-}, or Bi_4^{2-} [93,94] have square-planar (*arachno*) structures. Illustrations of these and other *closo*-, *nido*-, or *arachno*-systems held together by six or seven skeletal pairs are found in Figure 2-11.[95–100]

Among larger naked main group clusters, the 11 skeletal pair family is represented by *closo*-, *nido*-, and *arachno*-clusters with shapes based on the bicapped Archimedean antiprism, which is complete in the case of such *closo*-systems as Ge_{10}^{2-} [101] and $TlSn_9^{3-}$ [102] (Figure 2-11). Interestingly, the *nido*-clusters M_9^{4-} (M = Ge, Sn, or Pb) [103–105] have monocapped antiprismatic structures, leaving a low-connectivity site vacant, and the *arachno*-Bi_8^{2+} has a square antiprismatic structure, leaving both low-k sites vacant.[85,106]

Another category of cluster already well represented in naked metal cluster chemistry is the nine-atom cluster with a regular or slightly distorted D_{3h} tricapped trigonal prismatic geometry (Figure 2-11). Ideally, drawing on the analogy with $B_9H_9^{2-}$, such clusters should be held together by 10 skeletal electron pairs, though more such clusters have been characterized with more than 10 skeletal pairs. Examples include Bi_9^{5+}, with 11 skeletal pairs,[103] and the odd-

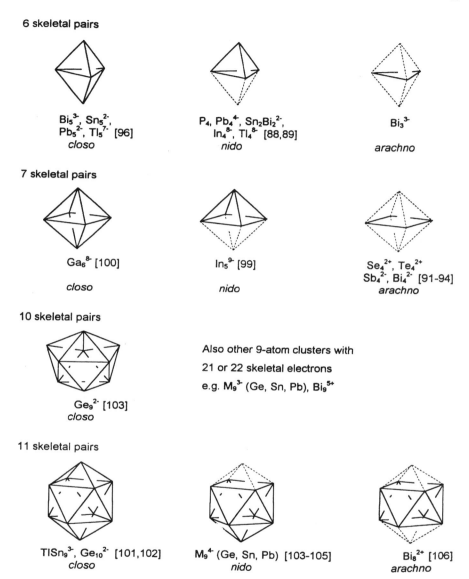

6 skeletal pairs

Bi_5^{3-}, Sn_5^{2-}, Pb_5^{2-}, Tl_5^{7-} [96]
closo

P_4, Pb_4^{4-}, $Sn_2Bi_2^{2-}$, In_4^{8-}, Tl_4^{8-} [88,89]
nido

Bi_3^{3-}
arachno

7 skeletal pairs

Ga_6^{8-} [100]
closo

In_5^{9-} [99]
nido

Se_4^{2+}, Te_4^{2+}, Sb_4^{2-}, Bi_4^{2-} [91–94]
arachno

10 skeletal pairs

Ge_9^{2-} [103]
closo

Also other 9-atom clusters with 21 or 22 skeletal electrons e.g. M_9^{3-} (Ge, Sn, Pb), Bi_9^{5+}

11 skeletal pairs

$TlSn_9^{3-}$, Ge_{10}^{2-} [101,102]
closo

M_9^{4-} (Ge, Sn, Pb) [103–105]
nido

Bi_8^{2+} [106]
arachno

Figure 2-11 Examples of naked ionic clusters formed by main group elements.[82,85]

electron systems M_9^{3-} [103] (M = Ge, Sn, or Pb), with 21 electrons formally available for skeletal bonding.[105,107–109] It was noted earlier that the non-degeneracy of the frontier MOs of $B_9H_9^{2-}$ (see Figure 2-4) allow the D_{3h} tricapped trigonal prismatic "*closo*" shape to tolerate from 18 to 22 skeletal electrons, and the existence of these relatively electron-rich members of this cluster family has allowed the distortion caused by the extra electrons (an elongation of the trigonal prism) to be probed in some detail.[85] Some *"hypercloso"* 11-atom *pentacapped* trigonal prismatic clusters In_{11}^{7-} and Tl_{11}^{7-} have been identified in alloys of these Group 13 elements with alkali metals.[110]

Many other Zintl-anionic cluster systems are known than are shown in Figure 2-11, including numerous systems containing M_{12} icosahedra held together by 13 skeletal bond pairs (see, e.g., Ref. 111), and various fused icosahedral systems [112] as examples of types as yet unknown in borane cluster chemistry except in one case in the structure of boron itself, in the β-rhombohedral form of which are sets of icosahedra sharing common faces, linking together the B84 "samson polyhedral" units therein.[112] It should be stressed, however, that most of the "naked" main group clusters discussed here have been found as constituents of alloy lattices, or low-oxidation-state halides, and relatively few have been shown to be capable of independent existence, as in the case of the original anions prepared and characterized in liquid ammonia solution by Zintl himself.[86,87]

2.10   CONCLUDING COMMENTS

In this chapter, we have attempted to show the range of cluster systems that have structures similar to or closely related to, those familiar to boron chemists in the structural patterns first documented by R. E. Williams. We have drawn attention to ways in which structures can be interpreted either by MO or localized bond treatments, to systems or features that may appear anomalous, and to the importance of ligands, particularly of *endo*-hydrogen atoms, in influencing the skeletal shapes of borane clusters. Knowledge of borane patterns can prove helpful to metal cluster chemists, to organometallic chemists studying metal–hydrocarbon π-complexes, and to solid-state chemists probing the intricacies of alloy structures. Though the core and capping atoms common in metal cluster chemistry provide reminders of the significant respects in which cluster systems may differ, the skeletons in the borane closet, originally viewed as rule-breakers, have long been widely recognized as patternmakers for that important area of chemistry where electrons are in relatively short supply.[113]

REFERENCES

1. Williams, R. E., *Adv. Organometal. Chem.*, **36**, 1, 1994.
2. Williams, R. E., *Inorg. Chem.*, **10**, 210, 1971.
3. Williams, R. E., *Adv. Inorg. Chem. Radiochem.*, **18**, 67, 1976.
4. Williams, R. E., *Chem. Rev.*, **92**, 177, 1992.
5. Wade, K., *Chem. Commun.*, 792, 1971.
6. Wade, K., *Adv. Inorg. Chem. Radiochem.*, **18**, 1, 1976.

7. Longuet-Higgins, H. C., *Quart. Rev. Chem. Soc.*, **11**, 121, 1957.

8. Lipscomb, W. N., *Boron Hydrides*, Benjamin, New York, 1963.

9. Muetterties, E. L. and Knoth, W. H., *Polyhedral Boranes*, Dekker, New York, 1968.

10. Grimes, R. N., *Carboranes*, Academic Press, New York, 1970.

11. Muetterties, E. L. (ed.), *Boron Hydride Chemistry*, Academic Press, New York, 1975.

12. Elian, M., Chen, M. M.-L., Mingos, D. M. P., and Hoffmann, R., *Inorg. Chem.*, **15**, 1148, 1976.

13. Mingos, D. M. P., *Adv. Organometal. Chem.*, **15**, 1, 1977.

14. Hoffmann, R., *Angew. Chem. Internat. Edn. Engl.*, **21**, 711, 1982.

15. Callahan, K. P., Evans, W. J., and Hawthorne, M. F., *Ann. N.Y. Acad. Sci.*, **239**, 88, 1974.

16. Dunks, G. B. and Hawthorne, M. F., p. 383 of Ref. 11.

17. Grimes, R. N., *Metal Interactions with Boron Clusters*, Plenum Press, New York, 1982.

18. Greenwood, N. N., *Chem. Soc. Rev.*, **13**, 353, 1984.

19. Kennedy, J. D., *Progr. Inorg. Chem.*, **32**, 519, 1984; **34**, 211, 1986.

20. Stone, F. G. A., *Adv. Organometal. Chem.*, **31**, 53, 1990.

21. Greenwood, N. N., *Chem. Soc. Rev.*, **21**, 49, 1992.

22. Siebert, W., *Adv. Organometal. Chem.*, **35**, 187, 1993.

23. Callahan, K. P., Evans, W. J., Lo, F. Y., Strouse, C. E., and Hawthorne, M. F., *J. Am. Chem. Soc.*, **97**, 296, 1975.

24. Johnston, R. L., Mingos, D. M. P., and Sherwood, P., *New J. Chem.*, **15**, 831, 1991.

25. Kabalka, G. W. (ed.), *Current Topics in the Chemistry of Boron*, Royal Society of Chemistry, London, 1994.

26. Stone, A. J., *Mol. Phys.*, **40**, 1339, 1980; *Inorg. Chem.*, **20**, 563, 1981; *Polyhedron*, **3**, 1299, 1984.

27. O'Neill, M. E. and Wade, K., *Polyhedron*, **3**, 199, 1984.

28. Mulvey, R. E., O'Neill, M. E., Wade, K., and Snaith, R., *Polyhedron*, **5**, 1437, 1986.

29. Porterfield, W. W., Jones, M. E., Gill, W., and Wade, K., *Inorg. Chem.*, **29**, 2914, 1990.

30. Atoji, M. and Lipscomb, W. N., *Acta Cryst.*, **6**, 547, 1953; *J. Chem. Phys.*, **21**, 172, 1953.

31. Massey, A. G., *Adv. Inorg. Chem. Radiochem.*, **26**, 1, 1983.

32. Maier, G., *Angew. Chem. Internat. Edn. Engl.*, **27**, 309, 1988.

33. O'Neill, M.E. and Wade, K., *Inorg. Chem.*, **21**, 461, 1982; *J. Mol. Struct.*, **103**, 259, 1983.

34. Corbett, J. D., *Progr. Inorg. Chem.*, **21**, 140, 1976.

35. Cavanaugh, M. A., Fehlner, T. P., Stamel, R., O'Neill, M. E., and Wade, K., *Polyhedron*, **4**, 687, 1985.

36. Porterfield, W. W., Jones, M. E., and Wade, K., *Inorg. Chem.*, **29**, 2919, 2923, 1990.

37. Porterfield, W. W., Jones, M. E., and Wade, K., *Inorg. Chem.*, **29**, 2927, 1990.

38. Greenwood, N. N. and Earnshaw, A., *Chemistry of the Elements*, *2nd. Edn.*, Butterworth-Heinemann, Woburn, U.K., 1997.

39. Hermánek, S., *Chem. Rev.*, **92**, 325, 1992.

40. Brint, P. and Sangchakr, B., *J. Chem. Soc.*, *Dalton Trans.*, 105, 1988.

41. Gillespie, R. J., Porterfield, W. W., and Wade, K., *Polyhedron*, **6**, 2129, 1987.

42. Wade, K., in *Electron Deficient Boron and Carbon Clusters* (eds Olah, G.A., Wade, K., and Williams, R. E.), Wiley, New York, 1991, p. 95.

43. Johnson, B. F. G. (ed.), *Transition Metal Clusters*, Wiley, Chichester, 1980.

44. Moskovits M. (ed.), *Metal Clusters*, Wiley, New York, 1986.

45. Olah, G. A., Prakash, G. K. S., Williams, R. E., Field, L. D., and Wade, K., *Hypercarbon Chemistry*, Wiley, New York, 1987.

46. Mingos, D. M. P. and Wales, D. J., *Introduction to Cluster Chemistry*, Prentice Hall, Englewood Cliffs, N J, 1990.

47. Gonzalez-Moraga G., *Cluster Chemistry*, Springer, Berlin, 1993.

48. Moore, M. J. and Brint, P., *J. Chem. Soc.*, *Dalton Trans.*, 427, 1993.

49. Housecroft, C. E. and Wade, K., *Tetrahedron Letters,* **34**, 3175, 1979.

50. Hawthorne, M. F., Young, D.C., and Wegner, P.A., *J. Am. Chem. Soc.*, **87**, 1818, 1965.

51. Hawthorne, M. F. and Pilling, R. L., *J. Am. Chem. Soc.*, **87**, 3987, 1965.

52. Hawthorne, M. F. and Andrews, T. D., *Chem. Commun.*, 443, 1965; *J. Am. Chem. Soc.*, **87**, 2496, 1965.

53. Sirigu, A., Bianchi, M., and Benedetti, E., *Chem. Commun.*, 596, 1969.

54. Braye, E. H., Dahl, L. F., Hübel, W., and Wampler, D. L., *J. Am. Chem. Soc.,* **84**, 4633, 1962.

55. Holt, E. M., Whitmire, K. H., and Shriver, D. F., *J. Organometal. Chem.*, **213**, 125, 1981.

56. Housecroft, C. E., *Advan. Organometal. Chem.*, **33**, 1, 1991.

57. Albano, V. G., Sansoni, M., Chini, P., and Martinengo, S., *J. Chem. Soc.*, *Dalton Trans.*, 651, 1973.

58. Albano, V. G., Chini, P., Ciani, G., Martinengo, S., and Sansoni, M., *J. Chem. Soc.*, *Dalton Trans.*, 463, 1978.

59. Vidal, J. L., Walker, W. E., Pruett, R. L., and Schoening, R. C., *Inorg. Chem.* **18**, 129, 1979.

60. Ciani, G., Garlaschelli, L., Sironi A., and Martinengo, S., *J. Chem. Soc.*, *Chem. Commun.*, 536, 1981.

61. Ciani, G., Sironi, A., and Martinengo, S., *J. Chem. Soc.*, *Dalton Trans.*, 519, 1981.

62. Tachikawa, M. and Muetterties, E. L., *Progr. Inorg. Chem.*, **28**, 203, 1981.

63. Herrmann, W. A., *Angew. Chem. Internat. Edn. Engl.*, **25**, 56, 1986.

64. Blake, A. J., Johnson, B. F. G., and Nairn, J. G. M., *Acta Crystallogr. Sect. C*, **50**, 1052, 1994.

65. Eady, C. R., Johnson, B. F. G., Lewis, J., Mason, R., Hitchcock, P. B., and Thomas, K. M., *J. Chem. Soc.*, *Chem. Commun.*, 385, 1977.

66. Doi, J. A., Teller R. G., and Hawthorne, M. F., *J. Chem. Soc., Chem. Commun.*, 80, 1980.

67. McPartlin, M., *Polyhedron*, **3**, 1279, 1984.

68. Hughes, A. K., Peat, K. L., and Wade, K., *J. Chem. Soc. Dalton Trans.*, 4639, 1996; 2139, 1997.

69. Wade, K., *Electron Deficient Compounds*, Nelson, London, 1971.

70. Cotton, F. A. and Haas, T. E., *Inorg. Chem.* **3**, 10, 1964.

71. Kettle, S. F. A., *Theor. Chim. Acta*, **3**, 211, 1965.

72. Abel, E. W., Stone, F. G. A., and Wilkinson, G. (eds), *Comprehensive Organometallic Chemistry II*, Pergamon, Oxford, 1995.

73. Raithby, P. R. and Rosales, M. J., *Adv. Inorg. Chem. Radiochem.*, **29**, 169, 1985.

74. Blount, J. F., Dahl, L. F., Hoogzand, C., and Hübel, W., *J. Am. Chem. Soc.*, **88**, 292, 1966.

75. Tachikawa, M., Shapley, J. R., and Pierpont, C.G., *J. Am. Chem. Soc.*, **97**, 7174, 1975.

76. Pierpont, C. G., *Inorg. Chem.*, **16**, 636, 1977.

77. Bradley, J. S., Chap. 5, p. 105 in Ref. 44.

78. Bradley, J. S., Ansell, G. B., and Hill, E. W., *J. Am. Chem. Soc.*, **101**, 7417, 1979.

79. Bradley, J. S., Ansell, G. B., Leonowicz, M. E., and Hill, E. W., *J. Am. Chem. Soc.*, **103**, 4968, 1981.

80. Simon, A., *Angew. Chem. Internat. Edn. Engl.*, **20**, 1, 1981.

81. von Schnering, H. G., *Angew. Chem. Internat. Edn. Engl.*, **20**, 33, 1981.

82. Corbett, J. D., *Chem. Rev.*, **85**, 383, 1985.

83. Mingos, D. M. P., Slee, T., and Lin, Z.-Y., *Chem. Rev.*, **90**, 383, 1990.

84. Nesper, R., *Angew. Chem. Internat. Edn. Engl.*, **30**, 789, 1991.

85. Corbett, J. D., *Structure and Bonding*, **87**, 157, 1997.

86. Zintl, E. and Kaiser, H. Z., *Z. Anorg. Allgem. Chem.*, **211**, 113, 1933.

87. Zintl E., *Angew. Chem.*, **52**, 1, 1939.

88. Schäfer, H., *Ann. Rev. Mat. Sci.*, **15**, 1, 1985.

89. Sevov, S. C. and Corbett, J. D., *J. Solid State Chem.*, **103**, 114, 1993.

90. Blase, W. and Cordier, G., *Z. Kristallogr.*, **196**, 207, 1991.

91. Prince, D. J., Corbett, J. D., and Garbisch, B., *Inorg. Chem.* **9**, 2731, 1970.

92. Couch, T. W., Lokken, D. A., and Corbett, J. D., *Inorg. Chem.*, **11**, 357, 1972.

93. Cisar, A. and Corbett, J. D., *Inorg. Chem.*, **16**, 2482, 1977.

94. Critchlow, S. C. and Corbett, J. D., *Inorg. Chem.*, **23**, 770, 1984.

95. Corbett, J. D., *Progr. Inorg. Chem.*, **21**, 129, 1976.

96. Krebs, B., Mummert, M., and Brendel, C., *J. Less-Common Met.*, **116**, 159, 1986.

97. Edwards, P. A. and Corbett, J. D., *Inorg. Chem.*, **16**, 903, 1977.

98. Dong, Z.-C. and Corbett, J. D., *J. Am. Chem. Soc.*, **116**, 3429, 1994; *Inorg. Chem.*, **35**, 3107, 1996; *Angew. Chem., Internat. Edn. Engl.*, **35**, 1006, 1996.

99. Zhao, J.-T. and Corbett, J. D., *Inorg. Chem.*, **34**, 378, 1995.

100. Liu, Q., Hoffmann, R., and Corbett, J. D., *J. Phys. Chem.*, **98**, 9360, 1994.

101. Belin, C., Mercier, H., and Angilella, V., *New J. Chem.*, **15**, 931, 1991.

102. Burns, R. C. and Corbett, J. D., *J. Am. Chem. Soc.*, **104**, 2804, 1982.

103. Belin, C., Corbett, J. D., and Cisar, A. J., *J. Am. Chem. Soc.*, **99**, 7163, 1977.

104. Burns, R. C. and Corbett, J. D., *Inorg. Chem.*, **24**, 1489, 1985.

105. Campbell, J., Dixon, D. A., Mercier H. P., and Schrobilgen, G. J., *Inorg. Chem.*, **34**, 5798, 1995.

106. Krebs, B., Hucke, M., and Brendel, C. J., *Angew. Chem.*, **94**, 453, 1982.

107. Angilella, V. and Belin, C. J., *J. Chem. Soc., Faraday Trans.*, **87**, 203, 1991.

108. Fässler, T. and Hunziker, M., *Inorg. Chem.*, **33**, 5380, 1994.

109. Critchlow, S. C. and Corbett, J. D., *J. Am. Chem. Soc.*, **105**, 5715, 1983.

110. Blase, W., Cordier, G., Müller, V., Häussermann, U., Nesper, R., and Somer, M., *Z. Naturforsch.*, **48b**, 754, 1993.

111. Blase, W. and Cordier, G., *Z. Naturforsch.*, **44b**, 1011, 1479, 1989.

112. Cordier, G. and Müller V., *Z. Naturforsch.*, **48b**, 1035, 1993; **49b**, 721, 1994.

113. Wade, K., *New Scientist*, **62**, 615, 1974.

3

DISOBEDIENT SKELETONS

JOHN D. KENNEDY

School of Chemistry, University of Leeds, Leeds LS2 9JT, U.K.

Bob Williams' perception that the cluster shapes of molecular polyhedral boron-containing cluster compounds could be interpreted as fragments of a series of regular closed polyhedra, rather than as fragments of an icosahedron, was ultimately published early in 1971.[1] This was followed closely by the perception [2–4] that if the cluster bonding electrons were counted up according to some simple rules, then a coherent relationship between structure, electron count, and formula could be discerned. The combination of these rules with an isolobal and isoelectronic cluster-fragment replacement principle thence permits the extension of the rules to heteroboranes, metallaboranes, metallaheteroboranes, etc.[2,4] The cluster-geometry/electron-counting formalism thereby generated (Figure 3-1) has been exceedingly helpful in the development of the increasingly expanding area of boron-containing cluster chemistry. This is generally because of its usefulness for the prediction and rationalization of many of the structures encountered. The detailed considerations were summarized by Bob Williams and Ken Wade some 20 years ago.[5,6] They are often incorporated into borane, carbaborane, and heteroborane cluster treatments in many standard general textbooks [7] in inorganic and structural chemistry. They are often referred to as "Wade's Rules," or the Williams/Wade formalism, although, as just mentioned, others have, of course, contributed to their development and application.[3,4]

However, even when the rules were first formulated, exceptions were recognized, and many more are now known. The exceptions arise either structurally, or in terms of electron count. Many of these exceptions, but by no means all of them, occur when transition-elements, with their varieties of oxidation states and bonding modes, are incorporated into the boron hydride matrices. They can involve subtle variations within the general pattern, or completely new geometries and systematics. There are now many of these. It is pertinent within the

| No. of vertices | closo | nido | arachno |

context of this book to delineate and categorize some of these ways in which boron-containing cluster compounds deviate, or apparently deviate, from the Williams/Wade structure-geometry/electron-counting paradigm.

In the first instance, it is useful to consider that, there has always been a recognition that a number of boron-containing cluster compounds do not fit into the Williams/Wade $closo/$ $(2n + 2)$-electron, $nido/$ $(2n + 4)$-electron, $arachno/$ $(2n + 6)$-electron, etc., parallels in the sense that, although they adopt structures that adhere to the $closo/$, $nido/$, $arachno$ structural patterns [8] of Figure 3-1, they do not always have the accompanying respective $(2n + 2)$, $(2n + 4)$, $(2n + 6)$ cluster electron counts. Additional confusion then often arises when it is not clear whether the descriptors $closo$, $nido$, $arachno$, etc. describe the geometry according to Williams,[1] or the electron counts of $(2n + 2)$, $(2n + 4)$, $(2n + 6)$, etc., respectively. Often it is not clear precisely how the electrons 'should' be counted in these ambiguous cases. As far as the experimental chemist is concerned, a geometry, for example as determined from a good diffraction analysis, is pretty unambiguous. The positions of individual electrons, however, are anybody's guess, and consequently it can be most difficult to decide where they are at all, let alone whether they might be considered as involved in the cluster bonding for counting purposes, or, indeed, what 'cluster bonding' might precisely mean in these circumstances, or what the significance of the resulting total count might be.

An example that it is useful to invoke at this point is the simple binary borane B_8H_{12}. Although this compound is of formal $nido$ B_nH_{n+4} constitution, it has an $arachno$ geometry (structure **1**) generated by the removal of two adjacent vertices from a 10-vertex $closo$ skeletal shape,[9] as has the more legitimately B_nH_{n+6} $arachno$ species B_8H_{14} (structure **2**).[10] Here, there are no special heteroatom bonding perturbations, and, as with all experimentally determined structural observations, the observer has, of course, to accept that this is the way the molecule wants to be, irrespective of any "rules" that she/he may think dictate otherwise. It is important to recognize that any such rules have to be tailored, constrained, modified, or compromised to fit the compounds. The converse is artificial if the compounds are disobedient and do not want to fit our rules.

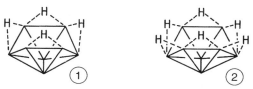

Figure 3-1 Representation of the classical Williams/Wade structural pattern (compare, e.g., Reference 3). The $(n-1)$-vertex $nido$ geometries are derived by the removal of one vertex of high connectivity from the n-vertex $closo$ geometries, and the $(n-2)$-vertex $arachno$-geometries by the removal of a second, adjacent, vertex. Horizontal progression $closo \rightarrow nido$ and thence $nido \rightarrow arachno$ for the same number of vertices is achieved by the successive addition of two pairs of electrons. In this formalism, the n-vertex $closo/$, $nido/$ and $arachno$ structures have, respectively, $(2n + 2)$, $(2n + 4)$ and $(2n + 6)$ cluster-bonding electrons associated with them.

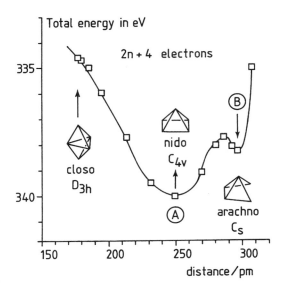

Figure 3-2   Calculated total energy as a function of geometry for the $\{B_5H_5\}^{4-}$ system which has a five-vertex *nido* electron count of 14 electrons (data taken from Figure 2 of Ref. 3). For the $D_{3h} \rightarrow C_{4v}$ *closo* $\rightarrow$ *nido* geometrical deformation, one equatorial distance was extended to 250 pm, and the axial distance shortened to 250 pm, with the others interboron distances being maintained at 177 pm. For the $C_{4v} \rightarrow C_s$ *nido* $\rightarrow$ *arachno* geometrical deformation, one basal interboron distance was extended, with the other maintained at 177 pm. The minimum A corresponds to the familiar square-pyramidal structure recognized for *nido*-B_5H_9, and the minimum B approximates to the more open structure observed in *arachno*-B_5H_{11}.

It is helpful to consider the approach exemplified initially by Ralph Rudolph in 1972.[3] This involved a consideration of the computed energies for various geometries of the $\{B_5H_5\}^{4-}$ fragment when they are associated with various numbers of skeletal bonding electrons. Although this was investigated in terms of the variation of essentially one parameter (Figure 3-2), and did not involve bridging hydrogen atoms, the calculation for $\{B_5H_5\}^{4-}$ clearly demonstrated the minimum A that corresponds to the familiar square-pyramidal B_5H_9 structure. There are other minima, e.g., minimum B, which is not as stable as A for this formulation, and which corresponds to a more open *arachno*-type structure. Obviously, if the same approach is used for B_8H_{12}, and if it is correctly computed, it will find that the *arachno*-structured minimum is the more stable and any *nido*-structured ones are less stable (see Chapter 9, this volume). A generalization from this type of consideration suggests a certain fortuitousness that most minima in many combinations of boron with carbon and hydrogen are such that (a) the classical *closo/nido/arachno* main-sequence structures of Figure 3-1 are the most stable and (b) they correspond, respectively, to the Wadian $(2n + 2)/(2n + 4)/(2n + 6)$ electron counts. In Williams' initial 1971 paper,[1] it was, in fact, noted that this sequence could well be an unlikely result of quite a number of coincidences. In terms of the sort of treatment summarized in Figure 3-2, there are, of course, many structural parameters to vary, and so the structure-energy surface must be regarded as multidimensional, and not just a function of essentially one distortion, as in Rudolph's initial treatment.[3]

Some consequences and corollaries arise out of these simple considerations. One, of course, is that minima not in accord with the Williams/Wade sequence in Figure 3-1 may be the most stable. This could involve a "mismatch" between electron count and structure, as in B_8H_{12} just discussed. Perhaps more interestingly, structures that are not part of the classical sequence of Figure 3-1 may be stabilized. This leads to a second corollary: by changing the cluster constituents and substituents, alternative minima might be stabilized.[11] This might be achievable quite easily in some cases, e.g., by deprotonation.[12] For a given formulation, if two or more minima are of sufficiently low energy relative to the general profile of the geometry-energy surface, and if the minima are isolated from each other by sufficiently high passes, then isomers of different skeletal geometries can be isolatable. It may be that some combinations stabilize potential wells that correspond to structures significantly outside the initial Williams/Wade sequences of Figure 3-1. If passes among the wells and valleys are low, then interconversions among similar or equivalent structures will be able to occur readily, so that isomerizations or fluxionalities can then be readily observable.

This last consideration introduces a somewhat loose, and, at this stage, necessarily vague concept of "fragility" that can be superposed upon any general structural patterns and/or rules that may be devised to account for them. For example, many of the established polyhedral shapes are not comfortable in a number of known compounds that otherwise obey the Williams/Wade formalism.[12] Thus the *closo* $[B_{11}H_{11}]^{2-}$ anion (structure **3**) is fluxional,[14,15] as is *closo* $[B_8H_8]^{2-}$ (structure **4**).[15,16] Thus, the *closo* $[B_{11}H_{11}]^{2-}$ anion (structure **3**) is fluxional,[13,14] as is *closo*-$B_8H_8^{2-}$ (structure **4**).[14,15] In each of these cases, there must be fluxional intermediates that have open-faced structures very close in energy to that of conventional closed geometry.[11] When it is recog-

(3)

(4)

nized how close these open and closed structures must be energetically, then it reinforces the perception about how fortuitous it is that there is, in fact, a coherent series of stable closed $[B_nH_n]^{2-}$ polyhedra, and, correspondingly, how fortuitous it is that this $[B_nH_n]^{2-}$ sequence does not contain a number of open structures. In contrast to this consistent sequence of closed structures, it has become increasingly apparent, since the 1970s, that the initially reasonably coherent set of open *nido*-type skeletons (the set in Figure 3-1) has become much more diverse, with many alternative structures, as discussed by Williams in Chapter 1 of this symposium volume. Within the more open *arachno* series, a greater diversity has always been generally recognized, ranging from the poly-hedral binary borane structures that are generally in accord with the *arachno* column of Figure 3-1, to the polygonal planar structures of the aromatic hydro-carbons at the other end of the borane–carbaborane–carbane continuum. Aspects of this last type of variation are dealt with more thoroughly by Fox and Wade in Chapter 2 of this book and are also mentioned further at various points in the following text.

Somewhat allied with the concept of cluster fragility are the "long" connec-tivities in some cluster structures. Thus, the nine-vertex *closo* cluster structure type typified by $[B_9H_9]^{2-}$ is static, but in many nine-vertex *closo* species, some of the cluster connectivities are long (structure **5**). This has long been recog-nized.[16] Another well recognized example consists of the long 5,10 and 7,8 linkages in *nido*-$B_{10}H_{14}$ (structure **6**).

In all of these fragile structures, it may often not require a particularly large per-turbation to distort and/or stabilize alternatives.[11] The small energies required can be induced by a variety of electronic and/or steric means, often within the context of the same formal Wadian electron count. In fluxional species the flux-ional intermediates may be preferentially stabilizable if the constituents or sub-stituents are changed. Conversely, of course, a robust cluster species may be rendered fragile by change of cluster constituents or substituents. In clusters with "long" connectivities, the connections may similarly be easily stretched to breaking point so that stable, more open structures may be generated. Some-times, there may be a continuum of progressively increasingly stretched struc-tures.[17,18] Alternatively, it could be that structure types may seem to opt for one or other of two or more extremes. This can happen, for example, with a symmetrical species, when the "fragility" can be distributed equally over more than one connectivity. A change in cluster constituents off the symmetry axes or planes can then localize the fragility, and, often, this type of effect can be quite dramatic. This may well occur in the 1,2,3 and 1,2,4 isomers of $[(\eta^6$-$MeC_6H_4CHMe_2)RuC_2Me_2B_8H_8]$ also mentioned below, which have different geometries. These are *closo*-type (structure **7**) [19] and a so-called *isonido*-type

(structure **8**, see also structure **36**),[20] respectively. Often, from an experimental view, a sufficiently coherent set of examples from which to make meaningful generalizations may not yet have been established.

To summarize thus far, there is often a tendency for skeletal structures of many boron-containing cluster compounds to disobey simplistic applications of the original 1971 Williams/Wade formalisms. It is of interest, in the context of this book, to discuss a few of these "disobedient skeletons" that have emerged over the intervening years. One objective is to help engender perceptions about how the Williams/Wade formalism might be modified or adapted to apply to the new structural phenomena observed. It is not practicable to mention all the disobediencies here. There is an emphasis on particular systems that have been the subject of interest in the Leeds laboratories. This means that some interesting areas of disobediency are not dealt with specifically in this account, e.g., the intriguing variety of 12-vertex *nido*-type systems.[21] It must also be recognized that there will be many more of these disobedient skeletons still hiding in the cupboard. They will presumably spring out upon us as progress in the subdiscipline moves away from the simple boranes and boron-rich carbaboranes.

One common apparent disobedience in the application of the Williams/Wade formalism is that typified by the structures of compounds that incorporate a 16-electron square-planar transition-element center as part of the cluster. For example, bis(ligand) platinum(II) units, such as $\{Pt(PMe_2Ph)_2\}$, which are generally held to be two-electron contributors, in accord with their platinum(II) character, are frequently found in cluster systems that may be two (or, in some cases, four) electrons short of the formal Wadian electron count associated with that particular cluster geometry. This type of dichotomy arises because simplistic applications of the skeletal electron-counting schemes generally presume 18-electron transition-element character and so do not accommodate deviations from this behavior.

Although the reasons for this are readily appreciated,[21] and are, indeed, included in some U.K. undergraduate chemistry courses,[22] compounds that display this type of behavior are often treated with surprise in contemporary research literature. In this context, for example, a series of $\{RhSB_9\}$ cluster compounds containing square-planar rhodium centres has received recent attention.[23,24] Thus, the *nido*-shaped 11-vertex clusters in compounds such as $[(PPh_3)_2RhSB_9H_{10}]$ (schematic **9**) are taken to have *closo* electron counts. Because of this apparent *closo*-count/*nido*-structure anomaly, these particular skeletons have been specifically presented as disobedient.[24,25] Further, some interesting and unusual coordinative phenomena have been postulated on the presumption that these disobedient molecules are, in fact, striving to obtain an obedient and thereby "correct" $(2n + 4)$-electron count.[25] Conversely, many

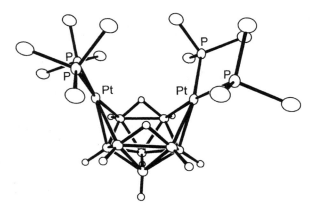

Figure 3-3 Molecular structure of the 10-vertex *arachno*-shaped diplatinadecaborane [{6,6,9,9-(PMe$_2$Ph)$_2$-6,9-Pt$_2$B$_8$H$_{10}$] (data from Ref. 27) (see also schematic cluster structure **10**). With each BH(*exo*) unit contributing two electrons, with two bridging hydrogen atoms, and with two {Pt(PMe$_2$Ph)$_2$} units contributing two electrons each, the electron count is 22, corresponding to ($2n + 2$) 10-vertex *closo*. However, the geometry is clearly *not* 10-vertex *closo* (schematic **25**). Rather, it is closely analogous to that of the straightforward *arachno*-[B$_{10}$H$_{14}$]$^{2-}$ anion (schematic **11**).

structures containing square-planar transition-element centres have been categorically dubbed "*closo*" when they are, by observation, anything but closed geometrically. An example that has been discussed in this context is the *nido*-structured species [(PMe$_2$Ph)ClPtB$_5$H$_8$Os(CO)(PPh$_3$)$_2$].[26]

This type of interpretation is exemplified by the open-structured species [(PMe$_2$Ph)$_2$Pt B$_8$H$_{10}$Pt(PMe$_2$Ph)$_2$] (Figure 3-3),[27] which has a formal "*closo*" count. However, in terms of geometrical and electronic structure, it is patently *arachno*, and a direct [B$_{10}$H$_{14}$]$^{2-}$ analog. Structurally, this is evident not only in the open-cluster aspect, but also in the detail of the positioning of the open-face bridging hydrogen atoms (schematics **10** and **11**). Electronically, it is evident in terms of clear parallels of cluster NMR shielding properties.[27] The anomaly arises because it is the nature of the orbital contributions from the metal center to the cluster that dictate the relevant aspects of cluster character, rather than a simplistic totaling of numbers of electrons in an electron "count". The square-planar platinum(II) moiety (schematic **12**) partakes in cluster bonding in a manner exactly analogous to the {BH$_2$}$^-$ unit (structure **13**), since it has two orbitals and two electrons available to get involved with the cluster bonding. In Wadian electron-counting terms, the two electrons in an *endo* BH terminal bond are included in electron-counting procedures (xx in structures **11** and **13**), but a rigorously square-planar transition-element center has no electrons in such a position, either for bonding or for counting (position Ⓐ in structures **10** and **12**). Thus, in cluster compounds containing such a square-planar transition-element center, there is no need [28] whatsoever to invoke any additional electronic contribution to the cluster bonding, providing that the square-planar metal centre is comfortable in that state, as it often is.[28–30] Clusters containing such centers will therefore be two electrons short for every such unit, and will be apparently anomalous in terms of formal counting rules.[25]

A more fundamental origin of this anomaly is that the two electrons in the *endo* BH bond, although formally included in the count, are, in fact, not involved in the actual cluster bonding. They are not involved in bonds that actually hold the skeleton together. In this sense, and in the more general case, the addition of two electrons to a cluster often actually removes electrons from the cluster bonding proper. This occurs particularly in *nido-to-arachno* transitions that result in the generation of *endo* terminal hydrogen atoms.[31] Here, the generation of an *endo*-hydrogen position will remove electrons from the cluster bonding itself.[32] This is readily illustrated at its simplest in a comparison of neutral *nido*-B_2H_6 and the *arachno*-$[B_2H_6]^{2-}$ dianion. In B_2H_6, two three-center two-electron bonds hold the two-boron "cluster" together, whereas in the $[B_2H_6]^{2-}$ anion, with two extra electrons, there is only one two-electron bond holding the two boron atoms together (structures **14** and **15**).

This type of behavior can also occur within the context of an unchanged skeletal electron count, leading to "disobedient" structural variation. For example, at the other end of the borane-carbaborane-carbane continuum, the pentagonal planar $[C_5H_5]^-$ anion has 16 skeletal bonding electrons, corresponding to a $(2n + 6)$-electron *arachno* count. All eight electron pairs contribute to the cluster bonding proper. All eight pairs are directly involved in intercarbon bonding. Although formally *arachno*, the planar aspect is clearly disobedient in terms of the deltahedral binary borane structural sequence of Figure 3-1. In the $(2n + 6)$-electron binary borane counterpart B_5H_{11}, the three electron pairs in the three *endo*-BH bonds are not involved in holding the cluster together, and the three bridging hydrogen atoms localize electrons and their associated molecular orbitals at specific regions on the open face.[1] This leaves only two filled cluster-bonding molecular orbitals for direct interboron bonding, and there is a corresponding gross change in geometry compared with the $[C_5H_5]^-$ anion. In sum, the effect of open-face hydrogen nuclei is to localize electron pairs on the open face, thus, resulting in a quite different cluster molecular-orbital scheme, and thence a quite different geometry. This is quite general. The discussion of this type of phenomenon, namely, the control of cluster geometry via an effective protonation of the open face, is developed further in Chapter 2 and also, to some extent, below, near structures **42–44**.

An additional interesting aspect of 16-electron transition-element centers such as square-planar rhodium(I) and platinum(II) is that they are also, in principle, capable of accepting two-electron ligands to form 18-electron centers. Also, they can oxidize up to rhodium(III) or platinum(IV) if the bonding systems in which they are involved permit it. Some possible platinum(II) bonding configurations are given very schematically in structures **16–18**, and some platinum(IV) configurations in structures **19–21**. In these, the drawn orbitals will be completely filled when binding with the boron-containing cluster fragment: in boron-containing cluster chemistry, any of these possibilities can occur, as well as intermediate types of bonding.[29,33–35] Aspects of this behavior are addressed below for the 11-vertex {PtC$_2$B$_8$} system. It has to be recognized that there is no hard-and-fast blanket rule with such flexible transition-element centers, especially in combination with different redox-flexible boron-containing skeletons. The general point here is that each type of system has to be considered on its own merits.

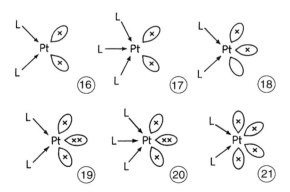

In terms of "fragility" of clusters, as introduced above, some skeletons with ostensibly classical Williamsian cluster shapes in fact exhibit lengthening of

---

[1]Note that one of the three *endo*-type BH hydrogen atoms is generally regarded as at least partially bridging. This does not affect the general point made.

some of their connectivities, to such an extent that new cluster shapes start to emerge. Examination of several formally "closed" geometries in the 9- and 10-vertex area reveals that many are, in fact, partially open. A complete opening of these 9-vertex *closo* (structure **22**) and 10-vertex *closo* (structure **25**) skeletons generates square-faced, more open, structures that have been called "*isonido*" (structures **23** and **26**).[17,18] Experimentally, it is found that there is a continuum of structure between formally *closo* (structures **22** and **25**) and *isonido* (structures **23** and **26**).[17,18] Nine-vertex and ten-vertex *isonido* compounds are exemplified by [(CO)(PPh$_3$)$_2$IrCB$_7$H$_8$] [21] and [(PPh$_3$)$_2$IrB$_9$H$_{10}$(PPh$_3$)],[17] respectively.

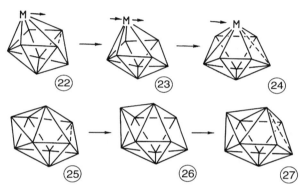

A continuation of the opening into a complete diamond-square-diamond (DSD) process across the relevant face-pair generates [36,37] the so-called *isocloso* geometries (and here it may be noted that the 9-vertex *closo* → *isocloso* transition can be accomplished by one DSD process, and that it does not need two as is sometimes maintained).[38,39] These *isocloso* geometries have their own symmetries, which are C_{2v} for the 9-vertex system (structure **24**), and C_{3v} for the 10-vertex system (structure **27**). It may be noted here that the 10-vertex *isocloso* system has a clear C_{3v}-based geometry and that these and other "*isocloso*" boron-containing skeletons do not "normally adopt capped polyhedral geometry" as is sometimes stated.[40] A second DSD process (other than a simple reversal of the first) can regenerate the *closo* cluster shape, although selected positions are now interchanged.[41] A process related to this occurs in the ready fluxional enantiomerization of [(η^5-C$_5$Me$_5$)RhC$_2$B$_7$H$_9$] and related compounds of structure **28** (Figure 3-4).[42] This implies an energetic similarity among all the structures along the fluxionality pathway for this particular combination and positioning of cluster constituent elements. It also demonstrates an increase in

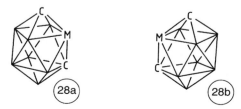

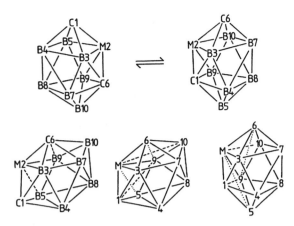

Figure 3-4 Diagram showing how the proposed fluxionality of compounds such as [$(C_5Me_5)RhC_2B_7H_9$] (after Refs. 41 and 42) interchanges enantiomers (structures **28a** and **28b**) and gives the positional changes consistent with assignments from NMR spectroscopy. The isomerization progresses through two simultaneous or consecutive diamond-square-diamond (DSD) processes involving the two face-pairs M(2)C(1)B(5)B(9) and M(2)C(6)B(9)B(10). One such DSD process would constitute a *closo* → *isonido* → *isocloso* transition, as in the schematic progression **25** → **27**.

fragility by change of cluster constituents, since the 10-vertex *closo* model [$B_{10}H_{10}$]$^{2-}$ itself is not fluxional. Such a fluxional process is not yet observed for the 9-vertex system, although it readily occurs in the 11-vertex system (see below),[14] and the 8-vertex system also is fluxional.[15]

For both the 9- and 10-vertex species, although a continuum of opening is observed from *closo* to *isonido* (sequences **22–24**, and **25–27**),[17,18] there are no examples yet of structures intermediate between *isonido* and *isocloso* (Figure 3-5). This may be because experimental chemistry has not yet happened upon appropriate systems to demonstrate this. If not, the observation may presumably be reflected in a smooth change in cluster molecular-orbital variation from *closo* → *isonido*, followed by a necessary jump to a completely different bonding scheme in *isocloso*. It may be noted here that, for compounds isolated so far, the *closo* and *isonido* species generally have formal *closo* Wadian electron counts, whereas the *isocloso* species have sub-*closo* totals when counted according to Wade's Rules. For the 9-vertex system, only one *isocloso*-species has been structurally characterized so far, that of [$(PMe_3)_2HIrB_8H_7Cl$].[43] It will be useful to have suitable synthetic systems to flesh out this area. Approaches based on B_8H_{12} or B_8H_{14} could well be useful. By contrast, several examples exist of the 10-vertex *isocloso* system, e.g., [$(\eta^6\text{-MeC}_6H_4CHMe_2)RuB_9H_9$][44] and the [$(CO)_3WB_9H_9$]$^{2-}$ anion.[45,46] It is becoming apparent that this is probably, in fact, the "normal" simple metallaborane closed 10-vertex cluster shape. Geometrically, each of the 9- and 10-vertex *isocloso* skeletal shapes is derived by

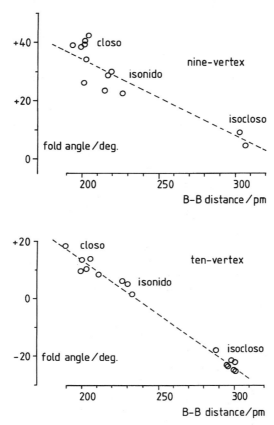

Figure 3-5 The changes from *closo* through *isonido* to *isocloso* are characterized by a DSD process involving two adjacent faces (sequences **22 → 24, 25 → 27,** and **54 → 56**). These are characterized by a change in the fold-angle θ between the original two faces, and a lengthening of the interboron edge *b* common to both original faces. The upper diagram plots fold-angle θ against the interboron distance *b* for known 9-vertex species (data from Ref. 18) and the lower diagram for known 10-vertex species (data from Ref. 17). In each case, there are several examples in the *closo → isonido* region, but none between *isonido* and *isocloso*.

the complete capping of the $(n-1)$-vertex *arachno*-geometry (structures **29** and **30**, respectively).[37] Conventional *closo*-shapes derive from the complete capping of the $(n-1)$-vertex *nido*-geometry.[1–6]

There is discussion about the relationship of these *isocloso* shapes to the Williams/Wade formalism.[36,38,39,47–49] It is clear that they have their own distinct symmetries [17,18,35,37] and are not merely slight distortions [38] of more classical *closo* shapes, or, indeed, of "capped" *closo* shapes, as is sometimes proposed.[40] Compounds of this type have been also called *hypercloso* (beyond *closo*), because a classical electron count gives a sub-*closo* total of $2n$

electrons.[38] However, a sub-*closo* electron count might imply that the term *hypocloso* (beneath *closo*) might be more appropriate. It is so used in this article. *Hypercloso* can then be used for closed structures with an excess of electrons over the $(2n + 2)$ count, as in eight-vertex $[(C_5H_5)_4Ni_4B_4H_4]$[50] and in certain of the 11-vertex $\{PtC_2B_8\}$ species discussed below. Total-cluster molecular-orbital schemes have been derived that show a compatibility with the *hypocloso* interpretation in that several molecular orbitals in the HOMO/LUMO region are generally nondegenerate in these *isocloso*-shapes. In terms of these schemes, these *isocloso* shapes could therefore, in principle, be compatible with a variety of even electron counts around $(2n + 2)$, with no tendency to adopt other structures via Jahn-Teller distortion.[38,48,49,51] In the total-cluster treatments reported so far, a three-orbital cluster contribution from the unique six-connectivity vertex is generally presumed.

Alternatively, it has been surmised that these *isocloso*-shaped clusters could, in principle, and may often, in fact, have a $(2n + 2)$ classically *closo* count.[36,47] This would require a metal center (so far, compounds of these types are all characterized by transition-element centers in the high-connectivity position) to contribute two more electrons to the cluster than are required by a strict application of the conventional electron-counting schemes. Such a $(2n + 2)$ count would then imply a four-orbital involvement of the metal center with the cluster, which would then reasonably generate geometries that are nonclassical.[36,47] If this four-orbital *isocloso* hypothesis holds, then the *isocloso* → *nido* → *arachno* sequence of cluster opening is characterized by a successive removal of metal-centered atomic orbitals from the cluster bonding proper (compare *arachno* 10-vertex discussion near structures **10–15** above). It would therefore represent an alternative fundamental sequence to the classical *closo* → *nido* → *arachno* progression.[52] This is well illustrated by the 9-vertex $\{IrB_8\}$ sequence represented in the structure progression **31** → **32** → **33**. In the *isocloso* structure **33**, as exemplified by $[(PMe_3)_2HIrB_8H_7Cl]$, the metal center would contribute four orbitals to the cluster, and in the classical *nido* structure **32**, as exemplified by $[(CO)(PMe_3)_2IrB_8H_{11}]$, the contribution would be three orbitals. In the *arachno* structure **31**, as exemplified by $[(CO)(PMe_3)_2HIrB_8H_{11}Cl]$, two orbitals are contributed as discussed above for platinum(II) square-planar species (structures **10–13**). See also the $B_2H_6 / [B_2H_6]^{2-}$ comparison (structures **14** and **15**). It may also be noted here that, in the *arachno*-species $[(CO)(PMe_3)_2HIrB_8H_{11}Cl]$, the octahedral iridium center does, in fact, have an *endo* Ir-H two-electron bond, which may be used for electron-counting purposes (if the cluster-pattern recognition chemist is really desperate).

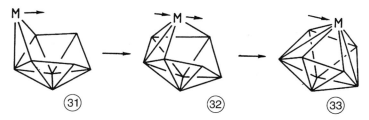

As mentioned above, most contiguous-cluster molecular-orbital schemes that have been set up to attempt to account for the *isocloso* skeletal types discussed here, and which have often generally been invoked to support the *hypocloso* hypothesis,[48] have presumed a classic three-orbital participation in the cluster bonding from the central high-connectivity atom on the central symmetry axis. Such schemes, of course, a priori preclude a four-orbital cluster involvement. It would therefore be instructive to see schemes set up with a four-orbital participation from this high-connectivity center. In clusters of transition-element atoms, particularly the earlier ones, it is not unusual for a transition-element center to donate four orbitals for skeletal bonding,[53–55] and it would be reasonable to examine for this in metallaborane clusters. The four-orbital participation could, for example, take the form of four tangentially directed vectors, or one radial and three tangential. One reported frontier-orbital approach based on metal-centered fragments and cluster fragments (rather than a total-cluster approach) could be interpreted as going someway along this line.[39] However, this was carried out at the extended Hückel level, and with fixed boron-fragment geometries. These are both perhaps rather severe constraints, both in view of the small energetic differences observed experimentally,[29] and in view of the magnitudes of the cluster-geometry flexing observed experimentally (see below). In practice, of course, a true picture of the bonding for a given cluster may sometimes reside as a hybrid somewhere between three-orbital *hypocloso* and four-orbital *isocloso*.[34,35] A recent perspective in this area alludes to the nominalistic maxim known as Ockham's Razor,[40] of which a common interpretation advocates the minimization of the number of assumptions used in an explanation. It is, in fact, instructive to apply nominalistic arguments to the Wade/Williams formalism, to the derivation of the formalism, and to its application to boron-containing cluster chemistry. At the simplest level, one of the first things that becomes apparent is that the Razor can cut both ways. Does the bewildered cluster-pattern recognition chemist discard the restrictive assumption that a vertex always contributes three orbitals to the skeletal bonding, or the liberating assumption that a vertex can sometimes contribute other than three orbitals to the scheme?

Some aspects of the interpretative problem between *isocloso* and *hypocloso* descriptions arise because there is a difficulty in establishing experimental criteria, other than shape, for these two forms. Here, some experimental progress has been made in the closed 11-vertex area.[56–59] Geometrically, this is interestingly different from the 9- and 10-vertex cases, since a DSD process involving the six-connectivity position on the 11-vertex *closo* cluster regenerates the same

C_{2v} geometry. Parallels with the 9- and 10-vertex cases above thence suggest that *closo* and *isocloso* 11-vertex structures have the same gross geometries.[37,60] This is supported by the similar gross geometries observed for $[(\eta^6\text{-MeC}_6\text{H}_4\text{CHMe}_2)\text{-}1,2,3\text{-RuC}_2\text{Me}_2\text{B}_8\text{H}_8]$,[19] which is classically *closo* (schematic **34**), and also for $[(\eta^6\text{-MeC}_6\text{H}_4\text{CHMe}_2)\text{-RuB}_{10}\text{H}_{10}]$ (schematic **35**),[19] which has to be interpreted as *isocloso* [47] or *hypocloso*.[38] This is also consistent with the geometrical approach that derives *isocloso* structures by the complete capping of *arachno* structures, since the 10-vertex *nido* and 10-vertex *arachno* clusters both have the same basic gross C_{2v} geometry.[38,60] Note, however, that the metal atom in the polar position is more tightly bound in the *isocloso* species than in the *closo* species,[19] and there are other significant differences in the detail of interatomic dimensions. Another parallel with the 9- and 10-vertex systems is that an intermediate 11-vertex *isonido* structure can be found, as exemplified by the 1,2,4 isomer of $[(\eta^6\text{-MeC}_6\text{H}_4\text{CHMe}_2)\text{Ru-}C_2\text{Me}_2\text{B}_8\text{H}_8]$ mentioned above (schematics **8** and **36**).[20] This latter open-faced compound is of classically *closo* electron count (as also are the known *isonido* species in the 9- and 10-vertex systems).

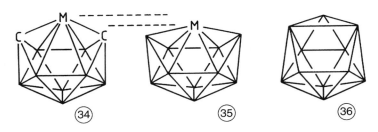

The geometrical *closo/isocloso* similarities involving the 11-vertex twofold C_{2v} symmetry are not disadvantageous in the present context. They permit, by NMR spectroscopy, a comparative assessment of any fundamental electronic differences associated with any cluster electronic changes. This is because the basic closed C_{2v} cluster geometry remains unchanged, and any gross changes in NMR shieldings must arise from gross electronic changes. With the 9- and 10-vertex clusters, however, both the geometry and the electronic structure change in the *isocloso* → *closo* transition. This inhibits comparative analysis of *closo* and *isocloso* metal-to-cluster bonding modes because NMR shieldings will change with gross geometry, as well as electronic structure, and so the two effects cannot simply be distinguished by NMR spectroscopy. Additionally, for the 11-vertex system, the NMR properties and geometries of the 10-vertex C_{2v} borane and carbaborane model fragments $[nido\text{-}6,9\text{-}C_2\text{B}_8\text{H}_{10}]^{2-}$, $[arachno\text{-}6,9\text{-}C_2\text{B}_8\text{H}_{14}]$, $[nido\text{-B}_{10}\text{H}_{14}]$, $[arachno\text{-B}_{10}\text{H}_{14}]^{2-}$, and the 11-vertex carbaborane model $[closo\text{-}2,3\text{-}C_2\text{B}_9\text{H}_{11}]$ are known, and permit direct comparison with the corresponding fragments in the 11-vertex metal-capped species,[57,59] whereas a set of corresponding 9-vertex fragment models for the 10-vertex system is lacking. Further, the twofold symmetry is more compatible with the orthogonal transition-element bonding axes, so that bonding vectors in the frontier area are easier to relate.

Within this basic C_{2v} 11-vertex geometry, it is possible to distinguish five structural types (structures **37** to **41**).[58] Initially, these progress from a bridged *arachno* 10-vertex structure, **37**, exemplified by the $[Ph_2PB_{10}H_{12}]^-$ anion,[12] to a bridged *nido* 10-vertex structure, **38**, exemplified by $[(Et_2O)EtAlC_2B_8H_{10}]$ and $[Me_2SnC_2B_8H_{10}]$.[61,62] In both of these types, the polar atom is joined to the cluster by two two-electron two-centre bonds to the "prow" (6) and (9) positions (*nido/arachno* 10-vertex numbering) of the respective *arachno* and *nido* 10-vertex subclusters. Types **37** and **38** are followed by a "true" *closo* type (represented by structure **39**), in which the atom M at the polar position is joined to the cluster by three multicenter bonds,[63] and two, more compact, types, structures **40** and **41**, in which the polar vertex probably has a four-orbital involvement with the cluster, as discussed below. In structure type **40**, represented by $[(PMe_2Ph)_2PtC_2B_8H_{10}]$ and its C-phenyl derivative, this four-orbital involvement generates a closed structure with a formal *nido* $(2n + 4)$ electron count.[58] In structure type **41** (see also structure **35**), represented by $[(\eta^6\text{-}MeC_6H_4CHMe_2)\text{-}RuB_{10}H_{10}]$,[19] it generates a formal *closo* $(2n + 2)$ count.[47] Alternatively, in the latter case, the *hypocloso* approach [38] would favor a $2n$ electron count and a polar vertex with a traditional three-orbital contribution.[39] Many compounds might well, in practice, exhibit electronic structures intermediate between the *hypocloso* and *isocloso* extremes. These may best be revealed as such by detailed rigorous molecular-orbital work.

The bridged structures *arachno* **37** and *nido* **38** are straightforwardly interpreted. The $[Ph_2PB_{10}H_{12}]^-$ anion of structure type **37** does, however, merit some

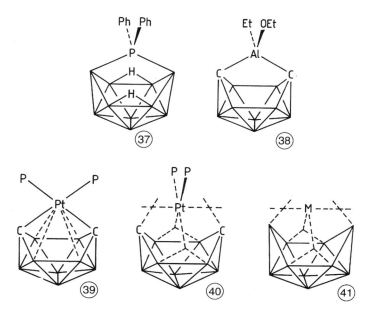

comment in the context of skeletal disobediency. It can be regarded either as a basic 10-vertex *arachno*-type in which the (6) and (9) *endo* positions are occupied by the {PPh$_2$} bridge,[12] or as an 11-vertex *arachno* type in which non-adjacent (remote) vertices have been removed from a 13-vertex *closo* shape. The classical Williams/Wade formalism [1–6] removes two adjacent vertices. This is a mild form of disobediency (boron chemists always gleefully describe the remote, planar, benzene as an *arachno* compound, see also discussion about the [C$_5$H$_5$]$^-$ anion above, near structures **14** and **15**); see Chapter 2. The crossover between "remote" and "adjacent" *arachno* structures depends upon the number of bridging hydrogen atoms (see Chapter 2, this book, and Ref. 63). The crossover occurs at two to three open-face bridging and/or *endo* hydrogen atoms, which is in accord with the adoption of a non-remote shape [64] by its protonated precursor, neutral Ph$_2$PB$_{10}$H$_{13}$ (schematic **42**). The latter has three bridging hydrogen atoms, rather than the two of the anion.[12] Interestingly, although the neutral species **42** is formally *arachno*, both from the 10-vertex and 11-vertex points of view, its cluster bonding now has an approximation to *nido*-decaboranyl 10-vertex cluster bonding. It entails a {PPh$_2$} moiety that bridges two adjacent boron sites, B(5) and B(6), by the use of two localized two-electron two-center bonds, rather than by the use of one bridging two-electron three-center bond, as to hydrogen in *nido*-B$_{10}$H$_{14}$ itself.[58] The rest of the structural electronics (schematic **43**), however, are very similar to those in *nido*-B$_{10}$H$_{14}$ (e.g., schematic **44**), as judged by NMR properties [12] and structural features such as bridging hydrogen-atom positioning and interboron distances.[64] This illustrates a general point that a designation of *closo, nido, arachno*, etc., does not necessarily imply only one overall structure that adheres to the appropriate *closo, nido, arachno*, etc., formalism of Figure 3-1 (or, indeed, any other total-cluster formalisms, such as that encompassed in Figure 3-6), but may sometimes be better treated in terms of electronic or structural subconfigurations appropriate to another cluster type in the *closo, nido, arachno*, etc., paradigm. This is somewhat related to the square-planar two-orbital arguments presented above (near structures **10–15**), which, in a sense, represent the converse presentation of this point. In extreme cases, local concentrations of several main-group heteroatoms in cluster compounds may engender disobediencies based on quite large domains of electron-precise bridging and polycyclic features that are conjoined to more contiguous deltahedral boron-based clusters.

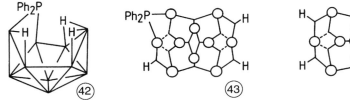

A related dichotomy is to be anticipated in macropolyhedral species of which the geometries are derived from the fusion of two subclusters with two or more boron atoms held in common. Thus, though each of the two known isomers of $B_{18}H_{22}$ consists of two *nido*-decaboranyl skeletons fused with two boron atoms in common,[65–68] the $B_{18}H_{22}$ formulation could alternatively accommodate one *closo* and one *arachno* skeleton fused also with two boron atoms in common. This has not yet been synthetically realized for an all-boron skeleton, although aspects of the analysis of the *closo:nido* 12-vertex:8-vertex species $[(PMe_3)_2(CO)IrB_{16}H_{16}Ir(PMe_3)_2(CO)]$ have been taken to illustrate the principle.[69] Also, within the context of a particular macropolyhedral formula, a change in the intimacy of the intercluster linkage can change the character of one or other of the subclusters by units of two electrons. It is perhaps worth emphasizing, therefore, that the "*nido*" total $(2n + 4)$ electron count of $B_{18}H_{22}$ does not necessarily dictate an exclusive *nido:nido* nature for a compound of this formulation. Here, it is also convenient to note that in fused cluster compounds, such as $B_{18}H_{22}$, it is necessary to count certain electrons twice to ensure compatibility of the individual subclusters with the Williams/Wade formalism. This is contrary to assertions in some textbooks.[70]

In the discussion of the sequence **37–41**, a particularly interesting and informative comparison is afforded by $\{L_2PtC_2B_8H_{10}\}$ species in the categories of structures **39** and **40**. If the ligand L is $P(OMe)_3$, or PPh_3, or SEt_2, then the thermodynamically stable configuration is that of structure **39**, with the ligands *trans* to the carbon atoms in the (2) and (3) positions (closed 11-vertex numbering now). The compounds then correspond to reasonably straightforward *closo* entities, both in terms of classical Wadian electron count and in terms of classical *closo* structure.[57,58,62] The metal atoms in the polar positions are more tightly bound than in the bridged *nido*-species of structure **38**, and the NMR properties of the $\{C_2B_8H_{10}\}$ unit are similar to those of the classic 11-vertex *closo*-model 2,3-$C_2B_9H_{11}$.[58,62] The $\{PtL_2\}$ moiety is behaving essentially as a Wadian $\{BH\}$ unit, i.e., as a 16-electron platinum(II) three-orbital cluster constituent (structure **18** above). By contrast, if the ligand L is PMe_2Ph (structural representation **40**), then the thermodynamically more stable ligand dispositions are orthogonal to those in the phosphite compound. The platinum atom is even more tightly bound with the cluster, which itself adopts a somewhat squatter geometry (see structures **39a** and **40a** below). The cluster NMR pattern also is now quite different, being similar to that of *arachno*-6,9-$C_2B_8H_{14}$.[57,58] It is therefore reasonable to invoke a bonding scheme in which the four *endo* and bridging hydrogen atoms of classically *arachno*-6,9-$C_2B_8H_{14}$ (schematic **45**) have been replaced by four two-electron bonding vectors to an octahedral platinum(IV) center (schematic **46**).[58] The $\{PtL_2\}$ moiety is now behaving as an 18-electron octahedral platinum(IV) four-orbital cluster constituent (schematic **21** above). This is interesting, as it implies a formal $(2n + 4)$ *nido* count for this closed 11-vertex cluster. Perhaps this cluster type should be the type best designated as *hypercloso* (see earlier discussion, near structures **29–33**), as it has a *closo*-structure but an electron count in *excess* of the formal *closo* $(2n + 2)$ total.

The PMe_2Ph compound (structure **40**) and its C-methyl and C-phenyl derivatives are fluxional.[56,58] The fluxionality involves a rotation of the

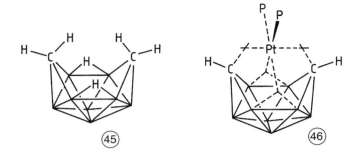

{Pt(PMe$_2$Ph)$_2$} moiety relative to the {C$_2$B$_8$H$_{10}$} unit about the C_2 axis. This is as represented in schematics **39a** and **40a** (note the cluster flexing). The fluxionality undergone by compounds of the type in structure **40** therefore implies an intermediate of structure the type in schematic **39**, and thence very similar energies for the two configurations of structures **39** and **40**. In turn, this implies very similar energies for the formal platinum(II) and (IV) valence states in this sort of system.[57] Conversely, the P(OMe)$_3$ species of the platinum(II) configuration **39** can be sterically forced into the platinum(IV) in configuration **40** by the incorporation of a C-phenyl cluster substituent.[57,59]

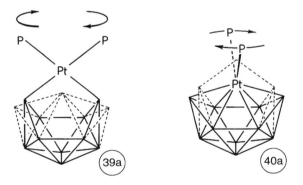

The final closed 11-vertex classification represented in structure **41**, may also have a four-orbital contribution to the cluster,[47] but now with an effective ($2n$ + 2) *closo* count, rather than the ($2n$ + 4) *nido* electron count of structure type **40**.[19,37] To support this hypothesis, it is possible to invoke the closed 11-vertex rhodaborane species [(PPh$_3$)$_2$HRhB$_{10}$H$_8$(OMe)Cl] which has a Rh–H–B bridge to the B(2) atom (schematic **47**).[71] If there was a "normal" three-orbital involvement of the rhodium atom with the cluster, together with a $2n$ electron count as required by the alternative [42] *hypocloso* hypothesis, then this Rh–H unit would be terminal. Contemporary results show that this Rh–H unit is, in fact, terminal in the corresponding "true *closo*" dicarbametallaborane species [1,1,1-H(PPh$_3$)$_2$-1,2,3-RhC$_2$B$_8$H$_{10}$] (schematic **48**).[72] In the absence of the Rh–H–B bridge, the 10-boron compound [(PPh$_3$)$_2$HRhB$_{10}$H$_8$(OMe)Cl] would have two electrons short of a $(2n + 2)$ total. However, there *is* a bridge: the two-electron Rh—H bond is added to the cluster, necessarily bringing with it its rhodium bonding orbital. It is therefore reasonable to postulate a four-orbital rhodium-to-borane interaction and a $(2n + 2)$ electron count within the constraints of an overall six-orbital "octahedral" rhodium(III) configuration.[71,72] As in the four-orbital structure type of schematic **40**, as discussed in the previous paragraph, the metal atom in this *isocloso* type, structure **41**, is also more tightly bound than in the "true" *closo* type of structure **39**.

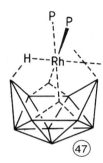

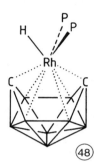

These considerations are more difficult to assess in the threefold 10-vertex *isocloso* system of structure **27**. The twofold symmetry (structure **24**) of the 9-vertex *isocloso* system would be more amenable, but synthetic breakthroughs are needed here. The above observations, however, constitute reasonable experimental evidence for four-orbital contributions from these late transition-element centers in two of the five 11-vertex cluster-bonding schemes. Similar four-orbital contributions to the *isocloso*-structured 9- and 10-vertex species are not excludable on present experimental evidence. An additional perception from the closed 11-vertex considerations is that the three-orbital "true" *closo* structure may be inherently less stable than the other 11-vertex closed forms. Thus, true 11-vertex *closo* structures easily either convert (a) with other, more open forms (as, e.g., in the fluxionality of the $[B_{11}H_{11}]^{2-}$ anion),[13,14] or (b) convert into the four-orbital *nido*-count closed *hypercloso* structure (as in the case of the $[(PR_3)_2PtC_2B_8H_{10}]$ species).[56–59]

In terms of cluster fragility and the stretching of intracluster linkages, as dealt with above for closed clusters in the *closo-isonido-isocloso* considerations (near structures **22–27**), there are also several incidences of stretchings of open clusters to breaking point, so that more extended open faces are generated. A good example here is $[1-(\eta^6-C_6Me_6)-1-RuB_9H_{13}]$ (structure **49**), in which one of the "long" $B(5)–B(10)/B(7)–B(8)$ interboron linkage of $B_{10}H_{14}$ (schematic **6** above) is stretched to 246 pm to give what might be termed an "*isoarachno*" configuration.[73] A second example is the isomer of the formally *closo* (at least in classic electron-counting terms) $[(\eta^6-C_6Me_6)OsC_2B_8Me_2]$ that has what, at first sight, looks like a conventional *nido* 11-vertex shape with a five-membered open face.[74] This latter *closo/nido* anomaly is, in itself, a stretching disobedience,

but the interboron distance in this open face is actually further stretched, to ca.
210 pm. This is not a dramatic stretching (although it does occur in an already
"stretched" cluster structure), but, again, it indicates an incipient *arachno* type
of geometry (structure **50**). This latter is now in the context of a formal $(2n + 2)$
closo count, with no 16-electron transition-element center to help rationalize the
anomaly. This *arachno* 11-vertex geometry is also exhibited by the classically
arachno $(2n + 6)$-electron $[S_2B_9H_{10}]^-$ anion,[75] as expected (structure **51**). It
also occurs in the $\{RhSB_9\}$ subclusters (structure **52**) of the interesting isomeric
macropolyhedral pair of formulation $[(C_5Me_5)_2Rh_2S_2B_{15}H_{16}(OH)]$,[76] in which
it has a formally *nido* $(2n + 4)$ electron count. In this latter case, it appears that
the *arachno* shape for the *nido*-count subclusters may well be dictated electronically. It also appears that the *arachno* shape for the 11-vertex *nido*-count *nido*-
shaped cluster may also be forceable sterically. This is illustrated, to some
extent, by a comparison of the geometry of the $\{C_2B_9\}$ residue in $Ph_2C_2B_9H_{11}$, in
which it is normal *nido*, and in $[(\eta^6\text{-}C_6H_6)RuC_2Ph_2B_9H_9]$ (schematic **53**), in
which the two cluster carbon atoms are separated by ca. 258 pm to give an
arachno-shaped 11-vertex fragment.[77] In sum, disobedient *closo* $(2n + 2)$ and
nido $(2n + 4)$ electron counts have both been found in this 11-vertex *arachno*-
geometry, as well as the obedient *arachno* $(2n + 6)$ count.

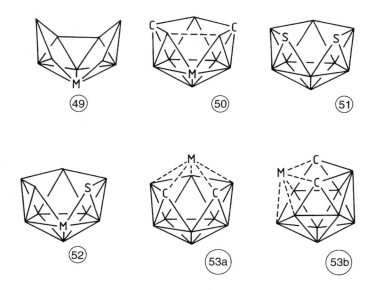

In terms of 12-vertex considerations, structure **53** is an *isonido* structure with
a four-membered open face (see also structure **55**). It derives from the 12-vertex
icosahedron of structure **54** via a diamond-to-square conversion. A notional
completion of the DSD process thence generates a 12-vertex *isocloso* geometry
(structure **56**). This has also been observed experimentally, e.g., as in the
$\{WC_2B_9\}$ subclusters in $[PtW(CO)_2(PEt_3)_2\text{-}(Me_2C_2B_9H_8)(CH_2C_6H_4Me)]$ and

related species (structure **57**).[78] It is therefore apparent that it is possible to synthesize *isonido* and *isocloso* compounds in the 12-vertex systems, to complement those in the 9-, 10- and 11-vertex systems discussed above.

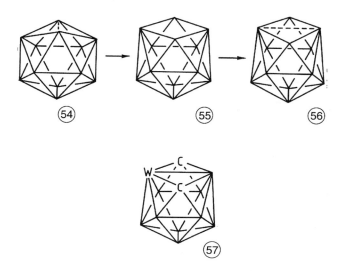

An interesting general perception is that these several alternative "disobedient" skeletal structures are perhaps starting to permit the construction of structural schemes of what could be termed, with the indulgence of the readers of these particular symposium proceedings, alternative *iso*-Williamsian structural sequences (e.g., Figure 3-6).[18] If there is then a desire that these be associated with corresponding "*iso*" electron-counting schemes, then it is clear that new subsections of the current general "rules" will have to be formulated, which would be an interesting exercise (see Chapter 1). Some of the results summarized in this chapter perhaps point the way to how this might be done in specific cases, although the large variety of factors that influence what appears to be an increasing variety of behavior may make this difficult in the general case. More experimental results to fill the gaps and address the uncertainties in the new sequences, such as the sequence in Figure 3-6, will obviously be helpful. Another general point is that it is interesting to speculate that, if the subdiscipline concerned with molecular boron-based cluster chemistry had developed initially by the examination of metallaboranes of the transition elements, rather than boranes themselves and carbaboranes, the accepted classical sequence might be based on that of Figure 3-6 and not that of Figure 3-1. It should be noted that any sequences of this kind link only selected minima on a multidimensional structure-energy surface, and so other sequences will also, in principle, be possible. Those of Figures 3-1 and 3-6 will be just two of these. Energetically, these several sequences can, in principle, cross, which accounts for many of the various anomalies within the sequences, and also accounts for fluxionalities that must involve intermediates with structures from other sequences.[11]

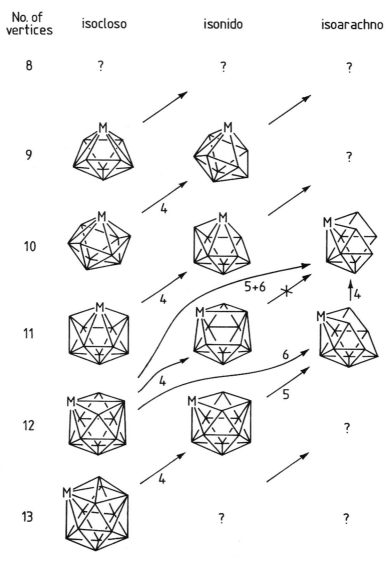

Figure 3-6 Schematic of a pattern of behavior that can be drawn for structural relationships among the *isocloso*, *isonido*, and "*isoarachno*" boron-containing cluster systems (after Ref. 18). Compare this with the classical *closo/nido/arachno* paradigm of Figure 3-1. The *isocloso*-system is generated from the classic *closo* system by a DSD process (see, e.g., sequences **22** → **24, 25** → **27,** and **54** → **56,** and also Figure 3-5). Removal of a vertex of low connectivity from an *n*-vertex *isocloso* geometry generates an (*n*–1)-vertex *isonido* geometry. Removal of two adjacent vertices generates a corresponding "*isoarachno*" geometry, although, in view of the more limited set of results available, it is not certain how general this might be. The numbers by the arrows indicate

There will be other disobedient skeletons in the cupboard waiting to pounce upon the unsuspecting cluster-pattern recognition chemist. What has been discussed above, so far, has often involved high-connectivity cluster vertices that involve metal centers for which high connectivities can be a comfortable concept. Boron atoms in such high-connectivity positions are rare. They are generally limited to awkward positions in supraicosahedral *closo*-type species.[79–81] They also, of course, exist at points of conjunction in *conjuncto* macropolyhedral species, but in this case they partake in bonding to two separate clusters, to which their individual connectivities are the usual three, four, or five.[69,82] When present in simpler closed species, such as $[B_{11}H_{11}]^{2-}$, their discomfort is manifested in that they are prone to fluxional instability via less-strained five-connectivity intermediates,[11,13,14] as mentioned above. Pervinder Kaur and Tom Jelínek have recently made neutral $[S_2B_{17}H_{17}(SMe_2)]$ and anionic $[S_2B_{18}H_{19}]^-$ (structures **58** and **59**).[83,84] Each of these has a six-connectivity boron in one of its subclusters, and in each case it is of interest that this subcluster is an open one. In the first, the subcluster is formally based on a novel *arachno* type of 10-vertex species $[SB_9H_{11}(SMe_2)]$ of configuration **60**. This configuration is quite different from the conventional *arachno* 10-vertex shape [85] of $[B_{10}H_{14}]^{2-}$ (structure **11**) or, perhaps more relevantly, $[6-SB_9H_{11}-9-(SMe_2)]$.[86] In the second, it is based on an *arachno* 11-vertex type of open subcluster (structure **61**), which is quite different from the conventional 11-vertex *arachno*-structure of the $[S_2B_9H_{10}]^-$ anion mentioned above (structure **51**).

It is not certain (a) whether these two new types of shape that involve boron atoms of cluster connectivity six are inherently stable, and (b) have not yet been observed in single clusters because reaction coordinates happen not yet to have led in those directions, or whether they are inherently unstable, and derive their stabilities from the constraints of the conjuncto two-boron linkages.[83] It is interesting in this regard that contemporary theoretical considerations are taken to suggest that the *p*-orbital overlap between a {BH} unit and a hexagonal planar arrangement of boron atoms is insufficient for stability in an isolated hexagonal bipyramidal unit.[87] Here, of course, the unit is monopyramidal rather than bipyramidal.

It is not clear how arguments that may selectively deemphasize selected observed boron-containing cluster structures, because they are perceived not be

---

the cluster connectivities of the removed vertices. Experimental results for the *isoarachno* column are, as yet, sparce. The eight-vertex area is, as yet, unexamined for "*iso*" behavior, although the very ready cluster isomerization of some eight-vertex *closo*-anions (Refs. 14 and 15) suggests that alternative open-cluster structures will be inherently isolatable by appropriate choice of cluster constituent. The *iso*-cluster structures are characterized by one or more vertices that have high cluster connectivities, and, in known examples, one or more of these is generally a transition-element center, designated here as M.

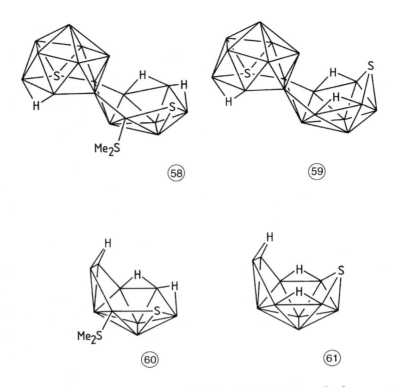

thermodynamically the most stable,[40,88] might fit in here. In fact, great care should be exercised in attempting to draw conclusions from any supposition [89] that the Williams–Wade formalism predicts the thermodynamically most stable structures, and that any other structures are kinetic artifacts.[82] A simple example here that is not in accord with such a supposition, is the eight-vertex *nido* system, as dealt with in the early part of this chapter and as dealt with more thoroughly elsewhere in this volume (Chapter 9). The supposition will, however, obviously apply to systems in which the Williams/Wade cluster structure *is* the most stable. On the other hand it, is also apparent from several of the examples dealt with in this chapter that thermodynamically favored conversions, both toward and away from the classical cluster structures of Figure 3-1, can occur. At the other end of the borane-carbaborane-carbane continuum, there is no difficulty about accepting the different geometric structures, bonding schemes, and chemistry of, say, $CH_2{=}CHCH_2OH$ and $(CH_3)_2C{=}O$, as two isomers of the same empirical formula, whichever is the thermodynamically more stable. In other systems, e.g., the two isomers $CH_3.CO.CH_2.CO.CH_3$ and $CH_3C(OH){=}CH.CO.CH_3$ of acetylacetone, different structures can have very similar thermodynamic stabilities indeed and be in dynamic kinetic equilibrium. All boranes and carbaboranes are, in any event, all endothermic species. Whatever the structure, a bonding scheme will be required to account for it.

Either way, the isolation of these two stable macropolyhedral thiaboranes means that the seven-boron hexagonal pyramid now has to be recognized as a viable structural building block. The classical sequence of Figure 3-1 is based on pentagonal pyramids and square pyramids. If, now, one or more boron-based hexagonal pyramids, such as those in structures **58** and **59**, can also be incorporated into single clusters, then still further sets of linked structural sequences that differ from those in Figures 3-1 and 3-6 can be envisaged. Whether these, and any other new types of skeletal behavior that may spring out upon the innocent hard-working boron-cluster chemist, will be ultimately regarded as obedient or disobedient, remains to be seen.

ACKNOWLEDGMENTS

This article is, of course, dedicated to Bob Williams on the occasion of his 70th birthday, and it is a pleasure and an honor to be invited to contribute to the symposium, and to this volume celebrating that event. Many of the perceptions in this chapter derive from earlier collaborative work with Norman Greenwood, and from more recent collaboration with the groups of Bob Štíbr and Trevor Spalding, and from numerous discussions with Jonathan Bould and Mark Thornton-Pett over many years, and it is a pleasure to acknowledge these various friendly scientific liaisons, together with the enthusiasm and productivity of the many other coworkers cited in the individual references. I should also like to thank Ken Wade and Tom Fehlner for helpful comments on this particular chapter. Much of the experimental work has been supported by the EPSRC (U.K.).

REFERENCES

1. Williams, R. E., *Inorg. Chem.*, 1971, **10**, 210–214.
2. Wade, K., *Chem., Commun.*, 1971, 792–793.
3. Rudolph, R. W. and Pretzer, W. P., *Inorg. Chem.*, 1972, **11**, 1974–1978.
4. Mingos, D. M. P., *Nature (Phys. Science)*, 1972, **236**, 99–102.
5. Williams, R. E., *Adv. Inorg. Chem. Radiochem.*, 1976, **18**, 67–142.
6. Wade, K., *Adv. Inorg. Chem. Radiochem.*, 1976, **18**, 1–66.
7. See, e.g., Owen, S. W. and Brooker, A. T., *A Guide to Modern Inorganic Chemistry*, Longman, Harlow, 1991; King, R. B., *Inorganic Chemistry of the Main Group Elements*, VCH, New York, Weinheim and Cambridge, 1994; Huheey, J. E., Keiter, E. A., and Keiter, R. L., *Inorganic Chemistry,* Harper Collins, New York, 1993; Spencer, J. L. in *Encyclopaedia of Inorganic Chemistry* (ed. King, R. B.), Wiley, Chichester, 1994, Vol. 1, pp. 338–357; Greenwood N. N. and Earnshaw, A. *The Chemistry of the Elements*, Pergamon, Oxford, 1987; Cotton, F. A. and Wilkinson, G., *Advanced Inorganic Chemistry*, Wiley, Chichester, 1988; Shriver, D. F., Atkins, P. W. and Langford, C. H., *Inorganic Chemistry*, Oxford University Press, Oxford, 1994.

8. Rudolph, R. W., *Accounts Chem. Res.*, 1976, **9**, 446–452.

9. Enrione, R. E., Boer, F. P., and Lipscomb, W. N., *J. Am. Chem. Soc.*, 1964, **86**, 1451–1452; *Inorg. Chem.*, 1964, **3**, 1659–1666; see also Pawley, G. S., *Acta Cryst.*, 1966, **20**, 631–638; and Rietz, R. R., Schaeffer, R., and Sneddon, L. G., *Inorg. Chem.*, 1972, **11**, 1242–1244.

10. Dobson, J. and Schaeffer, R., *Inorg. Chem.*, 1968, **7**, 402–408; Moody, D. C. and Schaeffer, R., *Inorg. Chem.*, 1976, **15**, 233–236.

11. Kennedy, J. D. and Štíbr, B., in *Current Topics in the Chemistry of Boron* (ed. Kabalka, G. W.), Royal Society of Chemistry, Cambridge, 1994, pp. 285–292.

12. Beckett, M. A. and Kennedy, J. D., *J. Chem. Soc. Chem. Commun.*, 1983, 575–576; Thornton-Pett, M., Beckett, M. A., and Kennedy, J. D., *J. Chem. Soc., Dalton Trans.*, 1986, 303–308.

13. Wiersema, R. E. and Hawthorne, M. F., *Inorg. Chem.*, 1973, **12**, 785–787; Tolpin E. I. and Lipscomb, W. N., *J. Am. Chem. Soc.*, 1973, **95**, 2384–2386; see also Porterfield, W. W., Jones, M. E., Gill, W. R. and Wade, K., *Inorg. Chem.*, 1990, **29**, 2914–2919.

14. Muetterties, E. L., Hoel, E. L., Salentine, C. G., and Hawthorne, M. F., *Inorg. Chem.*, 1975, **14**, 950–951.

15. Klanberg, F., Eaton, D. R., Guggenberger, L. J., and Muetterties, E. L., *Inorg. Chem.*, 1967, **6**, 1271–1281; Muetterties, E. L., Wiersema, R. J., and Hawthorne, M. F., *J. Am. Chem. Soc.*, 1973, **95**, 7520–7522; Gimarc, B. M. and Ott, J. J., *J. Am. Chem. Soc.*, 1987, **109**, 1388–1392; Bausch, Surya J. W., Prakash, G. K., and Williams, R. E., *Inorg. Chem.*, 1992, **31**, 3763–3768; see also Jelínek, T. Štíbr, B. Plešek, J. Kennedy, J. D., and Thornton-Pett, M., *J. Chem. Soc., Dalton Trans.*, 1995, 431–437, and other references cited therein.

16. O'Neill, M. E. and Wade, K., *Polyhedron*, 1983, **2**, 963–966.

17. Bould, J., Kennedy, J. D., and Thornton-Pett, M., *J. Chem. Soc., Dalton Trans.*, 1992, 563–576.

18. Štíbr, B., Kennedy, J. D., Drdáková, E., and Thornton-Pett, M., *J. Chem. Soc., Dalton Trans.*, 1994, 229–236.

19. Bown, M., Fontaine, X. L. R., Greenwood, N. N., Kennedy, J. D., and Thornton-Pett, M., *J. Chem. Soc., Dalton Trans.*, 1990, 3039–3049.

20. Bown, M., Fontaine, X. L. R., Greenwood, N. N., Kennedy, J. D., and Thornton-Pett, M., *Organometallics*, 1987, **6**, 2254–2255.

21. Pipal, J. R. and Grimes, R. N., *Inorg. Chem.*, 1979, **18**, 1936–1940; see also Grimes, R. N., in *Comprehensive Organometallic Chemistry* (eds. Wilkinson, G., Stone, F. G. A., and Abel, E.), Pergamon, Elmsford, 1982, Vol. 1, Chap. 5.5, pp. 459–542, and references cited therein.

22. See, e.g., question 4 in Examination CHEM 309001, February 1995, University of Leeds, Leeds, U.K.

23. Ferguson, G., Jennings, M. C., Lough, A. J., Coughlan, S., Spalding, T. R., Kennedy, J. D., Fontaine, X. L. R., and Štíbr, B., *J. Chem. Soc., Chem. Commun.*, 1990, 891-894.

24. Murphy, M., Spalding, T. R., Ferguson, G., and Gallagher, J. F., *Acta Cryst.*, *Sect. C*, 1992, **C48**, 638–641.

25. Adams, K. J., McGrath, T., and Welch, A. J., *Acta Cryst.*, *Sect. C*, 1995, **C51**, 401–403.

26. See, e.g., Kennedy, J. D., *Main Group Metal Chem.*, 1989, **12**, 149–154, and references cited therein; subsequently Barton, L., Bould, J., Rath, N. P., and Fang, H., *Inorg. Chem.*, 1996, **35**, 2062–2069.

27. Boocock, S. K., Greenwoood, N. N., Hails, M. J., Kennedy, J. D., and McDonald, W. S., *J. Chem. Soc., Dalton Trans.*, 1981, 1415–1429.

28. See, e.g., Macías, R., Synthesis, Structure and NMR Studies of Some Novel Metallaheteroboranes, Thesis, University of Leeds, Leeds, U.K., June 1996; Macías, R., Holub, J., Clegg, W., Thornton-Pett, M., Štíbr, B., and Kennedy, J. D., *J. Chem. Soc., Dalton Trans.*, 1997, 149–151.

29. O'Connell, D., Patterson, J. C., Spalding, T. R., Ferguson, G., Gallagher, J. F., Li, Y., Kennedy, J. D., Macías, R., Thornton-Pett, M., and Holub, J., *J. Chem. Soc., Dalton Trans.*, 1996, 3323–3333.

30. Murphy, M. P., Spalding, T. R., Cowey, C., Kennedy, J. D., Thornton-Pett, M., and Holub, J., *Organomet. Chem.*, in press.

31. Bould, J., Metallaboranes of the Platinum Group Metals, Thesis, University of Leeds, Leeds, U.K., 1983.

32. Moore, M. J. and Brint, P., *J. Chem. Soc., Dalton Trans.*, 1993, 427–432.

33. Colquhoun, H. M., Greenhough, T. J., and Wallbridge, M. G. H., *J. Chem. Soc., Dalton Trans.*, 1979, 619–628.

34. See, e.g., pp. 214–217 in Ref. 35.

35. Kennedy, J. D., *Prog. Inorg. Chem.*, 1986, **34**, 211–434.

36. Bould, J., Greenwood, N. N., and Kennedy, J. D., *J. Chem. Soc., Dalton Trans.*, 1990, 1451–1458.

37. Crook, J. E., Elrington, M., Greenwood, N. N., Kennedy, J. D., and Woollins, J. D., *Polyhedron*, 1984, **3**, 901–904.

38. Baker, R. T., *Inorg. Chem.*, 1986, **25**, 109–112.

39. Johnston, R. L., Mingos, D. M. P., and Sherwood, P., *New J. Chem.*, 1991, **15**, 831–841.

40. See p. 398 in Grimes, R. N., *Comprehensive Organometallic Chemistry*, 2nd Edition (eds Wilkinson, G., Stone, F. G. A., and Abel, E.), Pergamon, Elmsford, 1995, Vol. 1, Chap. 9, pp. 373–430, and references cited therein.

41. Bown, M., Jelínek, T., Štíbr, B., Heřmánek, S., Fontaine, X. L. R., Greenwood, N. N., Kennedy, J. D., and Thornton-Pett, M., *J. Chem. Soc., Chem. Commun.*, 1988, 974–975.

42. Nestor, K., Murphy, M., Štíbr, B., Spalding, T. R., Fontaine, X. L. R., Kennedy, J. D., and Thornton-Pett, M., *Collect. Czech. Chem. Commun.*, 1993, **58**, 1555–1568.

43. Bould, J., Crook, J. E., Greenwoood, N. N., Kennedy, J. D., and McDonald, W. S., *J. Chem. Soc., Chem. Commun.*, 1982, 346–348.

44. Kim, Y.-H., New Chemistry of Mono-, Di- and Trimetallaboranes of some Platinum Metals, Thesis, University of Leeds, Leeds, U.K., 1995–1996.

45. Macpherson, I., Some New Metallaborane Chemistry of Molybdenum and Tungsten, Thesis, University of Leeds, Leeds, U.K., 1987; and as cited in Ref. 43.

46. Kennedy, J. D., in *Boron Chemistry (IMEBORON VI)* (ed. S. Heřmánek), World Scientific, Singapore, 1987, pp. 207–243.

47. Kennedy, J. D., *Inorg. Chem.*, 1986, **25**, 111–112.

48. Johnston R. L. and Mingos, D. M. P., *Inorg. Chem.*, 1986, **25**, 3321–3323.

49. Johnston R. L. and Mingos, D. M. P., *J. Chem. Soc., Dalton Trans.*, 1987, 647–656.

50. Bowser, J. L., Bonny, A., Pipal, J. R., and Grimes, R. N., *J. Am. Chem. Soc.*, 1979, **101**, 6229–6236; see also O'Neill M. and Wade, K. *Inorg. Chem.*, 1982, **21**, 461–464.

51. Johnston, R. L. and Mingos, D. M. P., *Polyhedron*, 1986, 2059–2061.

52. See, e.g., pp. 88–89 in Ref. 70.

53. Johnston, R. L. and Mingos, D. M. P., *Inorg. Chem.*, 1986, **25**, 1661–1671, and as cited in Ref. 54.

54. Mingos, D. M. P., *Pure Appl. Chem.*, 1991, **63**, 807–872.

55. See, e.g., pp. 286–288 in Ref. 35.

56. Kennedy, J. D., Thornton-Pett, M., Štíbr, B., and Jelínek, T., *Inorg. Chem.*, 1991, **30**, 4481–4484.

57. Kennedy, J. D., Nestor, K., Štíbr, B., Thornton-Pett, M., and Zammitt, G. S. A., *J. Organomet. Chem.*, 1992, **477**, C1–C6.

58. Kennedy, J. D., Štíbr, B., Jelínek, T., Fontaine, X. L. R., and Thornton-Pett, M., *Collect. Czech. Chem. Commun.*, 1993, **58**, 2090–2120.

59. Nestor K., Kennedy, J. D., Štíbr, B., and Thornton-Pett, M., *J., Chem. Soc., Dalton Trans.*, in preparation for publication.

60. See, e.g., pp. 335–336 in Ref. 35.

61. Schubert, D. M., Knobler, C. B., Rees, W. S., and Hawthorne, M. F., *Organometallics*, 1987, **6**, 201–202 and 203–204.

62. Nestor, K., Štíbr, B., Jelínek, T., and Kennedy, J. D., *J. Chem. Soc., Dalton Trans.*, 1993, 1661–1663, and references cited therein.

63. Porterfield, W. W., paper presented to the Fourth National Working Meeting of British Inorganic Boron Chemists (INTRABORON 4), Durham, September 1984; see, subsequently, Gillespie, R. J., Porterfield, W. W., and Wade, K., *Polyhedron*, 1987, **6**, 2129–2135; Porterfield, W. W., Jones, M. E., and Wade, K., *Inorg. Chem.*, 1990, **29**, 2923–2927.

64. Friedman C. B. and Perry, S. L., *Inorg. Chem.*, 1973, **12**, 288–293.

65. Pittochelli A. R. and Hawthorne, M. F., *J. Am. Chem. Soc.*, 1962, **84**, 3218; Olsen, F. P., Vasavada, R. C., and Hawthorne, M. F., *J. Am. Chem. Soc.*, 1968, **90**, 3946–3951; Plešek, J., Heřmánek, S., and Hanousek, F., *Collect. Czech. Chem. Commun.*, 1968, **33**, 699–705.

66. Simpson P. G. and Lipscomb, W. N., *J. Chem. Phys.*, 1963, **39**, 26–34; Simpson, P. G., Folting, K., Dobrott, R. D., and Lipscomb, W. N., *J. Chem. Phys.*, 1963, **39**, 2339–2348.

67. Dixon, D. A., Kleier, D. A., Halgren, T. A., and Lipscomb, W. N., *J. Am. Chem. Soc.*, 1976, **98**, 2086–2096.

68. Barrett, S. A., Condick, P. N., Fontaine, X. L. R., Fox, M. A., Greatrex, R., Jelínek, T., Kaur, P., Kennedy, J. D., MacKinnon, P., and Štíbr, B., *J. Chem. Soc., Dalton Trans.*, to be submitted.

69. Barton, L., Bould, J., Kennedy, J. D., and Rath, N. P., *J. Chem. Soc., Dalton Trans.*, 1996, 3145–3149.

70. See, e.g., Housecroft, C. E., *Boranes and Metallaboranes* 2nd Edition, Ellis Horwood, Hemel Hempstead, U.K., 1994, pp. 113–114.

71. Fowkes, H. Greenwood, N. N., Kennedy, J. D., and Thornton-Pett, M., *J. Chem. Soc., Dalton Trans.*, 1986, 517–523.

72. Macías, R., Rath, N. P., Barton, L., and Kennedy, J. D., to be submitted, Structural study of $[1,1,1-(PPh_3)_2H-closo-1,2,3-RhC_2B_8H_{10}]$.

73. Bown, M., Fontaine, X. L. R., Greenwood, N. N., Kennedy, J. D., and MacKinnon, P., *J. Chem. Soc., Chem. Commun.*, 1987, 817–818.

74. Bown, M., Fontaine, X. L. R., Greenwood, N. N., Kennedy, J. D., and Thornton-Pett, M., *J. Chem. Soc., Chem. Commun.*, 1987, 1650–1651.

75. Holub, J., Wille, A. E., Štíbr, B., Carroll, P. J., and Sneddon, L. G., *Inorg. Chem.*, 1994, **33**, 4920–4926.

76. Kaur, P., Kennedy, J. D., Thornton-Pett, M., Jelínek, T., and Štíbr, B., *J. Chem. Soc., Dalton Trans.*, 1996, 1775–1777.

77. Lewis Z. G. and Welch, A. J., *J. Organomet. Chem.*, 1992, **430**, C45–C50; Cowie, J., Reid, B. D., Watmough, J. M. S., and Welch, A. J., *J. Organomet. Chem.*, 1994, **481**, 283–293; Brain, P. T., Bühl, M., Cowie, J., Lewis, Z. G., and Welch, A. J., *J. Chem. Soc., Dalton Trans.*, 1996, 231–237; see also Cowie, J., Donohue, D. J., Doueck, N. L., Kyd, G. O., Lewis, Z. G., McGrath, T. D., Watmough, J. M. S., and Welch, A. J., in *Current Topics in the Chemistry of Boron* (ed. Kabalka, G. W.), Royal Society of Chemistry, Cambridge, 1994, pp. 347–352; and also *Laboratory News* December, 1995, p. 8 and Thomas, R. L., Rosair, G. M., and Welch, A. J., *J. Chem. Soc., Chem. Commun.*, 1996, 1327–1328.

78. Attfield, M. J., Howard, J. A. K., Jelfs, A. N. de, M., Nunn, C. M., and Stone, F. G. A., *J. Chem. Soc., Dalton Trans.*, 1987, 2219–2233; Carr, N., Mullica, D. F., Sappenfield, E. L., and Stone, F. G. A., *Organometallics*, 1992, **11**, 3697–3704.

79. See, e.g., Evans W. J. and Hawthorne, M. F., *Inorg. Chem.*, 1974, **13**, 869–874; Churchill, M. R. and Deboer, B. G., *Inorg. Chem.*, 1974, **13**, 1411–1418; Evans, W. J. and Hawthorne, M. F., *J. Chem. Soc., Chem. Commun.*, 1974, 38–39; Lo, F. Y., Strouse, C. E., Callahan, K. P., Knobler, C. B., and Hawthorne, M. F., *J. Am. Chem. Soc.*, 1975, **97**, 428–429.

80. See, e.g., Maxwell, W. N., Weiss, E., Sinn, E., and Grimes, R. N., *J. Am. Chem. Soc.*, 1977, **99**, 4016–4029; Pipal, J. R. and Grimes, R. N., *Inorg. Chem.*, 1978, **17**, 6–10; Grimes, R. N., Sinn, E., and Pipal, J. R., *Inorg. Chem.*, 1980, **19**, 2087–2095.

81. See, e.g., Carr, N., Fernandez, J. R., and Stone, F. G. A., *Organometallics*, 1991, **10**, 2718-2725; Carr, N., Mullica, D. F., Sappenfield, E. L., and Stone, F. G. A., *Organometallics*, 1993, **12**, 1131–1139; Crennell, S. J., Devore, D. D., Henderson, S. J. B., Howard, J. A. K., and Stone, F. G. A., *J. Chem. Soc., Dalton Trans.*, 1989, 1363–1374; Devore, D. D., Henderson, S. J. B., Howard, J. A. K., and Stone, F. G. A., *J. Organomet. Chem.*, 1988, **358**, C6–C10; Jeffery, J. C., Jelliss, P. A., and Stone, F. G. A., *Inorg. Chem.*, 1993, **32**, 3382–3388.

82. Barton, L., *Topics in Current Chemistry. 100. New Trends in Chemistry* (ed. Boschke F. L.), Springer-Verlag, Berlin, Heidelberg and New York, 1982, pp. 169–206, and references therein.

83. Kaur, P., Holub, J., Rath, N. P., Bould, J., Barton, L., Štíbr, B., and Kennedy, J. D., *J. Chem. Soc., Chem. Commun.*, 1996, 273–275.

84. Jelínek, T., Cisařová, I., Thornton-Pett, M., and Štíbr, B., unpublished results 1995–1996, to be submitted; see also Jelínek, T., Štíbr, B., Kennedy, J. D., and Thornton-Pett, M., in *Advances in Boron Chemistry*, Royal Society of Chemistry, Cambridge, 1997, 426–429.

85. Kendall, D. S. and Lipscomb, W. N., *Inorg. Chem.*, 1973, **12**, 546–551.

86. Štíbr, B., Holub, J., Jelínek, T., Fontaine, X. L. R., Fusek, J., Kennedy, J. D., and Thornton-Pett, M., *J. Chem. Soc., Dalton Trans.*, 1996, 1741–1751.

87. Jemmis, E. D., Subramanian, G., and McKee, M. L., *J. Phys. Chem.*, 1991, **100**, 7014–7017.

88. See, e.g., Wang, X., Sabat, M., and Grimes, R. N., *Inorg. Chem.*, 1995, **34**, 6509–6513; Grimes, R. N., *Coord. Chem. Rev.*, 1995, **143**, 71–96.

89. Piepgrass, K. W., Curtis, M. A., Wang, X., Meng, X., Sabat, M., and Grimes, R. N., *Inorg. Chem.*, 1993, **32**, 2156–2163.

4

SECO-SYSTEMIZATION OF BORANES AND HETEROBORANES

STANISLAV HERMÁNEK

Institute of Inorganic Chemistry, Academy of Sciences of the Czech Republic, 250 68 Rez near Prague, Czech Republic

4.1 INTRODUCTION

In 1971, a basic paper appeared [1] in which R. E. Williams described a simple geometrical approach to the systemization of the title compounds, based on the debor principle, i.e, on a successive elimination of BH vertices from the basic *closo*-boranes $B_nH_n^{2-}$. In this way, *nido-, arachno-,* and *hypho*-compounds were formed stepwise, following the removal of one, two, and three B atoms, respectively, from the initial *closo*-compounds. This systematization, later expanded, gave borane chemists an excellent tool for predicting an immense number of possible new compounds.[2–4]

Almost simultaneously, a systemization, based on the number of the skeletal electron pairs $n + x$ (n = number of skeletal atoms, x = number of skeletal electron pairs, for *closo, x* = 1; *nido, x* = 2; *arachno, x* = 3; and *hypho, x* = 4), was reported by Wade.[5–7] His basic idea was slightly modified by Rudolf and Pretzer,[8] who suggested that these classes have $2n + 2$, $2n + 4$, $2n + 6$, and $2n + 8$ skeletal electrons, respectively. This shows that in analogy to organic chemistry, where the addition of two hydrogens converts acetylenes into olefins and two more hydrogens converts olefins into paraffins, in borane chemistry a similar reduction, i.e., the stepwise additions of three electron pairs, successively transforms *closo-* to *nido-*, nido- to *arachno-,* and finally *arachno-* to *hypho*-compounds.

117

4.2   SECO PRINCIPLE

By combining the above two approaches, we have proposed another systemization, namely, the seco concept,[9,10] which is based on the idea that the addition of two electrons lowers the number of three-center bonds in the given molecule by two, and, consequently, lowers the number of connections among skeletal atoms. According to this scheme, the successive opening of *closo-* to *nido-,* *arachno-,* and *hypho-*structures— caused by the addition of two, four, and six electrons, respectively—requires the stepwise elimination of two, three, and four edges, respectively (Figure 4-1), concurrent with the formation of one or two **Open Faces** (OF = four, five, or six-membered apertures in the clusters):

$$B_nH_n^{2-} \xrightarrow[-2E]{+2e} B_nH_n^{4-} \xrightarrow[-E]{+2e} B_nH_n^{6-} \xrightarrow[-E]{+2e} B_nH_n^{8-}$$

$$\textit{closo} \qquad\qquad \textit{nido} \qquad\qquad \textit{arachno} \qquad\qquad \textit{hypho}$$

$$e = \text{electron} \quad E = \text{edge}$$

4.2.1   Basic Prototypes

Figure 4-1 indicates that the prototypical skeletons in all classes belong to species that are composed of BH vertices only. The *closo-*$B_nH_n^{2-}$ compounds do not easily add protons but this is not the case for the *nido-*$B_nH_n^{4-}$ and arachno-$B_nH_n^{6-}$ anions, wherein the edges of the open face can successively accept protons as hydrogen bridges. Since the formation of bridge hydrogens is not a redox reaction, the class of the "protonized" compound is the same as that of the starting prototype. In those cases where the addition of another bridge hydrogen violates the μ–H Williams' Rule,[1,4] the proton either becomes an endo-H (in a BH_2 group), or the μ–H bridge opens and moves to another position in a new and larger open face. This can be demonstrated in the transformation of the hypothetical *nido-*$B_{10}H_{10}^{4-}$ anion (structure **19**), first to (5ω)*-*nido-*$B_{10}H_{14}$ (structure **29**), which must rearrange to (6ω)-*nido-*$B_{10}H_{14}$ (structure **30**) [*(5ω) = Five-membered aperture]. The intermediate (Structure **29**) was formerly considered in the alkaline conversion of 6-Cl$B_{10}H_{13}$ to $B_{10}H_{10}^{2-}$.[11] In this way, the five-membered open face of the *nido-*$B_{10}H_{10}^{4-}$ prototype (structure **19**) transforms into *nido-*$B_{10}H_{14}$ (structure **30**), which shows a six-membered open face and seemingly belongs, therefore, to the *arachno*-class (see Section 4.2.2).

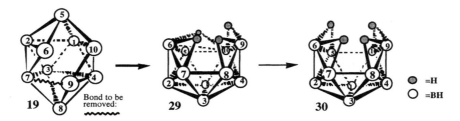

19	Bond to be removed:	**29**

30   ● =H   ○ =BH

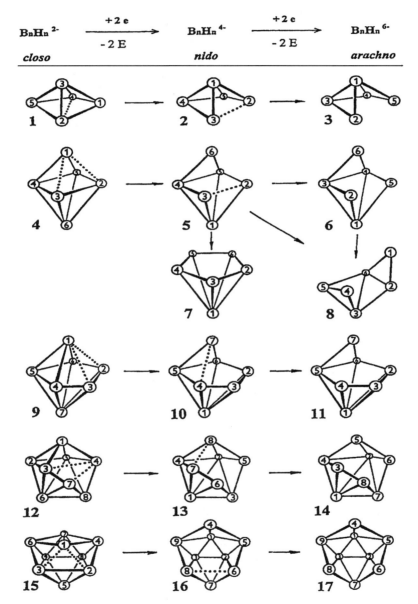

Figure 4-1 Interrelation between borane classes—seco principle. Dashed lines identify edges to be removed from individual skeletons by the addition of 2e.

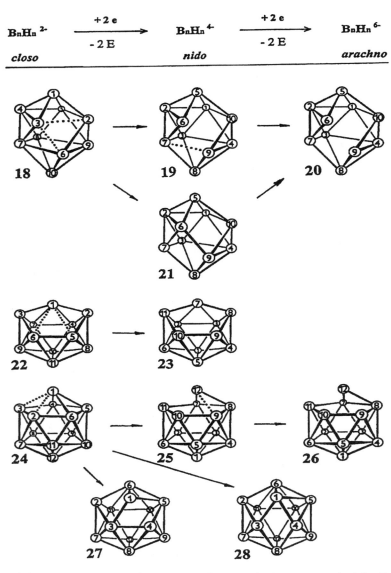

Figure 4-1 (*contd.*) Interrelation between borane classes—seco principle. Dashed lines identify edges to be removed from individual skeletons by the addition of 2e. As will be shown in the following text, the seco concept is not competitive, but rather is complementary to the debor concept in that it allows the deduction of additional skeletal types and implies that an explanation of skeletons with some weakened or almost missing skeletal bonds or connections may be possible.

By the same process, the six-membered open face of the hypothetical *arachno*-prototype (6ω)-$B_{10}H_{10}^{6-}$ (structure **20**) can theoretically be protonated to the (6ω)-$B_{10}H_{14}^{2-}$ anion (structure **31**):

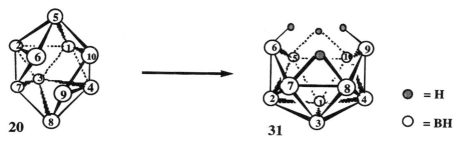

20　　　　　**31**　　　⬤ = H
　　　　　　　　　　　　　　○ = BH

These examples suggest that the 10-vertex *nido-* and *arachno*-prototypes are mutually exclusive and do not have identical skeletons as follows from the debor-concept.[1,2] This allows one to state that ***n*-vertex *closo-*, *nido-*, *arachno-*, and *hypho*-prototypes differ mutually in their number of edges**. And, vice versa, **the skeletons with the same number of vertices and edges belong to the same class** (see, e.g., structures **25**, **27**, and **28**).

4.2.2 The Size of Open Faces

If geometric relations among the redox classes shown in Figure 4-1 are valid, then *nido*-prototypes exhibit usually pentagonal open faces while *arachno*- prototypes show the hexagonal ones. The author of the debor concept has recently arrived at similar conclusions.[12]

A removal of edges produces further skeletons that are not readily accessible by the debor route, e.g. structures **10** and **11** (Figure 4-1).

In addition, the seco principle allows the removal of two nonadjacent edges from higher *closo*-skeletons, with the formation of two adjacent (structures **27** and **28**) or separated-squares open faces (4ω) (structure **32**):

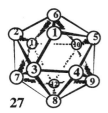
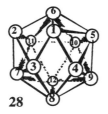
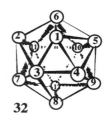

27　　　　　　**28**　　　　　　**32**

In this way, more than one species can be deduced with *nido*-skeletons that do not differ in the number of skeletal atoms n or the number of edges but do differ in the number of open faces. Which of these skeletons is preferred depends upon the character and number of the heteroatoms present, and their mutual locations can be inferred from Williams' Rule 3.[2] Nice examples can be found

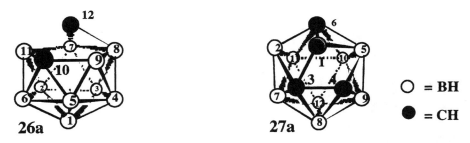

within the 12-vertex *nido*-series, namely, the dicarba anion **26a** [13] and the tetracarbadodecaboranes **27a**. [14]

4.2.3   Edge Tautomerism

The seco principle is compatable with the possibility that "edge tautomerism can take place," e.g., an interconversion of structure **27** to structure **28**, and further −2E 12-vertex analogs, all of which might result from the elimination of two edges from the *closo*-precursor. Such tautomerism can cause averaging of certain NMR signals and, consequently, mimic a skeleton of higher symmetry in some cases.

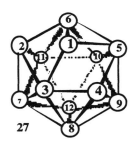

(1:2;1:4-diseco-)dodecahydrodo-
decaborate(2-) **27**

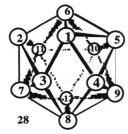

(1:2;3:4-diseco-)dodecahydrodo-
decaborate(2-) **28**

4.3   SECO NOMENCLATURE

In borane clusters, as in polycyclo-organic compounds (e.g., steroids), the position and size of an aperture can be described by means of the seco principle, which indicates the positions and number of eliminated edges, e.g., (1,2:3,4-diseco-)dodecahydrododecaborate(2−) (structure **28**). This "seco" nomenclature concept is not new [16a] and can replace the less suitable "debor" nomenclature description [16b] in which the number is sometimes higher than the number of actual vertices and which can confuse the reader.

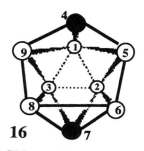

16   **33**

IUPAC Nomenclature [15]:	4,7-dicarba-*nido*-**nonaborate**(2–) (structure **16**)
DEBOR Nomenclature [16b]:	Nonahydro-1,10-dicarba-9-debor-*closo*-**decaborate** (2–) see (structure **33**)
SECO Nomenclature [16a]:	(1:2;1:3-diseco)-1,5-dicarbanonahydro-**nonaborate** (2–) see (structure **15**) in Figure 4-1

4.4   PARTIALLY OPEN SKELETONS AT METALLAHETEROBORANES

Further extension of the "seco" systematization accommodates the idea that not all of the electrons, available in the open face of the boron-cluster ligand, may be used in coordinating with the metal atoms located above the open face. An illustrative example is the $C_2B_9H_{11}^{2-}$ anion (dicarbollide = **dc**) representing a pentahapto (η^5) ligand. Despite the fact that this anion is one of the most electron-rich ligands, not all metal atoms are able to accommodate all six open-face electrons and thus form a fully symmetrically capped pentagon, i.e., an icosahedron.

Metal elements M, which incorporate nine atomic orbitals (i.e., $5d + s + 3p$), need 18 electrons to get a stable closed electron shell. These electrons originate from both the M^{n+} cation and from the free electron pairs associated with the ligands bound to the metal atom. Which of the ligands can be used and their number is governed by several factors, namely:

1. The number and distribution of electrons (paired and unpaired) located in the various d, s, and p atomic orbitals at the M^{n+} cation (x), which determines the number of occupied (y) and unoccupied (z) atomic orbitals (AO) of the nine that are available, e.g.:

$x = 3$ Cr^{III} (3 unpaired electrons in 3 AO; 6 unoccupied AOs)
   6   Fe^{II}, Co^{III}, Ni^{IV} (6e in 3 AO; 6 unoccupied AOs)
   6   Mo^0 (6e in 3 AO; 6 unoccupied AOs)
   7   Ni^{III} (7e in 4 AO, 1 unpaired e; 5 unoccupied AOs)
   8   Ni^{II}, Cu^{III}, Au^{III} (8e in 4 AO; 5 unoccupied AOs)
   9   Cu^{II} (9e in 5 AO, 1 unpaired e; 4 unoccupied AOs)
   11   Au^0 (11e in 6 AO, 1 unpaired e; 3 unoccupied AOs)

TABLE 4-1 Ligand Characteristics

Ligand	Formula	Ligand Notation	Donation (electron pairs)	Arrangement
Dicarbollide	$(C_2B_9H_{11})^{2-}$	**dc**	3	η^5
Cyclopentadienyl	$(C_5H_5)^-$	**cp**	3	η^5
Benzene	C_6H_6	**b**	3	η^6
Cyclobutadiene	C_4H_4	**cb**	2	η^4
Thiourea	$S_2 > C = N$		2	$M < S_2 > C = N$
Trialkylphosphine	$P(R)_3$		1	$M - P(R)_3$
Carbon monoxide	CO		1	$M - CO$

dc, Dicarbollide; cp, Cyclopentadienyl; b, Benzene; cb, Cyclobutadiene.

2. The number of unoccupied AOs (z), which indicates how many free electron pairs might be accepted from the individual ligands L; and

3. The character of the various ligands, i.e., the number of electron pairs available, their relative inclination to donate electrons, and their spatial arrangements; see examples in Table 4-1.

Uncomplicated representative examples of such complexes are the $Co^{III}(\mathbf{dc})_2^-$ (structure **35**) and $Ni^{IV}(\mathbf{dc})_2$ (structure **36**) sandwich compounds.[19]

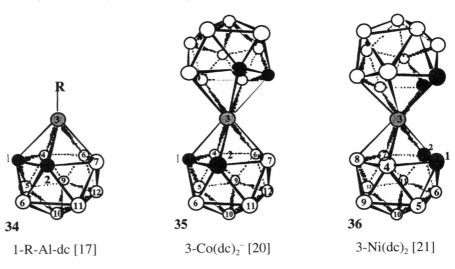

34
    1-R-Al-dc [17]

35
    3-Co(dc)$_2^-$ [20]

36
    3-Ni(dc)$_2$ [21]

In contrast to the above "symmetrical" compounds (structures **34**, **35**, and **36**), there are a number of bis-dicarbollide sandwich compounds in which the number of metal *dsp* electrons plus the 12 electrons supplied by the two dicarbollide anions surpass the total number of 18 electrons. In this case, any surplus electron cancels one three-center bond, i.e., one connection in the icosahedral parts of the sandwich. While the first surplus electron only weakens the bonds between the metal M and the **dc** ligands and keeps the B(10)-M-B(10′) axis intact, as in $Ni^{III}(dc)_2^-$,[22] the second and third surplus electrons cause lengthening of the M–C(1,2) and C(1′,2′) bonds leading to the formation of "slipped" sandwich structures, namely, in the $M(dc)_2$ complexes where M = Ni^{II},[19] Cu^{II},[23] Cu^{III},[24] and Au^{III}[25] (see structure types **37** and **38**).

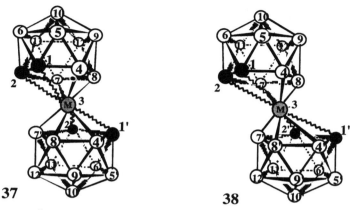

37 **38**

M = Ni^{II} [19], Cu^{II} [23], Cu^{III} [24], Au^{III} [25]

4. The hybridization that is typical for the various M^{n+} cations, e.g., sp for Hg^I, sp^3 for Al^{III}, dsp^2 for $Ni^{II}, Cu^{II}, Cu^{III}, Au^{II}, Au^{III}, Pd^{II}$, and d^2sp^3 for $Fe^{II}, Co^{III}, Ni^{IV}, Re^I, Mo^0, W^0, Au^0$, etc. This is probably a reason that sp^3 Al can form either the R–Al–**cb** complex (structure **34**) or the Al(**cb**)$_2$ sandwich structure (similar to structure **35** [28]); sp and dsp^2 metals can use only a part of one cluster-ligand and fill the opposite atomic orbitals by electron pairs from common mono- and bidentate ligands (see structure types **39** and **40**).

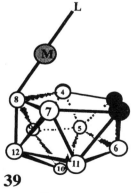

39

Me = Hg^{II}, L = PPh_3 [27]

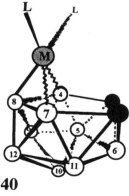

40

M = Pd^{II}, L = $P(Me_3)_2$ [29]
M = Pt^{II}, L = $P(Et_3)_2$ [30]
M = Au^{III}, L = S_2CNEt_2 [31]

5. The total charge of the resulting sandwich or coordination compound, the stability of which decreases sharply with increasing negative charge of the complex. A charge on the complex of (2–) seems to be the lower limit for stability. This is the reason that Mo^0, W^0, Mn^I, and Re^I (with six free atomic orbitals) cannot accommodate two dicarbollide(2–) cluster ligands, because the resulting sandwich compounds would have 4– and 3– charges, respectively. In these cases, the zero-charged CO ligands are the best counterparts (see structure **41** [19]).

REFERENCES

1. Williams, R. E., *Inorg. Chem.*, 1971, **10**, 210.
2. Williams, R. E., *Adv. Inorg. Chem. Radiochem.*, 1976, **18**, 67.
3. Williams, R. E., Prakash, G. K. S., Field, L. D., and Olah, G. A., in *Molecular Structure and Energetics, Vol. 5, Advances in Boron and the Boranes* (Liebman, J. F., Greenberg, A., and Williams, R. E., eds), VCH Publishers, New York, 1988, Chap. 9, p. 191.
4. Williams, R. E., in *Electron Deficient Boron and Carbon Clusters* (Olah, G. A., Wade, K., and Williams, R. E., eds), John Wiley & Sons, New York, 1991, Chap. 2.
5. Wade, K., *Chem. Commun.*, 1971, 792.
6. Wade, K., *Adv. Inorg. Chem. Radiochem.*, 1976, **18**, 1.
7. Wade, K., in *Electron Deficient Boron and Carbon Clusters* (Olah, G. A., Wade, K., and Williams, R. E. eds), John Wiley & Sons, New York, 1991, Chap. 3.
8. Rudolph R. W. and Pretzer, W. R., *Inorg. Chem.*, 1974, **13**, 248.
9. Hermánek, S., *Seco versus Debor Concept*, 23 May, 1985, The University of Munich, BRD.
10. Hermánek, S., *Chem. Revs*, 1992, **92**, 325.
11. Stíbr, B., Plesek, J., and Hermánek, S., *Collect. Czech. Chem. Commun.*, 1966, **34**, 194.
12. Williams, R. E., Chapter 1 in this volume.
13. Tolpin, E. I. and Lipscomb, W. N., *Inorg. Chem.*, 1973, **12**, 2257.
14. Freyberg, D. P., Weiss, R., Sinn, E., and Grimes, R. N., *Inorg. Chem.*, 1977, **16**, 1847.
15. Leigh G. J. (ed.), Nomenclature of Inorganic Chemistry, IUPAC, Recommendations 1990 Blackwell Scientific Publications, Oxford, 1990.
16. *Gmelin Handbuch der Anorganischen Chemie*, Springer, Berlin, Heidelberg, New York, 1974, Erg. zur 8. Aufg., Borver-bindungen 15, Teil 2, p. 62.
17. Churchill, M.R. and Reis, A.H., Jr, *J. Chem. Soc., Dalton Trans.*, 1972, 1317.
18. Ruhle, H. W. and Hawthorne, M. F., *Inorg. Chem.*, 1968, **7**, 2219.
19. Hawthorne, M. F., Young, D. C., Andrews, T. D., Howe, D. V., Pilling, R. L., Pitts, A. D., Reintjes, M., Warren, L. F., and Wegner, P. A., *J. Am. Chem. Soc.*, 1968, **90**, 879.
20. Zalkin, A., Hopkins, T. E., and Templeton, D. H., *Inorg. Chem.*, 1967, **6**, 191.

21. St. Clair, D., Zalkin, A., and Templeton, D. H., *J. Am. Chem. Soc.*, 1970, **92**, 1173.

22. Hansen, F. W., Hazell, R. G., Hyatt, C., and Stucky, G. D., *Acta Chem. Scand.*, 1973, **27**, 1210.

23. Wing, R. M., *J. Am. Chem. Soc.*, 1967, **89**, 5599.

24. Wing, R. M., *J. Am. Chem. Soc.*, 1968, **90**, 4828.

25. Colquhoun, H. M., Greenhough, T. J., and Wallbridge, M. G. H., *J. Chem. Soc. Chem. Commun.*, 1976, 1019.

26. Siedle, A. R., *J. Organometal. Chem.*, 1975, **90**, 249.

27. Colquhoun, H. M., Greenhough, T. J., and Wallbridge, M. G. H., *J. Chem. Soc., Dalton Trans.*, 1979, 619.

28. Bandman, M. A., Knobler, C. B., and Hawthorne, M. F., *Inorg. Chem.*, 1988, **27**, 2399.

29. Colquhoun, H. M., Greenhough, T. J., and Wallbridge, M. G. H., *J. Chem. Soc., Chem. Commun.*, 1978, 322.

30. Mingos, D. M. P., Forsyth, M. I., and Welch, A. J., *J. Chem. Soc., Dalton Trans.*, 1978, 1363.

31. Colquhoun, H. M., Greenhough, T. J., and Wallbridge, M. G. H., *J. Chem. Soc., Dalton Trans.*, 1978, 303.

PART II

THE CARBORANE–CARBOCATION CONTINUUM

5

BORON SUPERELECTROPHILES AND THEIR CARBOCATIONIC ANALOGS*

GEORGE A. OLAH

*Loker Hydrocarbon Research Institute, Department of Chemistry,
University of Southern California, University Park, Los Angeles, CA 90089-1661*

5.1 INTRODUCTION

Research observations accumulated over the years allowed me, in 1993 in a comprehensive paper,[1] to suggest the generalized concept of superelectrophiles. Protolytic (electrophilic) activation of electrophiles that are capable of further interaction (coordination) with strong Bronsted or Lewis acids results in their greatly enhanced reactivity. I also suggested that these activated electrophiles be named *superelectrophiles*. They include varied onium and carboxonium ion systems, showing that the concept of superelectrophilic activation is broad. Table 5-1 lists representative examples of superelectrophiles that have been thus-far investigated, and their parent electrophiles.

Despite the fact that oxonium ions are monopositive ions, it was found that their nonbonded electron pair is not inert and, with strong acids (Bronsted or Lewis), additional donor–acceptor interaction takes place; in the limiting case, gitonic dications are formed. Similar gitonic dications are formed by protonation (complexation, alkylation) of sulfur, nitrogen, halogen, *n*-donor oxonium, or carboxonium ions. Carbocations themselves can be similarly activated by protonation of their π- or σ-donor ligands.

Protolytic (electrophilic) activation can be considered to be the electrophilic equivalent of nucleophilic solvation. Nucleophilic solvation involves the inter-

---

*Dedicated to R. E. Williams with friendship and best wishes. For more years than either of us likes to remember, we have enjoyed comparing electron-deficient carbon with boron chemistry. I hope we will continue to do so for many more years.

TABLE 5-1 Examples of Superelectrophiles and the Electrophiles from which they are derived

Electrophile	Superelectrophile	Electrophile	Superelectrophile
$R_2\overset{+}{O}R$ (R, R, R on $\overset{+}{O}$)	$\overset{R}{\underset{R}{\,}}\overset{R}{O}^{2+}\overset{R}{\underset{R}{\,}}$	R_2X^+	R_3X^{2+}
		CX_3^+	$X_2\overset{+}{C}-XR^+$
$\overset{R}{\underset{R}{\,}}C=\overset{+}{O}R$	$\overset{R}{\underset{R}{\,}}\overset{+}{C}-\overset{+}{O}\overset{R}{\underset{R}{\,}}$	$R_2\overset{+}{C}NO_2$	$R_2\overset{+}{C}N\overset{+}{O}_2H$
$RC\equiv\overset{+}{O}$	$R\overset{+}{C}=OR$	$R_2\overset{+}{C}CN$	$R_2\overset{+}{C}C\overset{+}{N}H$
$R-\overset{+}{C}{\overset{OR}{\underset{OR}{\,}}}$	$R-C{\overset{\overset{+}{O}R}{\underset{\overset{+}{O}R_2}{\,}}}$	$R\overset{+}{C}=NH$	$R\overset{2+}{C}-NH_3$
HCO^+	$H\overset{+}{C}=\overset{+}{O}H$	$C^+(NH_2)_3$	$(H_2N)_2\overset{+}{C}-\overset{+}{N}H_3$
$HO\overset{+}{\underset{\overset{|}{O}H}{C}}OH$	$HO\,\overset{+}{\underset{HO}{C}}-\overset{+}{O}H_2$	$R_2C=\overset{+}{N}H_2$	$R_2\overset{+}{C}-\overset{+}{N}H_3$
R_3S^+	R_4S^{2+}	$H_2N\overset{+}{N}_2$	$H_2N\overset{2+}{N}_2H$
R_3Se^+	R_4Se^{2+}	NO_2^+	$NO_2\overset{2+}{H}$
R_3Te^+	R_4Te^{2+}		

A:R = H, alkyl, Lewis acid; B:X = Cl, Br; I:R = H, alkyl. In each group, the parent electrophiles are in the left column.

action of an electron-deficient species with a nucleophilic (i.e., electron-donor) solvent. In contrast, electrophilic solvation is the activation of an electron-deficient species that is capable of further coordination with strongly acidic media or complexing agents.

In any electron-deficient molecule, there is inevitably intramolecular electron flow from neighboring groups into the deficient center (Winstein's concept of neighboring group participation). At the same time, if electron-deficient species are put in a nucleophilic environment (solvent), intermolecular interaction (solvation) occurs. If the surrounding environment is only weakly nucleophilic, external interaction can be limited, but, of course, internal participation remains. With even stronger acids (or superacids), electrophilic solvation can set in. Such intermolecular protolytic interaction can decrease neighboring group participation by forcing the neighboring groups to interact intermolecularly with the external electrophile (acid) and thus decrease intramolecular participation.

Electrophilic solvation opens up new vistas in chemistry. Besides current investigations by active research groups in the area of superacid chemistry, some observations had been already reported in the literature, but were not recognized until recently and remained to be discovered.

5.2 COMPARISON OF CARBON AND BORON ELECTROPHILES AND SUPERELECTROPHILES

In 1979, Bach and Badger [2] found that carbon tetrachloride-aluminum chloride is a very reactive chlorinating agent for adamantane. In recent years, three groups (my group,[3a] Jean Sommer's group in Strasbourg,[3b] and Denis Sunko's group in Zagreb [3c]) have found that, in conversion of saturated hydrocarbons, the activity of protic superacids such as $HF-SbF_5$ or FSO_3H-SbF_5 is greatly enhanced in the presence of carbon tetrachloride, chloroform, or even, in certain cases, methylene chloride. This was surprising in view of the fact that we are dealing with the strongest protic acids known to mankind, and these extremely strong acids were found to be further activated by a chloroalkane such as carbon tetrachloride. Sommer et al. and Sunko et al. suggested the reasonable explanation that carbon tetrachloride ionizes to trichloromethyl cation in superacids acting as a reservoir for the initial carbocation. Trichloromethyl cation will start the reaction sequence by intermolecular hydride abstraction from the saturated hydrocarbons, which leads to subsequent carbocationic conversions. However, when my group, with my colleague Surya Prakash, prepared and studied long-lived trihalomethyl cations, including CCl_3^+, under stable ion conditions,[4.5a] we found that the C-13 chemical shifts of these carbocations are only relatively moderately deshielded compared with those corresponding trihalomethanes (CHX_3), indicating the highly delocalized resonance-stabilized nature of these ions [6] (Table 5-2).

Trichloromethyl, tribromomethyl, and triiodomethyl cations do not show highly deshielded carbenium centers as a consequence of obvious strong backdonation from the nonbonded halogen electron pairs into the carbenium center.[6] However, it should be pointed out that recent studies [5b] have shown that, particularly in the case of CBr_3^+ and CI_3^+, other factors such as spin-orbit coupling can affect chemical shifts, although the observed deshielding effects compared with trihalomethanes are not much different. Furthermore, theoretical calculations [5c] have revealed that the reaction of $^+CCl_3$ with propane to give isopropyl cation is exothermic by only -3.9 kcal/mol. In any case, these very strongly resonance-stabilized ions hardly could be highly electrophilic reactive species. Stability of intermediate carbocations and their reactivity are not parallel but are generally opposing properties. The more stable is a carbocation, generally the less reactive it is. This being the case, why is then carbon tetrachloride such a good activating agent for superacid systems? The reason suggested is that, in the presence of strong Bronsted or Lewis acids, the nonbonded halogen electron pairs will interact with them. This protolytic or electrophilic interaction can be depicted as a donor–acceptor interaction, which involves the halogen atoms, in a sense amounting to electrophilic solvation.[5c] This inter-

$$\text{CX}_4 \xrightarrow[-78°C]{n\text{SbF}_5/\text{SO}_2\text{ClF}} {}^+\text{CX}_3{}^-\text{Sb}_n\text{F}_{5n}\text{X}$$
$$X = C, Br, I$$

$$\text{CCl}_3\text{SO}_2\text{Cl} \xrightarrow[-78°C]{n\text{SbF}_5/\text{SO}_2\text{ClF}} {}^+\text{CCl}_3{}^-\text{Sb}_n\text{F}_{5n}\text{Cl} + \text{SO}_2$$

$$\text{CCl}_3\text{CO}_2\text{Cl} \xrightarrow[-78°C]{n\text{SbF}_5/\text{SO}_2\text{ClF}} {}^+\text{CCl}_3\text{Sb}_n\text{F}_{5n}\text{Cl}^- + \text{CO}$$

TABLE 5-2   Comparison of ^{13}C NMR Chemical Shifts of Trihalomethyl Cations and the Corresponding Trihalomethanes

Trihalomethyl cation	δ^{13}C	Trihalomethane	δ^{13}C	Δ^{13}C
$^+$CCl$_3$	236.3	HCCl$_3$	77.7	158.6
$^+$CBr$_3$	207	HCBr$_3$	+12.3	194.7
$^+$Cl$_3$	95	HCl$_3$	−139.7	234.7

action decreases back-donation (neighboring halogen participation) into the carbenium center and thus enhances the carbocationic reactivity.

Whereas the depicted interaction is only monodentate, all n-donor halogen atoms can be involved to some degree (i.e., the equivalent of electrophilic solvation). Further de facto ionization to the dichloromethyl dication can also be considered. A more detailed discussion of this question is, however, outside the scope of the present discussion (a comprehensive discussion is available elsewhere [5b]), except to note that the boron analog for CCl$_2^{2+}$ is BCl$_2^+$. Indeed, in our studies, when we generated the trichloromethyl cation from trichlorofluoromethane with antimony pentafluoride in the presence of a hydride donor isoalkane besides the expected chloroform, we also found that methylene chloride formed in the system. This could be indicative that a second ionization is indeed taking place, or at least that a highly polarized dicationic species is involved.

The energetics and structural parameters of the trichloromethyl cation, as well as its ^{13}C NMR chemical shift, were calculated using the ab initio/IGLO (individual gauge for localized orbitals) methods.[5a] The calculated and

$$CCl_3F \xrightarrow{SbF_5} \overset{+}{C}Cl_3 \underset{\rightleftharpoons}{\xrightarrow{H^+}} Cl_2\overset{+}{\underset{+}{C}}ClH \xrightarrow{RH} R^+ + CHCl_2 + H^+$$

with $\downarrow H^+$ branch:

$$H_2Cl^+ + [CCl_2^{2+}] \xrightarrow{RH} R^+ + \overset{+}{C}HCl_2$$

$$\downarrow {-Cl^-}$$

$$R^+ + \overset{+}{C}HCl_2$$

$$\downarrow RH$$

$$R^+ + CH_2Cl_2$$

experimental shift (in superacid) show a significant divergence.[5b] To rationalize this observation, we also calculated at the ^{13}C shift of the protiotrichloromethyl dication and found that the ^{13}C NMR shift moved in the direction of the experimentally observed value of $^+CCl_3$ in excess superacid. It is thus reasonable to suggest that the observed NMR shift of $^+CCl_3$ is affected by protolytic interaction (solvation) by the superacidic solvent.

3h

$C-Cl = 1.647$

$\delta^{13}C = 289.3$
exp. $\delta^{13}C = 236.0$

$E = -1416.50473$ hartrees

Cs

1.837
1.602   1.605

$\delta^{13}C = 262.1$

$E = -1416.50067$ hartrees

The close isoelectronic–isostructural relationship between trivalent carbenium ions and the corresponding neutral boron compounds has been frequently pointed out.[7] The trichloromethyl cation is the carbocation analog of boron trichloride.

The question thus arises as to how boron trichloride behaves under similar superelectrophilic activation. In 1959, the late Earl Muetterties reported [8] that if benzene is reacted with boron trichloride in the presence of aluminum trichloride and aluminum powder (to remove formed HCl), phenylborondichloride is obtained in good yield:

$$C_6H_6 + BCl_3 \xrightarrow{AlCl_3} C_6H_5BCl_2 + HCl$$

The reaction was used subsequently as a general method to prepare areneborondihalides. Onak has summarized these results [9] and Table 5-3 shows some of the compiled results.

TABLE 5-3   Summerized Results of Areneborondihalides

A reasonable mechanism for the reaction is suggested, in line with the fore-goings, to involve superelectrophilic activation of boron trichloride. Boron trichloride itself is only a relatively weak boron electrophile because of extensive $n–p$ back-donation. When boron trichloride is dissolved in benzene or other aromatic hydrocarbons, no reaction takes place; only the formation of weak π-complexes is observed. However, when $AlCl_3$ (a strong Lewis acid) is added, its complexation with the nonbonded chlorine electron pairs activates boron trichloride by decreasing back-donation, i.e., neighboring chlorine participation. Fast reaction with the aromatic hydrocarbon results in the formation of arenedichloroboranes:

Friedel-Crafts borylation of arenes

The Lewis acidity of boron trichloride (i.e., its boron electrophilicity) is enhanced by complexation with either Lewis or Bronsted acids. The close relationship of BCl_3 with the discussed carbon electrophilicity of CCl_3^+ (and other halomethyl cations) is apparent.

Whether there is also an ionization equilibrium involving the BCl_2^+ cation, which is the boron analog of CCl_2^{2+}, as mentioned, is still under investigation.

$$BCl_2 \cdot AlCl_3 \rightleftharpoons BCl_2^+ AlCl_4$$

Moreover, it is noteworthy that DePuy has recently obtained gas-phase data in his glowing discharge system for BH_2^+.[10] BH_2^+ is the boron analog of the parent CH_2^{2+} carbodication.

There are other relevant examples where the carbocation-borane or carbodication-boron cation analogy can be applied. One is CH_4^{2+}, the methane dication (or protomethyl dication).[11] Theoretical calculation of CH_4^{2+} shows it to be a planar species with an empty perpendicular p orbital. We have studied, its boron analog, BH_4^+, or the $BH_4^+ \cdot BH_3$ adduct both experimentally and theoretically, some years ago when we investigated the behavior of diborane in protic (deuterio) superacids.[12] Hydrogen–deuterium exchange was observed and we indicated involvement of the corresponding protonated ion(s).

Recently, CH_4^{2+} and higher homolog protoalkyl dications were studied, both theoretically and experimentally.[13]

Alkyl cations, such as the *tert*-butyl cation, are long lived in the superacidic media. This shows that no deprotonation equilibria with the corresponding olefins can be present. If this was the case, the olefin, a good π-base, and the

$$CH_4^{2+}$$

$$
\begin{array}{c}
H \\
\quad \diagdown \\
\qquad C \text{-}\text{-}\text{-}\!\!<^{H^{2+}}_{H} \\
\quad \diagup \\
H
\end{array}
$$

$$BH_4^{+}$$

$$
\begin{array}{c}
H \\
\quad \diagdown \\
\qquad B \text{-}\text{-}\text{-}\!\!<^{H^{+}}_{H} \\
\quad \diagup \\
H
\end{array}
$$

$$C(CH_3)_3H^{2+}$$

$$
\begin{array}{c}
CH_3 \\
\quad \diagdown \\
\qquad C\text{—}CH_2\text{-}\text{-}\!\!<^{H}_{H} \\
\quad \diagup \\
CH_3
\end{array}
\Bigg]^{2+}
$$

$$B(CH_3)_3H^{+}$$

$$
\begin{array}{c}
CH_3 \\
\quad \diagdown \\
\qquad B\text{—}CH_2\text{-}\text{-}\!\!<^{H}_{H} {}^{+} \\
\quad \diagup \\
CH_3
\end{array}
$$

alkyl cation, a very strong acid, would immediately react (giving oligomers, polymers, cyclized products, etc.). Thus, when observing the *tert*-butyl cation in superacid solution, we are assured that no isobutylene is simultaneously present in the system. Despite this, we have observed that, upon addition of strong deuterated superacid to a solution of the *tert*-butyl cation, slow hydrogen–deuterium exchange of the methyl hydrogens takes place. The only explanation for this is that, despite the fact that the *tert*-butyl cation is a monopositive relatively small ion, the C—H bonds of methyl groups involved in hypercon-jugative stabilization of the carbocationic center are still capable of acting as σ-donors in their intermolecular interactions with the protic superacid. In the limiting case, de facto deuteriation to deuterio, *tert*-butyl dication or equivalent deuterio (proto)solvation takes place. This interaction decreases the hypercon-jugative stabilization (σ-participation) and thus further activates the carboca-tion. Other alkyl cationic systems, such as the protoisopropyl cation, were also studied.[14]

The protolytic activation of alkyl cations by strong acid systems is consid-ered to be at least partly responsible for many facile hydrocarbon transformation processes in acid solutions, even at relatively low temperatures. Solid acid cata-lyst can also be effective when acidic sites are sufficiently close for similar activation.[15]

Trimethylboron in superacidic media also shows hydrogen–deuterium exchange, which can be similarly explained.[16]

$$
\begin{array}{c}
CH_3 \diagdown \diagup CH_3 \\
\qquad B \\
| \\
CH_3
\end{array}
\underset{\longleftarrow}{\overset{\text{"}D^+\text{"}}{\longrightarrow}}
\begin{array}{c}
CH_3 \diagdown \\
\qquad B\text{—}CH_2\text{-}\text{-}\!\!<^{H}_{D} \\
CH_3 \diagup
\end{array}\Bigg]^{+}
\xrightarrow{-\text{"}H^+\text{"}}
\begin{array}{c}
CH_3 \diagdown \\
\qquad B\text{—}CH_2D \\
CH_3 \diagup
\end{array}
$$

Hawthorne et al. reported[17] that protonation of a carborane dianion result-ed in a borane cation that subsequently inserted into hydrocarbons. This can also be considered to be an example of activation of a boron electrophile.

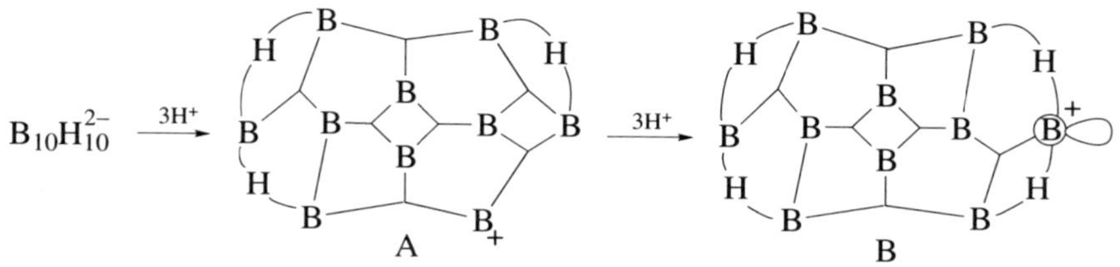

$$B_{10}H_{10}^{2-} \xrightarrow{3H^+} A \xrightarrow{3H^+} B$$

$B^+ = BH$ $\textcircled{B} = B$ Possible $B_{10}H_{13}^+$ Isomers

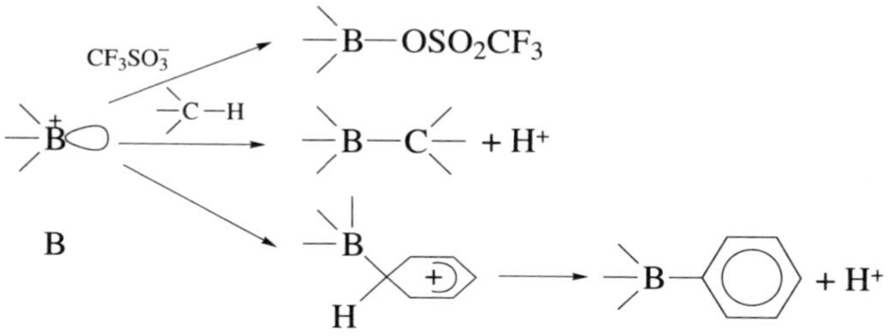

A further example of superelectrophilic boron activation is involved in the fluorine exchange reaction of boron trifluoride with antimony pentafluoride.[16] Studying the system with ^{19}F NMR, we observed fluoride scrambling, which indicates that BF_3 coordinates with the very strong Lewis acid SbF_5, in a similar way to the $BX_3 \cdot AlX_3$ systems.

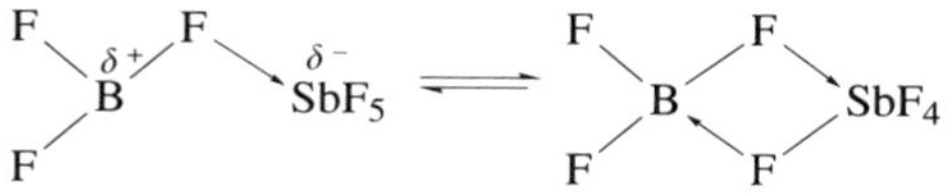

Antimony pentafluoride exists in a dimeric (trimeric) form and is a well-recognized Lewis "super" acid. This is the consequence of its remarkable mono-fluorine-bridged nature.[18] A very limited ionization equilibrium can also be involved:

Antimony pentafluoride was found to react readily with aromatics, such as benzene, to give phenylantimonytetrafluoride, and through further reaction with excess benzene to give triphenylantimony difluoride.[19] Antimony pentafluoride "self-activates" in its monofluorine-bridged dimer, which is superelectrophilic in nature.

In the aluminum-chloride-catalyzed Friedel-Crafts reaction of carbon tetrachloride with benzene (or other aromatics), the polarized $Cl_3CCl \rightarrow AlCl_3$ complex or the CCl_3^+ ion (even though it is resonance stabilized) can readily react with the π-electron-donor aromatics. The coordination of carbon tetrachloride with SbF_5 or Sb^V chloridefluorides is also the basis of the Swartz fluorine-exchange reaction giving CFCs:[20]

$$Cl_3C\diagdown\!\!\!\overset{Cl}{\underset{F}{\diamond}}\!\!\!\diagup SbCl_{4-x}F_x$$

Considering aluminum chloride, it has been known since Friedel and Crafts' studies 120 years ago that it is dimeric (except in donor solvent systems, where the complexed monomer is present). In the extensive literature of Friedel-Crafts chemistry,[20] it was frequently pointed out that the dimer, with its established doubly chlorine-bridged structure, itself cannot be the Lewis acid catalyst, because it is coordinatively saturated and thus lacks electron-deficient Lewis acid activity. It was therefore assumed that an equilibrium existed with the monomer, leading to the active Lewis acid. I would like to suggest, however, that if only one of the chlorine bridges is broken, the resulting equilibrium involves the monochlorine-bridged dimer, which, of course, is a very powerful superelectrophile for the same reasons discussed previously for the boron trichloride–aluminum chloride coordination complex. Again, a very limited ionization equilibrium is possible with $AlCl_2^+ AlCl_4^-$:

This explains why aluminum chloride could react, even in the absence of any proton sources giving strong conjugate Bronsted acids, as an aluminum super-electrophile with olefins, aromatics, etc. Under most conditions, however, protic impurities or co-acids are always present and lead to Bronsted acid catalysis.

$$ArH + Al_2Cl_6 \longrightarrow ArAlCl_2 + HAlCl_4$$

$$RCH{=}CH_2 + Al_2Cl_6 \longrightarrow R^+CH{-}CH_2AlCl_2\ \bar{A}lCl_4 \xrightarrow{RCH=CH_2} polymer$$

While discussing activating boron electrophiles, some comments on diborane itself are in order. Borane in noncoordinating solvents is always dimeric. The structure of the dimer is well established as a doubly hydrogen-bridged species. Thus, B_2H_6 should not be a boron electrophile. It is always assumed, however, that there is an equilibrium with monomeric BH_3, which is the de facto electrophile.

$$B_2H_6 \rightleftharpoons 2BH_3$$

H. C. Brown found [21] that complexation of borane with ether allows hydroboration with diborane in an efficient manner. In ether (or other *n*-donor solvent), borane is monomeric and can be transferred, as such, to an alkene. It seems, however, also feasible to consider that if only one of the hydrogen bridges of diborane opens up, this will also result in an activated very reactive boron electrophile:

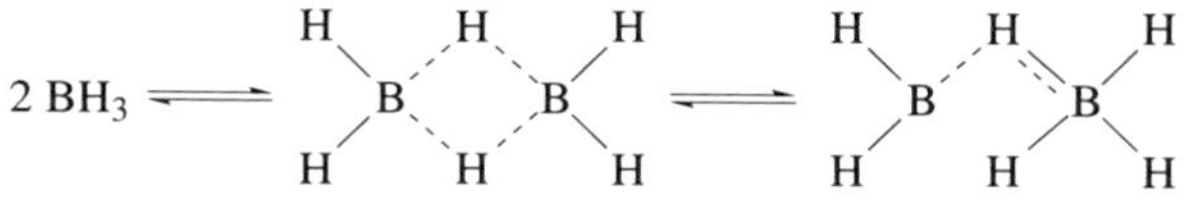

Monobridged diborane was mentioned before by Fehlner [22b] but has not received since much attention. Lipscomb also once discussed the unsymmetrical bridging of diborane:[23]

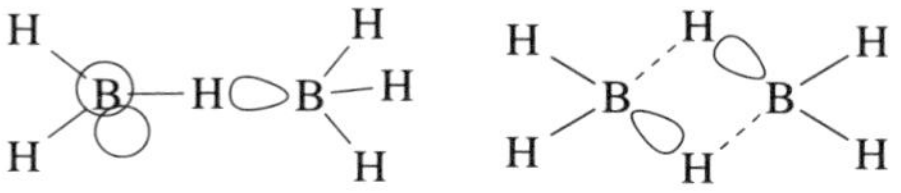

The foregoing consideration seems to give a reasonable explanation for the reactivity of diborane in electrophilic reactions involving apolar hydrocarbon systems, or even in gaseous reaction at moderate temperatures. In 1948, Hurd described reaction of diborane, not only with alkenes, but also with arenes and even with alkanes (including methane), to give derived boron compounds.[24] He carried out his reactions with diborane generally in the gas phase without any solvent. Under these conditions, monobridged reactive B_2H_6 could be involved as the reactive boron electrophile. Energetically, breaking two hydrogen bridges is certainly less favorable than breaking only one. The initial interaction of borane or activated diborane with alkenes is recognized as an electrophilic reaction. So too are the insertion reactions into C—H or C—C bonds initially reported by Hurd. The reaction takes place much more readily with silanes, as shown in studies by Olah et al.[25] It is, however, H. C. Brown's merit to have discovered the effect of ether (and related n-donor solvents) on the olefin hydroboration reaction and subsequently to have developed hydroboration to a general and most useful synthetic reaction.[21]

5.3 PROTONATED BORIC ACIDS

The last class of activated boron electrophiles to be discussed is that of protonated boric acid and its analogs. The carbon analog of orthoboric acid $B(OH)_3$ is protonated carbonic acid ($H_3CO_3^+$). Olah et al.[26] found that $H_3CO_3^+$ is a very stable species under superacid conditions. One of the first chemical experiments I ever did as a youngster was to take vinegar and drop some sodium bicarbonate or carbonate into it and watch the bubbles of carbon dioxide evolve. Many years later, using 4:1 $FSO_3H:SbF_5$ (Magic Acid) in repeating the experiment, to my initial surprise no bubbles appeared in the solution. Subsequently, using ^{13}C-labeled carbonate or bicarbonate, we were able to understand why there was no evolution of CO_2 at lower temperatures. The reason is that a stable solution of protonated carbonic acid was formed. It is stable because it is highly resonance stabilized and thus is a trioxa analog of the guadinium ion, which is one of most abundant ions in nature.

$$H_2O + CO_2 \rightleftharpoons H_2CO_3$$

$$\begin{array}{c} CO_3^{2-} \\ or\ HCO_3^- \end{array} \xrightarrow[-80°]{FSO_3H\text{-}SbF_5/SO_2} H_3CO_3^+ \xrightarrow[-10-0°C]{} CO_2 + H_3O^+$$

More recently, we reinvestigated the system and found that if we further increase the acidity (to 1:1 $FSO_3H:SbF_5$), then carbonic acid is protonated a second time. Seemingly, one charge is delocalized sufficiently on the carbocationic carbon and two of the neighboring hydroxyl groups, leaving the third hydroxyl group available for further protonation. The diprotonated gitonic carbonic acid

displays interesting cleavage chemistry, first into diprotonated CO_2, which then deprotonates into monoprotonated CO_2. The $CO_2H_2^{2+}$ and CO_2H^+ ions are of substantial interest, both theoretically and also from the practical point of view.

For diprotiocarbonic acid $H_4CO_3^{2+}$, the boron analog is protonated boric acid $B(OH)_2OH_2^+$:

Interestingly, protonated boric acid was only recently observed in the gas phase and by theoretical studies.[27] Its involvement in acidic solution chemistry, however, is now recognized.

Boric acid is a very weak boron electrophile because the three O—H groups strongly back-donate n electrons into the vacant boron p orbital. Consequently, boric acid does not react with nucleophiles, such as fluoride ion. However, if boric acid is reacted with HF or HSO_3F, it forms BF_3 quantitatively. It is suggested that the strong protic acids protonate (or protosolvate) boric acid, decreasing neighboring oxygen participation and thus enhancing the boron electrophilicity. Consequently, ready reaction of the activated boron center with fluoride ion takes place, thereby giving BF_3. More recently, similar reaction of trimethyl borate and other trialkyl(aryl)borates was also observed:

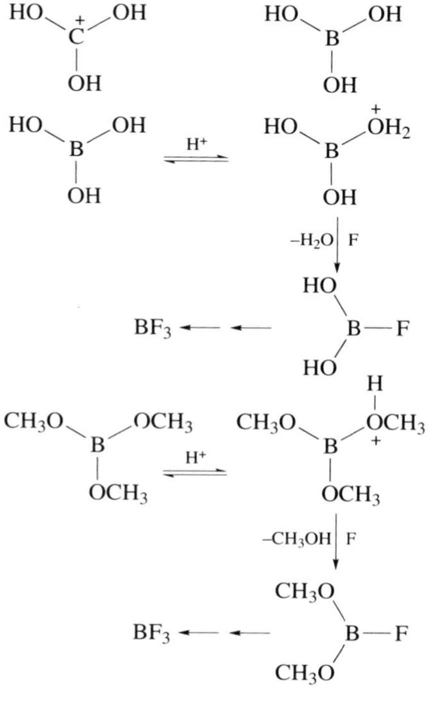

5.4   CONCLUSIONS

Superelectrophilic activation of electrophiles capable of further coordination with either Bronsted or Lewis acids is a general and effective way to facilitate a wide variety of chemical transformations. The discussed examples show that boron superelectrophiles, not unlike their carbocationic analogs, are readily generated and their chemistry is of substantial significance.

REFERENCES

1. Olah, G. A., *Angew. Chem. Int. Ed. English*, 1993, **32**, 761, and references therein.
2. Bach, R. D. and Badger, R. C., *Synthesis*, 1979, 529.
3. (a) Olah, G. A., et al., *J. Am. Chem. Soc.*, 1991, **113**, 3203; *J. Org. Chem.*, 1989, **54**, 14631; 1990, **55**, 1224; (b) Sommer, J., et al. *J. Chem. Soc. Chem. Comm.*, 1989, 1049; 1990, 1098; (c) Sunko, D. et al., *J. Am. Chem. Soc.*, 1990, **112**, 7418.
4. Olah, G. A., Heiliger, L., and Prakash, G. K. S., *J. Am. Chem. Soc.*, 1989, **111**, 8020.
5. (a) Olah, G. A., Rasul, G., Heiliger, L., and Prakash, G. K. S., *J. Am. Chem. Soc.*, 1996, **118**, 3580; (b) Kaupp, M., Malkina, O., and Malkin, V. G., personal communication, July 1996, *Chem. Phys. Lett.*, 1997, **265**, 55; (c) Olah, G. A., Rasul, G., Yudin, A. K., Burrichter, A., Prakash, G. K. S., Akhrem, I. S., Gambaryan, N. P., and Vol'pin, M. E., *J. Am. Chem. Soc.*, 1996, **118**, 1446.
6. Olah, G. A. and Schleyer, P. v. R. (eds), *Carbonium Ions*, Vol. V, Wiley-Interscience, New York, 1976, pp. 2135–2262.
7. (a) Olah, G. A., DeMember, J. R., Commeyras, A., and Bribes, J. L., *J. Am. Chem. Soc.*, 1971, **93**, 459; (b) Olah, G. A., Prakash, G. K. S., Williams, R. E., Field, L. D., and Wade, K., *Hypercarbon Chemistry*, Wiley-Interscience, New York, 1987, pp. 191–213.
8. Muetterties, E., *J. Am. Chem. Soc.*, 1959, **81**, 2597.
9. Onak, T., *Organoborane Chemistry*, Academic Press, New York, 1975, p. 36.
10. Ch. DePuy, personal communication, December 1994.
11. (a) Wong, M. W. and Radom, L., *J. Am. Chem. Soc.*, 1989, **111**, 1155; (b) Stahl, D., Maquin, F., Gäumann, T., Schwarz, H., Carrupt, P. A., and Vogel, P., *J. Am. Chem. Soc.*, 1985, **107**, 5049.
12. Olah, G. A., Aniszfeld, R., Prakash, G. K. S., Williams, R. E., Lammertsma, K., and Güner, O. F., *J. Am. Chem. Soc.*, 1988, **110**, 7885.
13. Olah, G. A., Hartz, N., Rasul, G., and Prakash, G. K. S., *J. Am. Chem. Soc.*, 1993, **115**, 6985.
14. Olah, G. A., Hartz, N., Rasul, G., Prakash, G. K. S., Burkhart, M. and Lammertsma, K., *J. Am. Chem. Soc.*, 1994, **116**, 3187.
15. Olah, G. A., and Molnar, A., *Hydrocarbon Chemistry*, Wiley-Interscience, New York, 1995.
16. Olah, G. A., et al., unpublished results.
17. Hawthorne, M. F., et al., *J. Am. Chem. Soc.*, 1992, **114**, 4427.
18. (a) Olah, G. A., *Friedel-Crafts Chemistry*, Wiley-Interscience, New York, 1973; (b) Cotton, F. A. and Wilkinson, G., *Advanced Inorganic Chemistry*, 5th Edition, Wiley-Interscience, New York, 1988, pp. 393–394.
19. Olah, G. A., Schilling, P., and Gross, I. M. *J., Am. Chem. Soc.*, 1974, **96**, 876.
20. (a) Olah, G. A., *Friedel-Crafts and Related Reactions*, Vol. I, Wiley-Interscience, New York, 1963, p. 150.
21. (a) Brown, H. C., *Boranes in Organic Synthesis*, Cornell University Press, Ithaca, NY, 1972; (b) Brown, H. C., *Organic Syntheses via Boranes*, Wiley, New York, 1975.

22. (a) Mappes, G. W., Fridmann, S. A., and Fehlner, T. P., *J. Phys. Chem.*, 1970, **74**, 337; (b) Fehlner, T. P., in *Advances in Boron and the Boranes* (eds Liebman, J. F., Greenberg, A., and Williams R. E.,), VCH Publishers, New York, Weinheim, 1988, p. 269.

23. Dixon, D. A., Repperberg, I. M., and Lipscomb, W. N., *J. Am. Chem. Soc.*, 1974, **96**, 1325.

24. Hurd, D. T., *J. Am. Chem. Soc.*, 1948, **70**, 2053.

25. Olah, G. A., Field, L. D., Pacquin, D., and Suemmerman, K., *Nuov. J. Chim.*, 1983, **7**, 279.

26. Rasul, G., Reddy, V. P., Zdunek, L. Z., Prakash, G. K. S., and Olah, G. A., *J. Am. Chem. Soc.*, 1993, **115**, 2236.

27. (a) Attina, M., Cacace, F., and Ricci, A., paper presented at the Congress of the Italian Chemical Society at San Benedetto del Tronto, 1990; (b) Attina, M., Cacace, F., Grandinetti, F., Occhiucci, G., and Ricci, A., *Int. J. Mass Spectrom.*, 1992, **117**, 47.

6

EXTENSION OF THE BORANE–CARBOCATION CONTINUUM TO CAGE SYSTEMS

G. K. Surya Prakash , Golam Rasul, Andrei K. Yudin, and Robert E. Williams

Loker Hydrocarbon Research Institute, Department of Chemistry, University of Southern California, University Park, Los Angeles, CA 90089-1661

6.1 INTRODUCTION

Boron and carbon are consecutive first-row elements. It follows that a neutral tetravalent carbon is isoelectronic with a tetravalent boron anion or a borane–Lewis base adduct.[1] Similarly, electron-deficient trivalent carbocations are isoelectronic with the corresponding neutral trivalent boron compounds. The structural similarity of trivalent boron compounds and trivalent carbocations was established experimentally by comparing their vibrational spectra. The infrared and Raman spectra of $(CH_3)_3C^+$ (structure **1**) in SO_2ClF solution were studied in 1971 by Olah et al.,[2] and vibrational frequencies, the number of lines, and the activity of those lines were correlated with the spectrum of $(CH_3)_3B$ (structure **2**). The similarity between the skeletal modes of vibration in $(CH_3)_3B$ (structure **2**) and $(CH_3)_3C^+$ (structure **1**), provided unambiguous evidence that the two species possessed analogous structures and bonding; i.e., the carbon skeleton in $(CH_3)_3C^+$ (structure **1**) is planar with C_{3v} symmetry. An X-ray structural analysis of (structure **1**) has recently been reported.[3] The infrared spectrum of CH_3^+ (structure **3**) in the gas phase was obtained by Oka and coworkers in 1985.[4] Analysis of the infrared bands strongly suggests a planar structure with D_{3h} symmetry, in agreement with theory.[5] The expected similar structure for the isoelectronic boron analog, BH_3 (structure **4**), is based on infrared analysis.[6]

In 1971, extrapolating from the square-pyramidal structure of pentaborane (structure **5**), Williams proposed [7] a similar square-pyramidal structure for $C_5H_5^+$ (structure **6**). Subsequent support for the $C_5H_5^+$ structure came from extended Hückel moleculer-orbital MO calculations by Stohrer and Hoffmann in 1972.[8] Masamune et al. prepared and characterized [9] the dimethyl analog $(CH_3)_2C_2H_3^+$ (structure **7**), whose structural and isoelectronic relationship with $1,2\text{-}(CH_3)_2B_5H_7^+$ (structure **8**) has been established.

Extending similar carbocation–borane analogies from trigonal to pentacoordinate species, Olah et al. proposed [10] in 1972 that the carbon analog of BH_5 (pentahydridoborane) (structure **9**) is the parent hypercoordinate carbocation, CH_5^+ (structure **10**). The structures of both (**9**) and (**10**) have been extensively studied by both theory and experiment. Current theoretical studies have focused on the analogy of CH_6^{2+} to BH_6^+ [11] and CH_7^{3+} to BH_7^{2+}.[11a, 12] Recently, DePuy et al. were able to generate BH_6^+ in the gas phase.[11b]

The analogy between carbocations and boranes is quite apparent in the ability of CH frangments to replace BH_2 groups in polyboranes, thus leading to carboranes. Based upon such an analogy, a *nido*-carbocation analog of B_6H_{10} (structure **11**), i.e., $Me_6C_6^{2+}$ (structure **12**), has been prepared and characterized, based upon ^1H and ^{13}C NMR spectroscopic data, by Hogeveen and coworkers.[13] The dication (structure **12**) can be depicted in several different ways. It can be represented as one of five possible canonical three-center two-electron (3c-2e) resonance forms.

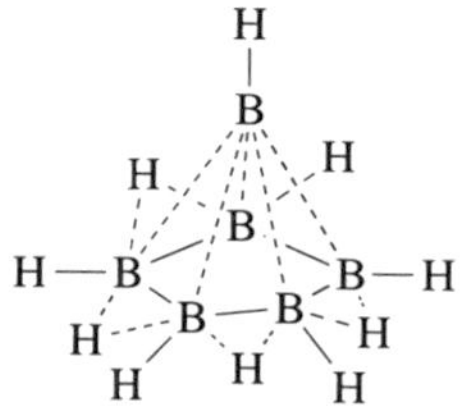

11 **12** etc.

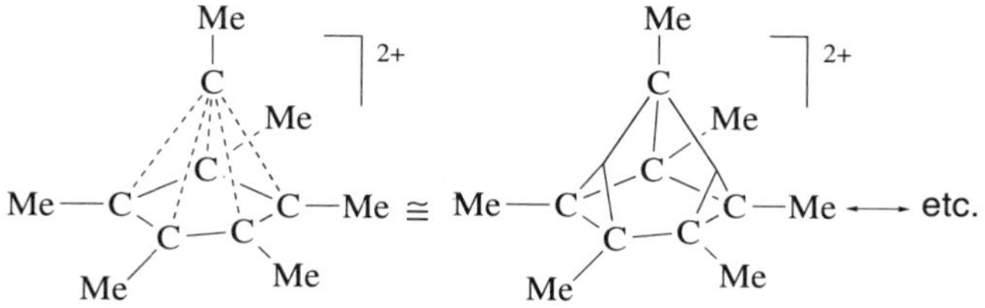

Similar analogies have been pointed out for many other carbocations and borane derivatives, which are shown below (structures **13** and **14**, **15** and **16**, **17** and **18**, and **19** and **20**).[14]

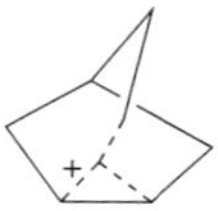

13 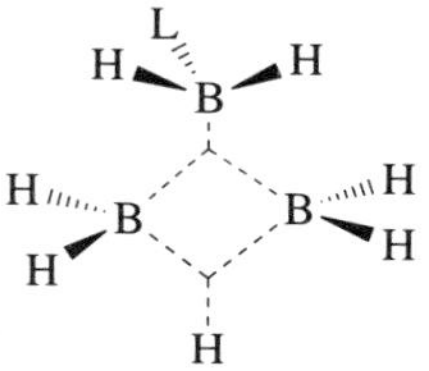 **14**

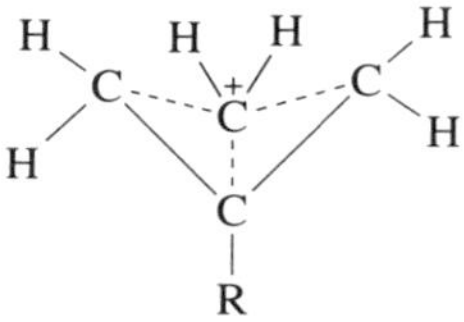

15 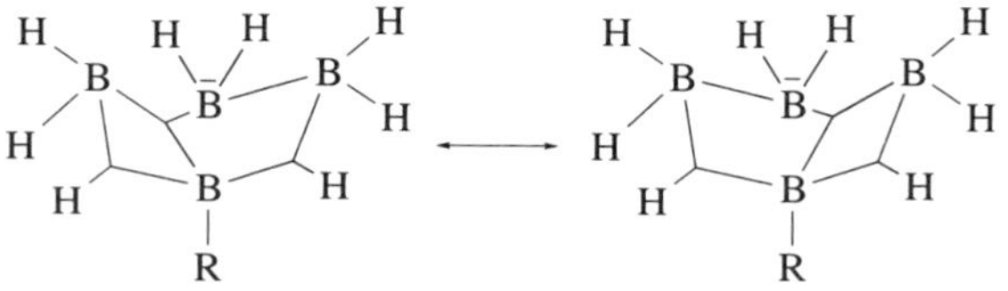 **16**

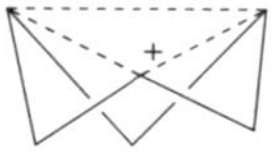

17 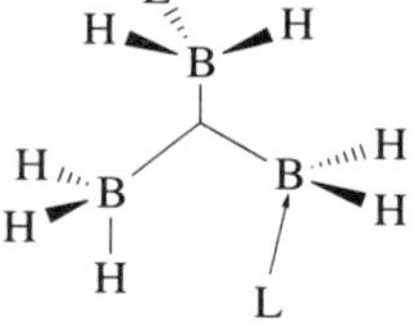 **18**

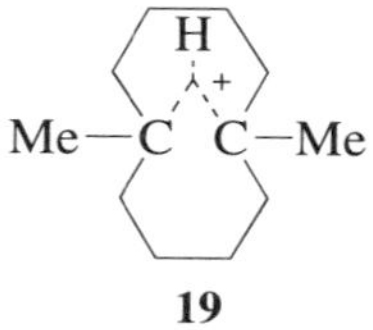

19 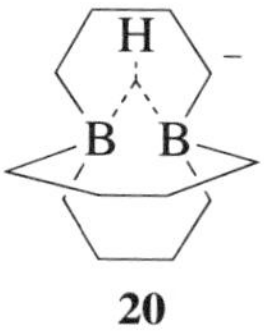 **20**

The close relationship between the ^{11}B NMR chemical shifts of the boron atoms in boron compounds and the ^{13}C NMR chemical shifts of the corresponding trigonal carbons in carbocations was first shown by Spielvogel and Purser [15a] and by Nöth and Wrackmeyer.[16] The general correlation equation for trigonal species is:

$$\delta^{11}B(BF_3:OEt_2) = 0.40 \, \delta \, ^{13}C_{(TMS)} - 46 \quad\quad (1)$$

In Eq. (1), the $\delta^{11}B$ is the chemical shift of the ^{11}B nucleus is in ppm with respect to the $BF_3:OEt_2$ absorption, while $\delta^{13}C$ is the chemical shift of the cationic carbon of the corresponding carbocation is with respect to the tetramethylsilane signal.

Trigonal compounds involving alkyl, halogen, oxygen, or hydrogen substituents correlate with Eq. (1), while trigonal compounds incorporating phenyl or cyclopropyl groups deviate from Eq. (1) to different degrees. Equation (1) is restricted to the comparison of neutral electron-deficient sp^2 boron compounds and sp^2 carbocations. However, if the ^{11}B chemical shifts of electron-precise tetracoordinate borate anions are compared with the ^{13}C chemical shifts of electron-precise neutral carbon analogs, they fall close to a similar line.

A general equation [Eq. (2)] has been derived by incorporating both electron-deficient trigonal and electron-precise tetrahedral ^{11}B and ^{13}C values in the analysis:

$$\delta^{11}B(BF_3:OEt_2) = 0.33 \, \delta \, ^{13}C_{(TMS)} - 30 \quad\quad (2)$$

Eqs (1) and (2) have been derived empirically. They are in good agreement with most of the available data that span some 600 ppm on the ^{13}C chemical shift scale. Combining both electron-deficient trigonal and nonclassic carbocations gave the following empirical equation [Eq. (3)][14]

$$\delta^{11}B(BF_3:OEt_2) = 0.41 \, \delta \, ^{13}C_{(TMS)} - 53 \quad\quad (3)$$

Clearly, the same factors that determine the chemical shifts of the boron nuclei also govern the chemical shifts of carbon nuclei. Now, we would like to extend this borane–carbocation continuum to some cage carbocationic systems and their isoelectronic and isostructural boron analogs.

6.2   RESULTS AND DISCUSSION

The structures of the cage carbocations (dications) and their boron analogs have been optimized at the DFT[17] B3LYP/6-31G* level using the GAUSSIAN 94[18] package of programs. The ^{13}C and ^{11}B NMR chemical shifts were calculated by IGLO methods. Calculations were performed according to the method reported [19] at the IGLO DZ level using B3LYP/6-31G* geometries. The ^{13}C NMR chemical shifts are referenced to TMS and the ^{11}B NMR chemical shifts are referenced to $BF_3:OEt_2$. Many of the cage carbocations (dications) have been observed experimentally. However, the corresponding boron analogs, with the exception of 1-boradamantane, have not been synthesized. Their structures and ^{11}B NMR chemical shifts are obtained solely from theoretical calculations.

6.2.1 1-Adamantyl Cation

Over several decades, extensive research has been carried out in investigating the so-called diamondoid molecules, of which adamantane, $C_{10}H_{14}$ (structure **21**), is of primary interest. The tight interlocking of cyclohexane rings into the rigid, relatively strain-free chair conformation in adamantane disallows easy formation of double bonds (e.g., through 1-adamantyl cation (structure **22**) and there is no possibility of back-side (nucleophilic or electrophilic) attack.

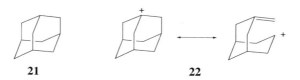

21   **22**

Of particular interest is the bridgehead 1-adamantyl cation (structure **22**), which has been prepared and characterized under long-lived stable ion conditions.[20] The X-ray structure of an adamantyl cation analog is also known.[21]

The ^{13}C NMR spectrum of 1-adamantyl cation, (structure **22**) shows four absorptions at δ^{13}C 300, 87, 66, and 34. The deshielded peak at δ^{13}C 300 due to the carbocationic center C^+, however, is 36 ppm more shielded than that of the *tert*-butyl cation. This reflects the extensive C—C hyperconjugative interaction of the neighboring three C—C bonds with the p orbital of the carbocationic center. This is also in agreement with our DFT calculated structure (compound **22a**) of 1-adamantyl cation (Figure 6-1). Thus, the C2—C3 bond distance is 1.629 Å, significantly longer than carbon–carbon single bonds. On the other hand, the C1—C2 bond distance, 1.453 Å, is between a carbon–carbon single and a carbon–carbon double bond. The structure of 1-adamantyl cation previously calculated by Schleyer et al. at the MP2(FU)/6-31G* level [22] agrees very well with our DFT results. In addition, the analogous 3,5,7-trimethyl-1-adamanyl cation has been calculated at the DFT B3LYP/6-31G* level [22] and, as mentioned earlier, the X-ray structure of this cation is also known.[21]

We have calculated the structure of compound **22b**, an isoelectronic boron analog of 1-adamantyl cation, at the B3LYP/6-31G* level. There is also significant C—C hyperconjugative interaction by three neighboring C—C bonds with the empty p orbital of the boron atom (Figure 6-1). The C2—C3 bond distance of 1.580 Å is about 0.04 Å longer than a carbon–carbon single bond.

We have calculated the ^{13}C and ^{11}B NMR chemical shifts of compound **22a** and **22b** at the IGLO/DZ//B3LYP/6-31G* level. The calculated δ^{13}C value of the C^+ atom in compound **22a** is 324.5 (w.r.t. TMS) and the calculated δ^{11}B of compound **22b** is 71.7 (w.r.t. BF_3:OEt_2), to be compared with the experimental

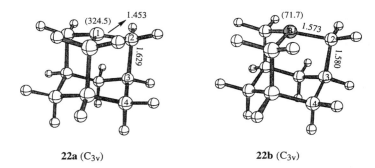

22a (C$_{3v}$)   22b (C$_{3v}$)

Figure 6-1   The calculated B3LYP/6-31G* structures and IGLO chemical shifts (in parentheses) of structures **22a** and **22b**.

values of 300 ppm and 73.2 ppm,[23] respectively. We also estimated the $\delta^{11}B$ values in compound **22b** by applying Eq. (3) and by using the experimental $\delta^{13}C$ value of 300 for C$^+$ in compound **22a**. This yields a $\delta^{11}B$ value of 69.9 for the neutral boron compound **22b**. The $\delta^{11}B$ of 69.9 is very close to those of the IGLO calculated value, 71.7, and the experimental value, 73.2. This good correlation reconfirms the close relationship between the ^{11}B and ^{13}C chemical shifts of isoelectronic carbon and boron analogs **22a** and **22b**, respectively.

6.2.2   1,3-Dehydro-5,7-adamantanediyl Dication

By exploiting the guidance furnished by the theoretical calculations, Schleyer et al. were able to prepare[24] and characterize the intriguing 1,3-dehydro-5,7-adamantanediyl dication (structure **23a**) under long-lived stable ion conditions. The dication **23a** is an example of a three-dimensional aromatic system in which the p orbitals of the four bridgehead carbons overlap in a tetrahedral fashion involving two electrons. The experimental $\delta^{13}C$ value of 6.6 for the four eqivalent bridgehead carbons clearly attests to the nonclassic nature of the ion. Schleyer et al. have also calculated [24a] the structures and the ^{13}C NMR chemical shift value of the dication **23a** and these agree very well with the experimental data.

We have now recalculated the structures of compound **23a** and its isoelectronic neutral boron analog **23b** at the DFT B3LYP/6-31G* level of theory. The calculated structure of compound **23a** clearly shows its three-dimensional aromatic character. Interestingly, the calculated structure of the boron analog **23b** also indicates the existence of three-dimensional aromaticity in which the p orbitals of the two carbons and the two bridgehead borons overlap in a pseudotetrahedral fashion involving two electrons (Figure 6-2). The calculated $\delta^{13}C$ values of the bridgehead carbon of compound **23a** are 7.0, which is very close to that of the experimental value of 6.6. The calculated $\delta^{11}B$ value of compound **23b** is –49.2. The estimated $\delta^{11}B$ value of compound **23b** obtained from apply-

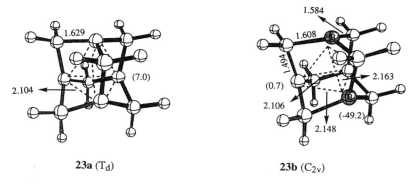

23a (T$_d$)   23b (C$_{2v}$)

Figure 6-2   The calculated B3LYP/6-31G* structures and IGLO chemical shifts (in parentheses) of structures **23a** and **23b**.

ing Eq. (3) and using a $\delta^{13}C$ value of 6.6 for compound **23a** is −50.4 ppm, which is similar to that of the IGLO calculated value of −49.2 ppm. This close correlation reconfirms the close relationship between ^{11}B and ^{13}C chemical shift values within the nonclassic isoelectronic carbon and boron analogs **23a** and **23b**, respectively.

Such homoconjugative interactions have been proposed earlier, based upon theoretical calculations, in the cases of 7-boranorbornene and 7-boranorbornadiene [24b]

6.2.3   Dodecahedryl Cation

Olah, Prakash, Paquette, and their coworkers were able to generate [25] dodecahedryl cation **24a** by ionizing chlorododecahedrane in a solution of SbF$_5$/SO$_2$ClF at −78°C. The ^{13}C NMR spectrum reveals six peaks at $\delta^{13}C$ values of 363.9, 81.1, 64.4, 64.1, 63.0, and 60.9. These data allow the species to be assigned to the C_{3v} symmetric dodecahedryl cation **24a** (Figure 6-3). This cation had been anticipated to undergo facile hydrogen scrambling through 1,2-hydride shifts similar to that observed in the related cyclopentyl cations, a process that would render all the carbon and hydrogen atoms equivalent with a 20-fold degeneracy. However, no such degenerate rearrangements were observed, as indicated by the lack of change in the ^1H NMR line shapes, even when a solution of compound **24a** was allowed to warm to 0°C. From this observation, the lower barrier for such a degenerate rearrangement in compound **24a** was estimated to be approximately 15 kcal/mol.

The ^{13}C NMR shifts of the positively charged center in compound **24a** ($\delta^{13}C$ value of 363.9) happens to be the one of the most deshielded signals ever observed for a monocarbocationic species, indicating the highly localized positive charge at this center. The application of the ^{13}C NMR shifts additivity criterion[26] estimated a net deshielding of 283 ppm, in agreement with the formation of a monocation. However, the magnitude of deshielding of the positive charge in compound **24a** is less than that observed in a typical tertiary carbocation, such as the 1-methyl-1-cyclopentyl cation ($\Sigma \Delta\delta$ = 374 ppm). This indicates the prevalence of some unique cage-shielding effect in the dodecahedryl cation. The origin of this effect is difficult to identify because of the complexity of the multiple bond connectivities.

We have fully optimized the structures of dodecahedryl cation **24a** and its unknown isoelectronic boron analog **24b** at the B3LYP/6-31G* level. The cation **24a** exhibits considerable deformation around the cationic center, as expected from the preference of a trigonal cation to adopt a planar sp^2 geometry (Figure 6-3). Complete planarization, however, is precluded within the spherical framework as a direct consequence of the simultaneous increase in bond-angle distortion at the adjacent tetragonal center. These forces are balanced in cation **24a** at a pyramidalization level of about 10°, expressed as the out-of-plane bending angle of the central atom relative to the plane defined by its three bonding partners. For comparison, the corresponding value for a regular sp^3-hybridized corner in dodecahedrane is about 21°. The situation calculated for cation **24a** compares nicely with the structure of the adamantyl cation **22a**, which has out-of-plane deviation of 9°. In contrast to the electronic structure of adamantyl cation **22a**, the β-bonds (1.569 Å) in the dodecahedryl cation **24a** are not optimally oriented to stabilize the empty p orbital by C—C hyperconjugation, as they have a dihedral angle of close to 90°.

Similarly, the isoelectronic and isostructural boron analog **24b** exhibits deformation around the boron center (Figure 6-3). The pyramidalization level about the boron atom in cation **24b** is about 14°, which is 4° more than that of the dodecahedryl cation **24a**. However, the β-bond (1.568 Å) in cation **24b** is

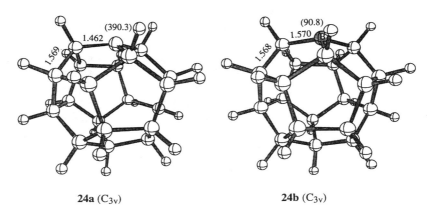

24a (C$_{3v}$)   **24b** (C$_{3v}$)

Figure 6-3   The calculated B3LYP/6-31G* structures and IGLO chemical shifts (in parentheses) of structures **24a** and **24b**.

almost same as that of the dodecahedryl cation **24a**, which argues for similar C—C hyperconjugation in both molecules.

The calculated $\delta^{13}C$ value (of C$^+$) in cation **24a** and the $\delta^{11}B$ value in cation **24b** are 390.3 and 90.8, respectively. The calculated $\delta^{13}C$ value of the C$^+$ in cation **24a** (390.3) is about 26 ppm more deshielded than that of the experimental value (363.9). The calculated $\delta^{11}B$ value in cation **24b**, derived from the application of Eq. (1) and using the experimental $\delta^{13}C$ value of C$^+$ in cation **24a** is 99.6, which is also close to that of the IGLO calculated value of 90.8. This reconfirms the close relationship between the ^{11}B and ^{13}C chemical shifts of the isoelectronic carbon and boron analogs **24a** and **24b**, respectively.

6.2.4   1,16-Dodecahedryl Dication

The 1,16-dodecahedryl dication **25a** has also been generated in superacid solutions.[25] The dodecahedryl cation **24a**, upon standing in the superacid solutions for 6–7 h at 50°C, is slowly and irreversibly rearranged into the 1,16-dodecahedryl dication **25a**. The ^{13}C NMR spectrum reveals three peaks at $\delta^{13}C$ values of 379.2, 78.8, and 59.8. This observation clearly indicates the formation of a static D_{3d} symmetric 1,16-dodecahedryl dication **25a**. The formation of this unique, highly electron-deficient dication can be rationalized by protolytic ionization of the C—H bond at position 16 (farthest from the cationic center at position 1) involving a hypercarbon intermediate.

The ^{13}C NMR shift of the positively charged center in cation **25a** ($\delta^{13}C$ value of 379.2) is 15.3 ppm more deshielded than that of the dodecahedryl cation **24a**. This is the most deshielded ^{13}C NMR shift ever observed for any carbocationic species, again indicating high localization of the positive charge at these centers. Application of the ^{13}C NMR shifts additivity criterion [26] reveals a net

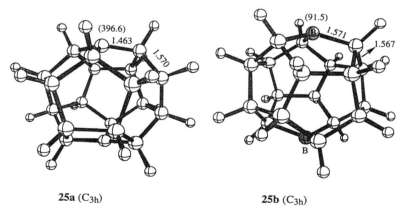

25a (C$_{3h}$)   **25b** (C$_{3h}$)

Figure 6-4   The calculated B3LYP/6-31G* structures and IGLO chemical shifts (in parenthese) of **25a** and **25b**.

deshielding of 610 ppm, in agreement with the formation of dication. However, the magnitude of deshielding per unit positive charge in cation **25a** is less than that observed in the 1-methyl-1-cyclopentyl cation ($\Sigma \, \Delta\delta = 374$ ppm), which, again, indicates the presence of some unique cage-shielding effect in the 1,16-dodecahedryl dication.

We have fully optimized the dodecahedryl dication **25a** and its isoelectronic boron analog **25b** at the B3LYP/6-31G* level. As expected, cation **25a** also exhibits substantial deformation around the cationic centers (Figure 6-4). The pyramidalization level about the boron atom in cation **25a**, about 10°, is almost the same as that in the monocation **24a**. The β-bond length (1.570 Å) in the 1,16-dodecahedryl dication **25a** is also the same as that in **24a**.

The calculated $\delta^{13}C$ value of C$^+$ in cation **25a** is 396.6 and the calculated $\delta^{11}B$ of cation **25b** is 91.5. The calculated $\delta^{13}C$ value of C$^+$ in cation **25a** (396.6) is about 17 ppm more deshielded than the experimental value (379.2 ppm). The estimated $\delta^{11}B$ value in cation **25b**, obtained by applying Eq. (1) (and by using the experimental $\delta^{13}C$ value of 105.6 for C$^+$ in cation **24a**, is somewhat close to that of the IGLO calculated value of 91.5.

6.2.5   [1.1.1.1]Pagodane Dication

During the attempted superacid-catalyzed isomerization of the undecacyclic hydrocarbon [1.1.1.1]pagodane compound **26** to the pentagonal dodecahedrane, the formation of the intriguing, stable [1.1.1.1] pagodane dication **27** was discovered.[27] Dication **27** may be considered to be a frozen Woodward-Hoffmann transition-state analog.[28]

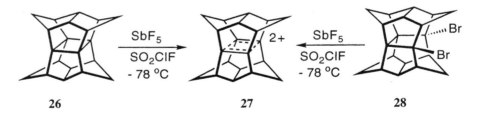

deshielding of 610 ppm, in agreement with the formation of dication. However,

When compound **26** was dissolved in an excess of SbF$_5$:SO$_2$ClF at –78°C, the solution turned yellow instantaneously. The proton spectrum, recorded immediately, was composed of complex broad signals in the aliphatic region, indicating the presence of paramagnetic radical cations.[29]

After standing for about 3 h at constant temperature, the proton and carbon spectra had simplified to a four-line pattern. The solution revealed a very clean ^1H NMR spectrum, δ^1H 3.37 (br, 8H), 3.68 and 2.72 (AX doublets, $J_{H-H} = 13.2$ Hz, 8H), and 2.39 (br, 4H). The 50-MHz ^{13}C NMR spectrum of the same solution at –80°C showed only four peaks at $\delta^{13}C$: 251.0 (singlet), 65.3 (triplet, J_{C-H} = 141.9 Hz). The observed symmetry and the extent of deshielding in both ^1H and ^{13}C NMR spectra of the species in SbF$_5$/SO$_2$ClF solution when compared with the progenitor pagodane (compound **26**) [δ^1H, 2.60 (4H, bridgehead), 2.24 (8H, bridgehead), 1.56 and 1.60 (8H, methylene, J_{AB}= 10 Hz); δ^{13}C, 62.9 (singlet), 59.6 (doublet), 42.7 (doublet), and 41.9 (triplet)] seem to imply that the species is ionic in nature and has the D_{2h} symmetry of the parent pagodane itself. The dication in solution was found to be surprisingly stable.

The same dicationic species were successfully obtained from different precursors. Under the same ionizing conditions employed for pagodane (structure **26**), the ionization of the dibromide (structure **28**) produced the same dication (structure **27**). The "closed" dication (structure **29**) (the "real pagodane dication") or the "open" dication (structure **30**), however, were excluded as possible structures.

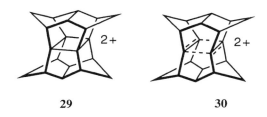

29   30

The dication **27** is considered to be the first representative of a novel class of 2e-"aromatic" pericyclic systems, topologically equivalent to the transition state for the Woodward-Hoffmann "allowed" cycloaddition of ethylene to ethylene dication.[28]

The pagodane dication **27** has also been subjected to MNDO and HF/3-21G level calculations.[27,30] Previous ^{13}C NMR chemical shift additivity analysis had ruled out the possibility of describing dication **27** as an average structure of several classic dications. In spite of the large strain imposed by the planar cyclobutane moiety in the polycyclic framework, some in-plane aromatic stabilization must occur, leading to bishomoaromatic 2e stabilization in dication **27**. Support for the bishomoaromatic dication **27** also comes from IGLO ^{13}C NMR chemical shift calculations.[30]

The B3LYP/6-31G* optimized structure of the pagodane dication **27a** and its isoelectronic neutral boron analog **27b** are given in Figure 6-5. The calculated structure of dication **27a** is consistent with the 2e-"aromatic" pericyclic systems. The calculated structure of the boron analog, structure **27b**, also supports the possibility of 2e-"aromatic" pericyclic systems in the molecule (Figure 6-5). The calculated δ^{13}C of aromatic carbon in dication **27a** is 248.4, which is very close to that of the experimental value (251.0 ppm). The calculated δ^{11}B value in dication **27b** is 86.5. The δ^{11}B value in dication **27b**, estimated by the application of Eq. (1) and by using the δ^{13}C value in dication **27a**, was assessed to be 53.4 ppm, which is at higher field by 33 ppm than the IGLO calculated value of 86.4 ppm. This unusual 33 ppm deshielding of boron-based relative IGLO calculated value can be rationalized by consideration of the strain in the molecular framework and lack of pericyclic aromatic character in the dibora analog **27b**.

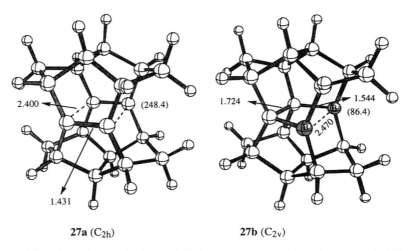

27a (C_{2h})   **27b** (C_{2v})

Figure 6-5   The calculated B3LYP/6-31G* structures and IGLO chemical shifts (in parentheses) of structures **27a** and **27b**.

6.2.6 [1.1.1.1]Isopagodane Dication

Recently, the highly symmetric [1.1.1.1]isopagodane (structure **31**) was synthesized [31] in order to explore the unusual cage radical cations and dications. Even though the interaction of ethylene and ethylene radical cation to form cyclobutane radical cation is symmetry-forbidden, such a 4c-3e radical cation was found stabilized in the rigid [1.1.1.1]pagodane and [1.1.1.1]isopagodane skeletal frameworks.[32,33] On the other hand, as explained above, the interaction of ethylene and ethylene dication is symmetry-allowed based on the Woodward-Hoffmann theory.[30,34]

Dissolution of [1.1.1.1]isopagodane (structure **31**) in SbF$_5$/SO$_2$ClF solution at −78°C (using a dry ice–acetone bath) gave a dark greenish blue solution, indicating initially the formation of the radical cation **32**. Upon prolonged vortex stirring, the dark greenish blue solution disappeared to give a light-yellow solution. The 75-MHz ^{13}C NMR spectrum of the solution suggested a species with lower C_2 symmetry than that of the parent D_{2d} isopagodane itself.[27] Accordingly, seven absorptions were observed at δ^{13}C 251.7 (s), 66.8 (d, J_{C-H} = 169 Hz), 66.5 (t, J_{C-H} = 139.6 Hz), 63.4 (d, J_{C-H} = 149.3 Hz), 51.2 (d, J_{C-H} = 163.0 Hz), 45.3 (d, J_{C-H} =154.3 Hz) and 41.7 (t, J_{C-H} = 148.8 Hz). The 300-MHz proton NMR spectrum showed six peaks, at δ^1H 3.35 (b, 4H), 3.2 (b, 4H), 2.76 (b, 2H), 2.27 (b, 2H), 1.84 (b, 2H) and 1.72 (b, 2H). Based on the observed NMR deshieldings of both carbons and protons as well as the behavior of isomeric [1.1.1.1]pagodane in SbF$_5$/SO$_2$ClF [34,] it was presumed to be a structurally new dication, i.e., the [1.1.1.1]isopagodane dication **33**, a novel four-center two-electron cyclobutane dication. The lowering of the symmetry of the dication compared with that of isopagodane itself indicates that facile degenerative

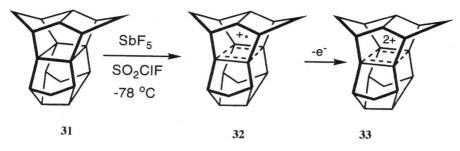

exchange between two identical structures has a high barrier (more than 18 kcal/mol).[35] The oxidation appears to occur stepwise through the intermediacy of the 4c-3e radical cation,[33] as indicated by the intial dark greenish blue color of the solution followed by subsequent oxidation to the dication.

We have also carried out a geometry optimization of the isopagodane dication **33a** and its isoelectronic boron analog **33b** at the B3LYP/6-31G* level. The bonding in structures **33a** and **33b** can be characterized as 4c-2e σ-bishomoaromatic systems.[30,34] They may also be depicted as frozen Woodward-Hoffmann transition-state analogs.[34]

The ^{13}C NMR chemical shifts were also calculated for [1.1.1.1]isopagodane dication **33a** and its isoelectronic neutral boron analog **33b** at the IGLO/DZ//B3LYP/6-31G* level of theory. The agreement is very good between the calculated (δ ^{13}C value of 257.5) and the observed chemical shifts (δ^{13}C value of 251.7) of aromatic carbon of dication **33a**. The ^{13}C NMR chemical shift additivity analysis[26] was also applied to structure **33**. The total chemical

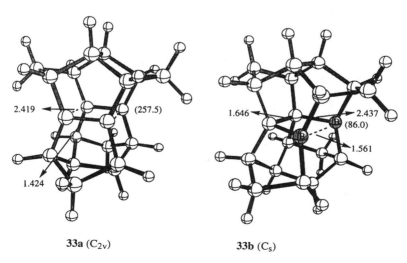

Figure 6-6　The calculated B3LYP/6-31G* structures and IGLO chemical shifts (in parentheses) of structures **33a** and **33b**.

shift difference of 910 ppm is observed for structure **33**. More than 450 ppm deshielding per unit positive charge seems to imply strain in the 4c-2e σ-bisomoaromatic dicationic frameworks. The calculated ^{11}B NMR chemical shift of analog **33b** is, again, unusually deshielded at 86 ppm (similar to analog **27b**).

6.2.7   [1.1.1.1]Secopagodane and [1.1.1.1]Isosecopagodane Cations

Dissolution of bissecododecahedradiene (structure **34**)[36] in the CF$_3$SO$_3$H-B(OSO$_2$CF$_3$)$_3$/SO$_2$ClF system at –78°C resulted in the desired monoprotonation to produce the 2-secopagodyl cation **35**. This was confirmed by independent generation of cation **35** by the ionization of 2-chlorosecopagodane (structure **36**) in SbF$_5$/SO$_2$ClF at –78°C. All attempts to protonate pagodane (structure **26**) to generate cation **35** with various oxidizing and nonoxidizing superacids were unsuccessful. Under these conditions, the dication **27** was formed exclusively by rapid two-electron oxidation. The 50-MHz ^{13}C NMR spectrum of each light-yellow-colored solution showed the same 12 ^{13}C resonances. The peaks were δ^{13}C 279.2 (C2), 140.6 (C11), 129.1 (C1), 56.4 (C3, C15), 55.6 (C16), 55.1 (C6, C17), 53.8 (C12), 52.3 (C5, C13), 47.3 (C10, C18), 40.0 (C4, C14), 37.5 (C8, C20), and 30.0 (C9, C19). These results are in accord with the formation of the monocation **35** with C_s symmetry. The secocation **35** was formed by protolytic ionization of secopagodane (structure **37**)[37] in SbF$_5$-FSO$_3$H/SO$_2$ClF. However, no protonation to bissecododecahedryl cation **38** was observed. In the superacidic medium, the cation **35** was found to be quite stable even at 0°C.

The relatively small, low field shift of the C2 carbon (δ^{13}C value of 279.2) in (structure **35**) indicates that the electron deficiency is largely alleviated by C—C hyperconjugation, i.e., by electron delocalization from the C1—C11 σ-bond, as is represented by resonance structures **35a′** and **35b′**. Such a phenomenon was reported,[20] e.g., for tertiary 1-adamantyl cation **22**, wherein the bridgehead γ-methine carbons are much more deshielded than the β-methylene carbons. This has been interpreted theoretically [38] and was substantiated by X-ray crystallographic studies.[39] From the magnitude of the relative chemical shift differences, it is concluded that the C—C hyperconjugation in (structure **35**) is much more pronounced than that in (structure **22**).

The additivity of ^{13}C NMR chemical shift analysis [26] for (structure **35**) gives a total chemical shift difference of 385 ppm relative to (structure **37**) (as reference for dication **35a**), which is well within the range of classic cations (350–400 ppm). This is in contrast to a comparison with bissecododecahedrene (as the hydrocarbon reference corresponding to dication **32b**) that resulted in a net difference of only 169 ppm.

Clearly, the intriguing C1—C11 bonding situation in structure **35** stems from two components: (1) molecular strain imposed by the special polyquinane framework and (2) electronic weakening by perfectly coplanar alignment with the empty C2 p-orbital.

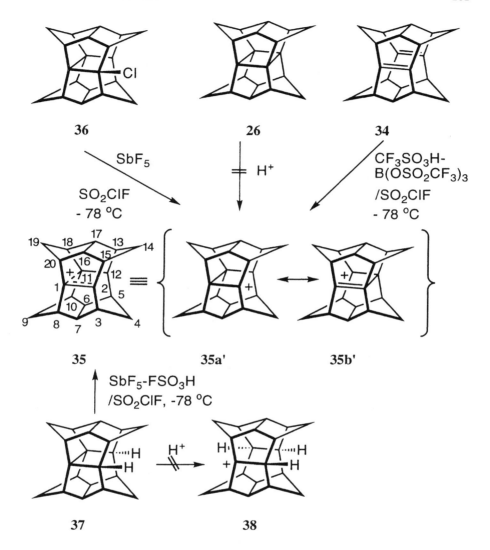

The B3LYP/6-31G* optimized structures of the secopagodyl cation **35a** and its isoelectronic neutral boron analog **35b** are given in Figure 6-7 and the unknown secoisopagodyl cation **39a** and its isoelectronic neutral boron analog **39b** are given in Figure 6-8. The calculated $\delta^{13}C$ value of C^+ in cation **35a** is 226.6 and the calculated $\delta^{13}C$ value of C^+ in cation **39a** is 338.8. The calculated $\delta^{11}B$ value of analogs **35b** and **39b** are 82.9 and 78.1, respectively. The bora analog **35b** of secopagodyl cation **35a** is, again, deshielded, indicating lack of homoconjugation. On the other hand, the unknown secoisopagodyl cation analog correlates well.

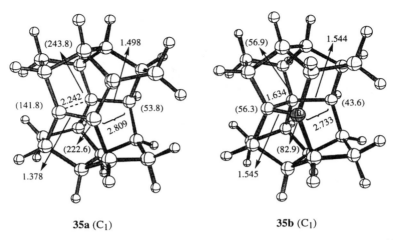

35a (C₁)   **35b** (C₁)

Figure 6-7   The calculated B3LYP/6-31G* structures and IGLO chemical shifts (in parentheses) of structures **35a** and **35b**.

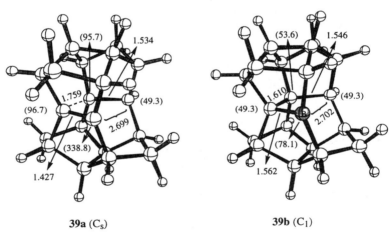

39a (Cₛ)   **39b** (C₁)

Figure 6-8   The calculated B3LYP/6-31G* structures and IGLO chemical shifts (in parentheses) of structures **39a** and **39b**.

6.3   CORRELATION OF ¹³C AND ¹¹B NMR CHEMICAL SHIFTS

The calculated ¹³C NMR chemical shifts of carbocations are in very good agreement with the experimental data (Figure 6-9a). The experimental ¹³C NMR chemical shifts of carbocations are also in reasonable agreement with the calculated ¹¹B NMR chemical shifts of the corresponding isoelectronic boron analogs (Figure 6-9b). With the exception of the pagodyl and isopagodyl systems, **27** and **33**, and secopagodyl analog **35**, the close relationship between the ^{11}B NMR chemical shifts of the boron atoms in boron compounds (see Table 6-1) and the ^{13}C NMR chemical shifts of the corresponding carbocations further demonstrate that even cage systems closely follow Eq. (3) derived for both trigonal and non-classical systems. We also found linear correlation between the calculated ^{13}C NMR chemical shifts and the calculated ^{11}B NMR chemical shifts (with the

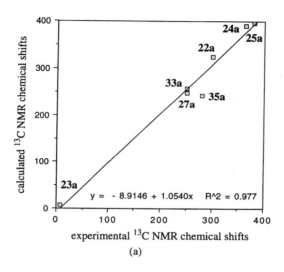

(a)

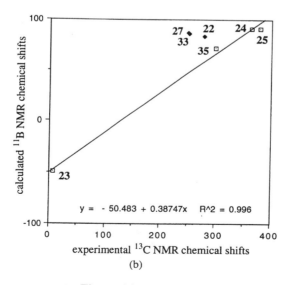

(b)

Figure 6-9 (*Continued*)

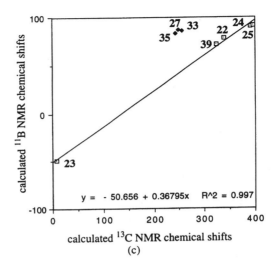

(c)

Figure 6-9   (a) Experimental vs calculated ^{13}C NMR chemical shifts, (b) experimental ^{13}C NMR chemical shifts vs calculated ^{11}B NMR chemical shifts, (c) calculated ^{13}C NMR chemical shifts vs calculated ^{11}B NMR chemical shifts.

TABLE 6-1   Experimental and calculated [13]C[a] and [11]B[b] NMR chemical shifts

no	IGLO	expt
	^{13}C NMR chemical shifts	
22a	324.5	300.0
23a	7.0	6.6
24a	390.3	363.9
25a	396.6	379.2
27a	248.4	251.0
33a	257.5	251.7
35a	243.8	279.2
	222.6	
39a	338.8	—
	^{11}B NMR chemical shifts	
22b	71.7	73.2
23b	−49.2	—
24b	90.8	—
25b	91.5	—
27b	86.4	—
33b	86.0	—
35b	82.9	—
39b	78.1	—

[a] [13]C shifts are referenced to TMS
[b] [11]B shifts are referenced to BF_3:OEt_2

exception of structures **27, 33** and **35**) of the corresponding isoelectronic boron analogs (Figure 6-9c). The equation of the correlation line of Eq.(4) derived from Figure 6-9c closely corresponds to the equation of the correlation line of Eq. (3).

$$\delta^{11}B(BF_3{:}OEt_2) = 0.37\ \delta\ {}^{13}C_{(TMS)} - 51 \tag{4}$$

ACKNOWLEDGMENTS

Support of our work by the Loker Hydrocarbon Research Institute is gratefully acknowledged. We also thank Professor George A. Olah for his encouragement and support and Professors Leo A. Paquette (Ohio State) and Horst Prinzbach (Freiburg) for their collaborative work with G.K.S.P.

REFERENCES

1. Olah, G. A., Prakash, G. K. S., Williams, R. E., Field, L. D., and Wade, K., *Hypercarbon Chemistry,* John Wiley & Sons, New York, 1987.
2. Olah, G. A., Demember, J. R., Commeyras, A., and Bribes, J. L., *J. Am. Chem. Soc.,* 1971, **93**, 459.
3. Hollenstein, S. and Laube, T., *J. Am. Chem. Soc.,* 1993, **115**, 7240.
4. Crofton, M. W., Kreiner, W. A., Jagod, M. F., Rehfuss, B. D., and Oka, T., *J. Chem. Phys.,* 1985, **83**, 3702.
5. (a) Lathan, W. A., Hehre, W. J., and Pople, J. A., *J. Am. Chem. Soc.,* 1971, **93**, 808; (b) Ragavachari, K., Whiteside, R. A., Pople, J. A., and Schleyer, P. v. R., *J. Am. Chem. Soc.,* 1981, **103**, 5649.
6. Kaldor, A. and Porter, R. F., *J. Am. Chem. Soc.,* 1971, **93**, 2140.
7. Williams, R. E., *Inorg. Chem.,* 1971, **10**, 210.
8. Stohrer, W. D. and Hoffmann, R., *J. Am. Chem. Soc.,* 1972, **94**, 1661.
9. Masamune, S., Sakai, M., Ona, H., and Jones, A. L., *J. Am. Chem. Soc.,* 1972, **94**, 8956.
10. Olah, G. A., Westerman, P. W., Mo, Y. K., and Klopman, G., *J. Am. Chem. Soc.,* 1972, **94**, 7859.
11. (a) Rasul, G. A. and Olah. G. A., *Inorg. Chem.* 1997, **36**, 1278; (b) DePuy, C. H., Gareyev, R., Hankin, J., and Davico, G. E., *J. Am. Chem. Soc.,* 1997, **119**, 427.
12. Olah, G. A. and Rasul, G., *J. Am. Chem. Soc.,* 1996, **118**, 8503.
13. Hogeveen, H. and Kwant, P. W., *J. Am. Chem. Soc.,* 1974, **96**, 2208.
14. Williams, R. E., Prakash, G. K. S., Field, L. D., and Olah, G. A., *Advances in Boron and Boranes* (eds) Liebman, J. F., Greenberg, A., and Williams, R. E., VCH, New York, 1988 and references therein.
15. (a) Spielvogel, B. F. and Purser, J. M., *J. Am. Chem. Soc.,* 1971, **93**, 4418. (b) Spielvogel, B. F., Nutt, W. R., and Izydore, R. A., *J. Am. Chem. Soc.,* 1975, **97**, 1609.
16. Nöth, H. and Wrackmeyer, B., *Chem. Ber.,* 1974, **107**, 3089.

17. Ziegler, T., *Chem. Rev.* 1991, **91**, 651.

18. Frisch, M. J., Trucks, G. W., Schlegel, H. B., Gill, P. M. W., Johnson, B. G., Robb, M. A., Cheeseman, J. R., Keith, T. A., Petersson, G. A., Montgomery, J. A., Raghavachari, K., Al-Laham, M. A., Zakrzewski, V. G., Ortiz, J. V., Foresman, J. B., Cioslowski, J., Stefanov, B. B., Nanayakkara, A., Challacombe, M., Peng, C. Y., Ayala, P. Y., Chen, W., Wong, M. W., Andres, J. L., Replogle, E. S., Gomperts, R., Martin, R. L., Fox, D. J., Binkley, J. S., Defrees, D. J., Baker, J., Stewart, J. P., Head-Gordon, M., Gonzalez, C., and Pople, J. A., Gaussian 94 (Revision A.1), Gaussian, Inc., Pittsburgh, PA, 1995.

19. (a) Schindler, M., *J. Am. Chem. Soc.*. 1987, **109**, 1020; (b) Kutzelnigg, W., Fleischer, U., Schindler, M., *NMR Basic Princ. Prog.* 1991, **23**, 165.

20. Olah. G. A., Prakash, G. K. S., Shih, J. G., Krishnamurthy, V. V., Mateescu, G. D., Liang, G., Sipos, G., Buss, V., Gund, T. M., Schleyer, P. v. R., *J. Am. Chem. Soc.*, 1985, **107**, 2764.

21. Laube, T., *Angew. Chem. Int. Ed. Engl.*, 1986, **25**, 349.

22. Schleyer, P. v. R. and Maerker, C., *Pure Appl. Chem.*, 1995, **67**, 755.

23. Aniszfeld, R., Ph.D. Dissertation, University of Southern California, 1990.

24. (a) Bremer, M., Schleyer, P. v. R., Schötz, K., Kuasch, M., and Schindler, M., *Angew. Chem. Int. Ed. Engl.*, 1987, **26**, 761; (b) Schulman, J. M., Disch, R. L., Schleyer, P. v. R., Bühl, M., Bremer, M., and Koch, W., *J. Am. Chem. Soc.*, 1992, **114**, 7897.

25. (a) Olah. G. A., Prakash, G. K. S., Kobayashi, T., and Paquette, L. A., *J. Am. Chem. Soc.*, 1988, **110**, 1304; (b) Olah, G. A., Prakash, G. K. S., Fessner, W.-D., Kobayashi, T., and Paquette, L. A., *J. Am. Chem. Soc.*, 1988, **110**, 8599.

26. Schleyer, P. v. R., Lenoir, D., Mison, P., Liang, G., Prakash, G. K. S., and Olah, G. A., *J. Am. Chem. Soc.*, 1980, **102**, 683.

27. Prakash, G. K. S., Krishnamurthy, V. V., Herges, R., Bau, R., Yuan, H., Olah, G. A., Fessner, W.-D., and Prinzbach, H., *J. Am. Chem. Soc.*, 1986, **108**, 836, *J. Am. Chem. Soc.*, 1988, **110**, 7764.

28. Goldstein, M. J., and Hoffmann, R., *J. Am. Chem. Soc.*, 1971, **93**, 6193.

29. Prinzbach, H., Murty, B. A., Fessner, W.-D., Mortensen, J., Heinze, J., Gescheidt, G., and Gerson, F., *Angew. Chem. Int. Ed. Engl.*, 1987, **26**, 457.

30. Herges, R., Schleyer, P. v. R., Schindler, M., and Fessner, W.-F., *J. Am. Chem. Soc.*, 1991, **113**, 3649.

31. Wollenweber, M., Pinkos, R., Leonhardt, J., and Prinzbach, H., *Angew. Chem. Int. Ed. Engl.*, 1994, **33**, 117.

32. Weber, K., Lutz, G., Knothe, L., Mortensen, J., Heinze, J., and Prinzbach, H., *J. Chem. Soc.*, *Perkin Trans. 2*, 1995, 1991.

33. Prinzbach, H., Wollenweber, M., Herges, R., Neumann, H., Gescheidt, G., and Schmidlin, R., *J. Am. Chem. Soc.*, 1995, **117**, 1439.

34. (a) Prakash, G. K. S., Krishnamurthy, V. V., Herges, R., Bau, R., Yuan, H., Olah, G. A., Fessner, W.-D., and Prinzbach, H., *J. Am. Chem. Soc.*, 1986, **108**, 836; (b) Prakash, G. K. S., Krishnamurthy, V. V., Herges, R., Bau, R., Yuan, H., Olah, G. A., Fessner, W.-D., and Prinzbach, H., *J. Am. Chem. Soc.*, 1988, **110**, 7764.

35. Prinzbach, H., Gescheidt, G., Martin, H.-D., Herges, R., Prakash, G. K. S., and Olah, G. A., *Pure Appl. Chem.*, 1995, **67**, 673.

36. Spurr, P. R., Murty, B. A. R. C., Fessner, W.-D., Fritz, H., and Prinzbach, H., *Angew. Chem., Int. Ed. Engl.* 1987, **26**, 455.

37. Fessner, W.-D., Ph.D., Dissertation, Universität Freiburg, 1986. Prakash, G. K. S., Fessner, W-D., Olah, G. A., Lutz, G., and Prinzbach, H., *J. Am. Chem. Soc.*, 1989, **111**, 746.

38. Sunko, D. E., Hirsl-Starcevic, S., Pollack, S. K., and Hehre, W. J., *J. Am. Chem. Soc.*, 1979, **101**, 6163.

39. (a) Laube, T., *Angew. Chem. Int. Ed. Engl.*, 1986, **25**, 349; (b) Olah, G. A., Prakash, G. K. S., Bau, R., et al., Unpublished results.

7

ARE POLYHEDRAL BORANES, CARBORANES, AND CARBOCATIONS AROMATIC?

PAUL VON RAGUÉ SCHLEYER and KATAYOUN NAJAFIAN

Computer Chemistry Center, Institut für Organische Chemie, Universität Erlangen-Nürnberg, Henkestrasse 42, D-91054 Erlangen, Germany

The "three-dimensional aromaticity of polyhedral boranes" was first proposed by Aihara in 1978 on the basis of a graph-theoretical treatment of aromaticity.[1] The idea that aromaticity can be extended beyond two dimensions was supported by our "six interstitial electron" analogies between the electron delocalization in the $(4n + 2)\pi$ electron Hückel [n]annulenes and the delocalized bonding in *nido*-carboranes, pyramidal nonclassic carbocations, and isostructural transition metal derivatives.[2]

Robert E. Williams,[3] whom we are honoring in this book, was the first to propose the pyramidal $C_5H_5^+$ in 1971 on the basis of the isoelectronic analogies with boranes and carboranes that stressed the relationships among these fascinating compounds. The title question of this chapter certainly is not new. The 12-vertex $C_2B_{10}H_{12}$ isomers have long been known to be exceptionally stable, to undergo electrophilic substitution reactions like benzene, and even to have "aromatic" odors.[4,5] However, other boranes and carboranes are highly unstable, inflame in air, and are unlikely candidates for electrophilic substitution experiments.[5] Chemical properties are poorly suited as criteria of aromaticity.

What is aromaticity? What are the relationships among the various kinds of systems that have been termed "aromatic"? Nineteenth-century chemists were content with Erlenmeyer's 1866 distinction that aromatic compounds undergo substitution, rather than addition.[6] The large C/H ratios were noted by M. Faraday, the discoverer of benzene, in 1825.[7] Despite the high degree of unsaturation, not only benzene, but also benzenoid hydrocarbons like naphthalene, phenanthrene, anthracene, etc., are unusually stable.[8] In 1926, Robert Robinson introduced his "electron sextet" concept, and extended it to heterocyclic systems, like pyrrole and thiophene.[9]

E. Hückel, in 1932, provided his famous theoretical framework.[10] In the late 1930s, M. G. Evans noted the relationship between the transition state for the Diels-Alder reaction and benzene.[11] This forerunner of the Woodward–Hoffmann theory pointed to cyclic delocalization as the essential characteristic of aromaticity.[12] Although most chemists still regard geometric [8,13] and energetic [8,13,14] features to be the main criteria of aromaticity, these have considerable drawbacks for general application. Instead, the abnormal magnetic properties of aromatic compounds are arguably better as defining characteristics.[15]

That ordinary organic compounds are repelled (slightly) by a magnetic field was known in the last century. In 1910, Pascal first devised an increment system that allowed such diamagnetic susceptibilities to be predicted with good accuracy.[16] However, aromatic compounds were abnormal; their magnetic susceptibilities were larger than estimated by Pascal's increments. Later work, particularly by H. J. Dauben[17] and his coworkers in the late 1960s, showed aromatic compounds to be the *only* exceptions among closed-shell molecules: "Diamagnetic susceptibility exaltation, defined as a difference between the measured molar susceptibility and the susceptibility estimated neglecting contribution of ring current, is found to be a property solely of aromatic compounds."[17b] Dauben's untimely death effectively ended this promising line of research, and recent organic textbooks do not discuss "magnetic susceptibility" at all.

Nevertheless, abnormal magnetic properties have long been associated with aromatic compounds by organic chemists. The proton chemical shifts of benzene and other aromatic molecules are deshielded relative to the NMR spectra of olefins. In contrast, protons above or (when geometrically possible) inside aromatic rings resonate at high fields.[8] Pople first applied "ring-current" theory to explain this behavior in 1962.[18] Much earlier, in 1936, Linus Pauling had attributed Pascal's magnetic susceptibility exaltations [16] to induced "ring currents" in aromatic compounds.[19] A year later, London [20] proposed a simple way to compute such contributions, and assessments of "London diamagnetism" were the forerunners of modern ab initio methods (e.g., IGLO [21] and GIAO [22]) of computing magnetic properties.

All William's categories of "electron-deficient" compounds, e.g., "*closo*," "*nido*," and "*arachno*," utilize delocalized multicenter bonding.[3] As a first answer to the title question of this chapter, we have chosen the central family of *closo*-borane dianions for detailed analysis. The aromaticity of these $B_nH_n^{2-}$ species, examined by employing energetic, structural, and magnetic criteria, once established, should set the patterns for polyhedral compounds in general.

7.1 INTRODUCTION

The *closo*-borane dianions, $B_nH_n^{2-}$ ($n = 5$–12), constitute a basic class of nonclassic compounds with nearly spherical structures.[3,5,23–26] The bonding is "electron deficient" since there are more valence orbitals than valence electrons.[24] Many experimental and theoretical investigations have been concerned with the nature of these polyhedral molecules.[3,5,23–33] Lipscomb's topological *styx* formalism[25] systematized the structure and bonding in boranes. Williams' *closo*-, *nido*-, *arachno*-, and *hypo*-polyhedron classification emphasized the structural and electronic analogies.[3] Wade's electron-counting rules[24,26] provided a better understanding of the interrelationship between the various types of cluster geometries. King's topological and group-theoretical approaches[27] predicted the rigidity and fluxionality in polyhedral cluster geometries. The stabilities of *closo*-borane dianions have also been investigated extensively,[28–31] but only qualitatively or by using lower level theoretical calculations. Lipscomb et al. (PRDDO),[28] Wade and colleagues (extended Hückel theory),[29] Zhao and Gimarc (three-dimensional Hückel theory and minimal basis set ab initio)[30,31] each computed the relative stabilities of *closo*-boranes. All these studies concluded that the relative stability of $B_nH_n^{2-}$ ($n = 5$–12) increases generally with cluster size, but differed somewhat regarding the relative stabilities of individual members of this class.

The three-dimensional delocalization in polyhedral boranes with n vertices and containing ($n + 1$) skeletal electron pairs were discussed by King and Rouvray.[32] In 1982, Jemmis and Schleyer[2] extended the planar ($4n + 2$) Hückel rule to the aromaticity in three-dimensions using the "six interstitial electron" concept. Aihara estimated (1978) the resonance energies associated with the "three-dimensional aromaticity" of *closo*-borane dianions.[1] The most symmetric $B_{12}H_{12}^{2-}$ (I_h) had the greatest resonance stabilization (1.763β), in agreement with its high stability, but $B_5H_5^{2-}$ (resonance energy of 0.0β) was classified as "nonaromatic".[1] However, our recent ab initio study,[33] which found $B_5H_5^{2-}$ to be stabilized by about 35 kcal/mol because of three-dimensional delocalization, contradicts the "nonaromatic" description offered by Aihara. One of the goals of this chapter is to reevaluate the stability of the *closo*-borane dianions at higher, electron-correlated theoretical levels than have been employed previously.[27–31]

The aromaticities of two-dimensional molecules have been examined using three principal criteria:[8,15] energetic (resonance energies, aromatic stabilization energies), geometric (bond length equalization, bond order indices, etc.), and magnetic (^1H NMR chemical shifts, magnetic susceptibility anisotropies and exaltations).[8,13,15] While direct relationships among these criteria for a wide-ranging set of five-membered heterocycles was demostrated using ab initio and density functional theory data,[13] such parallel behavior did not extend to more complex systems where other effects dominate.[34]

We will now apply these criteria (with special emphasis on the magnetic properties) to the question of aromaticity in the *closo*-borane clusters ($B_n H_n^{2-}$ where $n = 5$–12). Along with ^{1}H NMR chemical shifts,[35] magnetic susceptibility anisotropies (χ_{anis}),[36] and exaltations (Λ)[17] have been widely used as magnetic criteria of aromaticity in organic molecules.[8,13,15,33,34] However, χ_{anis} is small or zero for molecules with near-spherical symmetry, like the *closo*-boranes, and ^{11}B as well as ^{1}H NMR chemical shifts are not informative in this context. Diamagnetic susceptibility exaltations, Λ, are directly associated with aromaticity,[15,17] but depend on the square of the ring area [8b] and require suitable increment schemes for the evaluation of Λ.[17] The choice of reference molecules often is not straightforward. Chemical shifts of encapsulated [3]He atoms (computed, as well as experimental) have been used recently as measures of aromaticity in fullerenes and their derivatives.[37] Since most of the *closo*-systems are too small to accommodate [3]He or other elements, we have employed nucleus-independent chemical shifts (NICS)[38] similarly. These are based on the absolute chemical shieldings computed at the cage centers (to conform with the chemical shift convention, the signs of the computed shieldings are reversed to give NICS).

7.2   METHODS

The geometries of *closo*-borane dianions (Figure 7-1) were optimized within the symmetry restriction using the GAUSSIAN 94 program, first at HF/6-31G*, then at MP2/6-31G*, B3LYP/6-31G*, and finally at B3LYP/6-311+G**.[39] The frequency calculations at B3LYP/6-31G* characterized the stationary points as minima and provided the zero-point energies (ZPE).[40] The geometry and energy data discussed here (Table 7-1) were obtained at the electron-correlated B3LYP/6-311+G** optimized DFT level. The NICS were computed at GIAO-HF/6-31G* on the B3LYP/6-311+G** optimized geometries [22] (Table 7-1).

It is important to note that all of these dianions are unbound. Not only the HOMOs, but also six to eight of the highest-occupied orbitals have positive eigenvalues. The situation is not improved at the B3LYP/6-311+G** level (i.e., with the inclusion of diffuse functions). Indeed, these dianions have only been observed at condensed phases, e.g., in the stabilizing electronic field of the counterions.

7.3   RESULTS AND DISCUSSION

7.3.1   Energies of $B_n H_n^{2-}$ ($n = 5$–12)

Stabilization energies arising from multicenter bonding in the "electron-deficient" *closo*-boranes should be directly related to the three-dimensional delocalization. However, evaluations for most of the boron cages are not straightforward, since suitable classic reference structures are hard to define. Prior computational assessments of the relative stability of $B_nH_n^{2-}$ ($n = 5$–12) *closo*-boranes were reported by Lipscomb and coworkers,[28] and by Wade and coworkers [29] as well as by Zhao and Gimarc, but they came to somewhat different conclusions.[5,30,31] Using PRDDO data, Lipscomb and coworkers [28] evaluated the relative stabilities of *closo*-borane dianions from Eq. (1). The stabilities of *closo*-boranes were based on the average energy per BH group in the $B_nH_n^{2-}$ cluster compared with the BH group in the octahedral reference molecule,

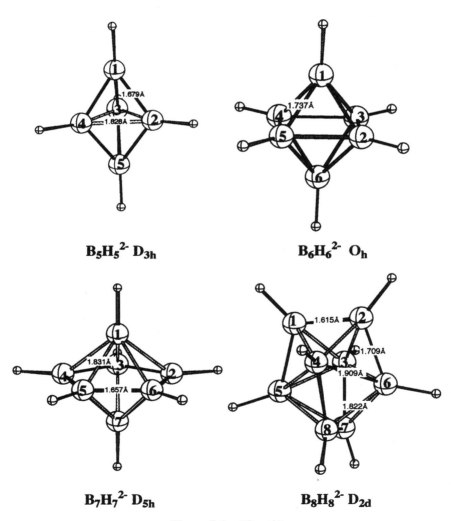

$B_5H_5^{2-}$ D_{3h}

$B_6H_6^{2-}$ O_h

$B_7H_7^{2-}$ D_{5h}

$B_8H_8^{2-}$ D_{2d}

Figure 7-1 (*Contd.*)

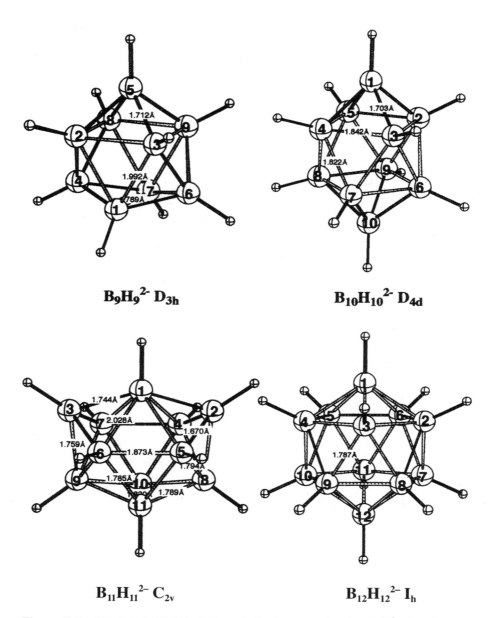

Figure 7-1   The B3LYP/6-311+G** optimized geometries for B$_n$H$_n^{2-}$ *closo*-borane dianions (n = 5–12).

TABLE 7-1 Data for *Closo*-borane Dianions, $B_nH_n^{2-}$ (n = 5–12)

Total Energies in a.u.; Other Energies in kcal/mol. Zero-Point Energies (ZPE); [a] Reaction Energies from Eq. (2) (ΔH); [b] Average BH Unit Energies (ΔH$_{ave}$); [c] Disproportionation Energies from Eq. (3) (ΔE$_{dis}$); [d] Bond Length Alternation (ΔR, Å); [e] Nucleus-Independent Chemical Shifts (NICS, ppm)

Molecule	Symmetry	B3LYP/6-311+G**	ZPE[a]	ΔH[b]	ΔH$_{ave}$[c]	ΔH$_{dis}$[d]	ΔR[e]	NICS[f]
$B_5H_5^{2-}$	D_{3h}	−127.17129	36.7	−156.82	−31.36		0.149	−26.44
$B_6H_6^{2-}$	O_h	−152.72451	47.1	−242.59	−40.43	+40.19	0.000	−34.22
$B_7H_7^{2-}$	D_{5h}	−178.21180	56.3	−288.18	−41.16	+5.98	0.174	−27.43
$B_8H_8^{2-}$	D_{2d}	−203.68894	65.1	−327.79	−40.97	−14.79	0.293	−24.15
$B_9H_9^{2-}$	D_{3h}	−229.19051	74.5	−382.18	−42.46	−12.63	0.281	−27.38
$B_{10}H_{10}^{2-}$	D_{4d}	−254.71361	84.7	−449.21	−44.92	+33.52	0.139	−33.43
$B_{11}H_{11}^{2-}$	C_{2v}	−280.18060	93.2	−482.72	−43.88	−69.37	0.359	−32.41
$B_{12}H_{12}^{2-}$	I_h	−305.76291	104.8	−585.60	−48.80		0.000	−35.73

[a] At B3LYP/6-31G*.

[b] $B_2H_2^{2-} + (n-2)BH_{inc} \longrightarrow B_nH_n^{2-}$ (n = 5–10) at B3LYP/6-311 + G**, with ZPE corrections (Ref. 40) scaled by 0.98 (and 0.89 for RHF/6-31G*) in kcal/mol.

[c] ΔH$_{ave}$ = ΔH/n.

[d] $2B_nH_n^{2-} \longrightarrow B_{n-1}H_{n-1}^{2-} + B_{n+1}H_{n+1}^{2-}$ (n = 6–11) at B3LYP/6-311 + G** with ZPE corrections scaled as in c.

[e] Difference (Å) between the longest and the shortest BB distances in each dianion at B3LYP/6-311 + G**.

[f] At GIAO-HF/6-31 + G*//B3LYP/6-311 + G**.

$B_6H_6^{2-}$ (this equation is an approximation, as the charges are not balanced). The stabilites of *closo*-borane dianions were found to increase generally as the cluster size increases (ΔH_L, Table 7-2). Thus, $B_{12}H_{12}^{2-}$ is the most stable and $B_5H_5^{2-}$ is the least stable. Only $B_{11}H_{11}^{2-}$ deviated from the general trend. Problems were noted in this analysis, as $B_7H_7^{2-}$ was indicated to be more stable than $B_6H_6^{2-}$.

$$[E(B_6H_6^{2-})/6] - [E(B_nH_n^{2-})/n] = \Delta H \qquad (1)$$

Wade and coworkers [29] evaluated the stability of the $B_nH_n^{2-}$ *closo*-borane dianions using the extended Hückel formalism. Overall, the electronic energy decreases as a function of cluster size. However, the average energy of the BH moieties (total electronic energy divided by the number of vertexes) plotted as a function of cluster size revealed that the "*closo*-borane dianions $B_nH_n^{2-}$ fall into two main categories; with $B_6H_6^{2-}$ and $B_{12}H_{12}^{2-}$ markedly more stable than the remainder, among which $B_{11}H_{11}^{2-}$ and perhaps $B_5H_5^{2-}$ make the least effective use of their skeletal electrons."

TABLE 7-2   Energetic trends in *Closo*-borane Dianions, $B_nH_n^{2-}$ (n = 5-12) Based on Various Treatments Relative to $B_{12}H_{12}^{2-}$ in kcal/mol. The PRDDO relative energies (ΔH_L) of Lipscomb and Coworkers;[a] STO-3G energy per BH Unit of Zhao and Gimarc (ΔH_G a.u.);[b] Graph-theoretical Resonance Energies of Aihara (RE, β);[c] Average BH Energy (ΔH_{ave}, from Table 7-1).[d] Degenerate Rearrangement Barriers ($E_{barrier}$) Taken From Various Sources

Molecule	$\Delta H_L{}^a$ (kcal/mol)	$\Delta H_G{}^b$ (kcal/mol)	RE^c (β)	$\Delta H_{ave}{}^d$ (kcal/mol)	$E_{barrier}$ (kcal/mol)
$B_5H_5^{2-}$	61.5	63.2	0.00	17.44	20^e
$B_6H_6^{2-}$	38.3	38.3	0.12	8.37	44.7^f
$B_7H_7^{2-}$	27.6	28.1	0.13	7.64	76.7^g
$B_8H_8^{2-}$	23.9	22.6	0.10	7.83	5^h
$B_9H_9^{2-}$	17.0	17.2	0.09	6.34	25.7^i
$B_{10}H_{10}^{2-}$	10.7	10.3	0.12	3.88	21.7^j
$B_{11}H_{11}^{2-}$	11.9	9.5	0.09	4.54	3^k
$B_{12}H_{12}^{2-}$	0.0	0.0	0.15	0.00	144.5^l

[a] From Eq. (1) (Ref. 28).
[b] $\Delta H_G = (E_{ab\ initio}/n)$ from Ref. 30.
[c] From Ref. 1.
[d] From Table 7-1.
[e] The rearrangement of 1,2-$C_2B_3H_5$ to the most stable 1,5-$C_2B_3H_5$ isomer from Ref. 61.
[f] Calculated barrier of 1,2-$C_2B_4H_6$ to the most stable 1,6-$C_2B_4H_6$ isomer from Ref. 43.
[g] Ref. 51.
[h] Ref. 47.
[i] Ref. 57.
[j] Ref. 49.
[k] Ref. 55.
[l] Refs 44 and 45.

Employing three-dimensional Hückel theory, as well as minimal basis set STO-3G ab initio calculations, Gimarc and coworkers [30] investigated the energy relationships of *closo*-borane dianions with cluster size. The negative total energy of *closo*-borane dianions divided by the number of vertices (ΔH_G, Table 7-2; note the correspondence with ΔH_L), showed increasing stability of *closo*-$B_nH_n^{2-}$ ($n = 5–12$) as a function of polyhedral size, in agreement with the earlier conclusions.[28,29,41]

Following King,[27] Zhao and Gimarc [31] related the stability of the polyhedra to the fluxionality or rigidity of the cage, based on the ease of intramolecular diamond-square-diamond (DSD)[42] rearrangements. Accordingly, the barriers for the DSD processes (evaluated by three-dimensional Hückel theory) of *closo*-borane dianions follow the order:[31] $B_6H_6^{2-} > B_{10}H_{10}^{2-} \sim B_7H_7^{2-} > B_8H_8^{2-} \sim B_{11}H_{11}^{2-}$. For $B_5H_5^{2-}$ and $B_9H_9^{2-}$, the energy barrier was anticipated to be high (in contradiction to King)[27] since a single DSD mechanism is forbidden by orbital symmetry.[31] However, the nonallowed pathways in higher symmetries assumed in DSD analyses need not be followed. Indeed, McKee [43] and Wales [44] showed that twist DSD mechanisms, with reduced symmetry, result in lower barriers.

We employed B3LYP/6-311+G** data and Eq. (2) to examine the energetic relationships in the *closo*-$B_nH_n^{2-}$ family. The acetylene-like $B_2H_2^{2-}$ was employed and the BH_{inc} increment energy is taken as the difference between B_3H_5 (C_{2v}, planar) and B_2H_4 (D_{2h}, ethylene-like). (Note that the computed BH_{inc} increment does not possess any inherent stabilization due to hyperconjugation or delocalization.)

$$B_2H_2^{2-} + (n - 2)\, BH_{inc} \rightarrow B_nH_n^{2-} \qquad (n = 5–12) \qquad \Delta H \qquad (2)$$

As summarized in Table 7-1, all the reaction energies [Eq. (2)] are exothermic. In order to evaluate the relative stability of boron clusters, the average BH bond energy ($\Delta H_{ave} = \Delta H/n$) is plotted as a function of cluster size in Figure 7-2. This emphasizes the variation among these dianions more clearly than the earlier treatments.[28–31] The trend towards increasing stability with cluster size is confirmed, but the magnitude is less (Table 7-2) at our much better theoretical levels.

In order to include this trend, we have employed the most regular borane dianions, $B_{12}H_{12}^{2-}$ and $B_6H_6^{2-}$ (which define the straight line in Figure 7-2), as the basis for the quantitative comparison of *closo*-borane dianion stabilities. Except for $B_5H_5^{2-}$, the deviations of ΔH_{ave} for the other *closo*-borane dianions from the reference line in Figure 7-2 are within about 4.2 kcal/mol. (However, note that the total deviation, i.e., ΔH_{ave} times the number of vertices, can be quite large). Compounds $B_{10}H_{10}^{2-}$ and $B_7H_7^{2-}$ have the smallest deviations, ΔH_{ave} 1.1 and 0.7 kcal/mol, respectively, while the $B_9H_9^{2-}$ and $B_8H_8^{2-}$ deviations from the $B_6H_6^{2-}$ and $B_{12}H_{12}^{2-}$ defined line (Figure 7-2) are larger (ΔH_{ave} = Both 2.2 kcal/mol). According to this energetic analysis, the stabilities of *closo*-boranes fall into four groups of generally decreasing stability: (1) $B_{12}H_{12}^{2-}$ and $B_6H_6^{2-}$; (2) $B_{10}H_{10}^{2-}$

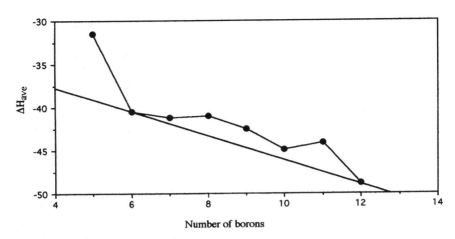

Figure 7-2   The ΔH_{ave} values (average energy per BH unit in kcal/mol from Table 7-1) vs number of boron atoms in $B_nH_n^{2-}$ *closo*-borane dianions. The trend to lower ΔH_{ave} values with increasing cluster size, as well as the variations for individual dianions, is indicated by the line defined by the 6- and the 12-vertex points.

and $B_7H_7^{2-}$; (3) $B_8H_8^{2-}$ and $B_9H_9^{2-}$; and (4) $B_{11}H_{11}^{2-}$ and $B_5H_5^{2-}$. The most highly symmetrical species, $B_{12}H_{12}^{2-}$ and $B_6H_6^{2-}$, also are the most stable, are the hardest to deform to other geometries, and are reluctant to undergo degenerate rearrangement. Both Wales [44] and Gimarc et al.[45] suggested that the DSD[42] mechanism for $B_{12}H_{12}^{2-}$ ($I_h \rightarrow D_{3h}$) should have a high energy barrier (around 145 kcal/mol). Single DSD mechanisms that are forbidden by orbital symmetry do not necessarily curtail the automerization, since alternative processes are possible. For example, $B_{12}H_{12}^{2-}$ may undergo a series of rearrangements [44] involving allowed lower-symmetry pathways, but these also are high in energy. Compound $B_6H_6^{2-}$ (O_h)[46] is predicted to be rigid according to topological and graph theory.[27,31] The triple DSD process [31] for the intramolecular rearrangement of $B_6H_6^{2-}$ would have to proceed via a high-energy trigonal prism (D_{3h}). The fluxional $B_nH_n^{2-}$ barrier is not known experimentally, but should be related to data for the isoelectronic $C_2B_{n-2}H_n$ species. Other pathways may be more favorable; e.g., the computed rearrangement barrier of isoelectronic 1,2-$C_2B_4H_6$ to 1,6-$C_2B_4H_6$ is 44.7 kcal/mol and has a C_{2v} transition structure.[43]

The stabilities of $B_{10}H_{10}^{2-}$ and $B_7H_7^{2-}$ [Eq. (2)] are somewhat lower than those of most symmetrical boranes. The D_{4d} structure of $B_{10}H_{10}^{2-}$ is the global minimum.[47a] The experimental activation energy for polyhedral rearrangements of derivatives is 37 kcal/mol.[48] Semiempirical MNDO [49] calculations on the parent $B_{10}H_{10}^{2-}$ suggested that the C_{2v} transition state is 21 kcal/mol higher than the D_{4d} TS. The global D_{5h} $B_7H_7^{2-}$ minimum has been characterized by ^{11}B NMR experimentally.[47b,50] Gimarc and coworkers found the energy barrier for the DSD rearrangement of $B_7H_7^{2-}$ to be 76.6 kcal/mol at the ab initio STO-3G level.[51] Graph theory [27] and preliminary ab initio computations [47b] suggest that $B_7H_7^{2-}$ should be rigid.

Compounds $B_8H_8^{2-}$ and $B_{11}H_{11}^{2-}$ are the most highly fluxional *closo*-borane dianions; they are also less stable thermodynamically according to Eq. (2). An X-ray [52] of the $B_8H_8^{2-}$ tetramino zinc salt showed D_{2d} symmetry, i.e., two sets of nonequivalent boron atoms. However, ^{11}B NMR study of $B_8H_8^{2-}$ indicates a 2:4:2 ratio of ^{11}B resonances (C_{2v} symmetry) at low temperatures (4°C).[50b] At higher temperatures (46°C), only one signal could be observed (i.e., effective D_{4d} symmetry). The boron scrambling ($D_{2d} \rightarrow C_{2v} \rightarrow D_{4d}$) provides striking evidence for the highly fluxional behavior of $B_8H_8^{2-}$. Lipscomb and coworkers [53] computed (PRDDO) the D_{2d} form of $B_8H_8^{2-}$ to be the most stable isomer; the activation energy for polyhedral rearrangement is less than 4 kcal/mol. Our RMP2(fc)/6-31G* calculations also indicate the DSD rearrangement barrier for this dianion to be only 5 kcal/mol.[47c]

Like $B_8H_8^{2-}$, the $B_{11}H_{11}^{2-}$ dianion also is fluxional in solution on the ^{11}B NMR time scale.[54] Although no experimental value has been reported, PRDDO calculations predict a < 3 kcal/mol barrier.[55]

Judging from Figure 7-2, the stability of *closo*-$B_9H_9^{2-}$ is between that of $B_{10}H_{10}^{2-}$ and $B_8H_8^{2-}$. Contrasting views on the stability of $B_9H_9^{2-}$ have been expressed.[56,58] Scrambling of the B atoms was not expected to be observable by NMR at ordinary temperatures.[56a] While topological and graph-theoretical studies concluded that $B_9H_9^{2-}$ should be an inherently fluxional molecule,[27] our B3LYP/6-311+G** DFT calculations [57] quantify Gimarc and Ott's [58] prediction that the intramolecular cluster rearrangement of *closo*-$B_9H_9^{2-}$ should proceed by a double DSD mechanism (21.3 kcal/mol) rather than the single DSD (28.4 kcal/mol) process.

Although $B_5H_5^{2-}$ is not known, Gimarc and coworkers [31,59] expected this species to be rigid since the single DSD mechanism is forbidden by orbital symmetry (HOMO–LUMO crossing). However, alternate mechanisms (turnstile, edge-cleavage) were probed by Fenske-Hall calculations for $B_5H_5^{2-}$ by Wales and Stone.[60] However, their attempts to find a concerted DSD pathway were unsuccessful. However, recent ab initio studies on isoelectronic *closo*-$C_2B_3H_5$ isomers found a relatively low barrier (20 kcal/mol) for the rearrangement of the 1,2-$C_2B_3H_5$ isomer into the much more stable 1,5-$C_2B_3H_5$.[61]

We conclude that the fluxionality of the *closo*-$B_nH_n^{2-}$ dianions provides only a rough indication of stability. The degree of perfection of the bonding of the most stable structure is not directly related to the barriers for rearrangement. According to our quantitative evaluation [Eq. (2) and Figure 7-2], the stability of *closo*-borane dianions, $B_nH_n^{2-}$, decreases in the following sequence:

$$B_{12}H_{12}^{2-} \approx B_6H_6^{2-} > B_7H_7^{2-} > B_{10}H_{10}^{2-} > B_9H_9^{2-} \approx B_8H_8^{2-} > B_{11}H_{11}^{2-} > B_5H_5^{2-}.$$

Besides Eq. (2) and Figure 7-2, the stability of an individual cluster (E_{dis}, B3LYP/6-311+G**), compared with its neighbors, can be estimated by the energy of the disproportionation reaction, Eq. (3):[29]

$$2B_nH_n^{2-} \rightarrow B_{n-1}H_{n-1}^{2-} + B_{n+1}H_{n+1}^{2-} \quad (n = 6\text{--}11) \tag{3}$$

As shown in Table 7-1, $B_6H_6^{2-}$, $B_7H_7^{2-}$, and $B_{10}H_{10}^{2-}$ are stable towards disproportionation [Eq. (3) is endothermic], whereas the disproportionations of $B_{11}H_{11}^{2-}$, $B_9H_9^{2-}$, and $B_8H_8^{2-}$ are exothermic. Known experimental results lend support to these results.[29] The cesium salt of $B_{11}H_{11}^{2-}$ disproportionates above 600°C into a equimolar mixture of $B_{10}H_{10}^{2-}$ and $B_{12}H_{12}^{2-}$.[56b] The air oxidation [50c] of $B_9H_9^{2-}$ generates $B_8H_8^{2-}$, and small amounts of other boranes, such as $B_6H_6^{2-}$, $B_7H_7^{2-}$, and $B_{10}H_{10}^{2-}$. The oxidation of the sodium salt of $B_8H_8^{2-}$ produced $B_7H_7^{2-}$ and large amounts of $B_6H_6^{2-}$; small amounts of $B_{10}H_{10}^{2-}$ and $B_{12}H_{12}^{2-}$ also formed.[50c]

7.3.2 Are the Increasing Stability and ΔH_{ave} Measures of Cage Delocalization in *Closo*-borane Dianions?

The generally increasing stability of the $B_nH_n^{2-}$ dianions with increasing size (Table 7-1, Figure 7-2, and Refs 28–31) is unusual. For comparison, we have evaluated the aromatic stabilization energies (ASE) of cyclic [n]annulenes ($n = 6$, 18, and 30, Figure 7-3) that obey the ($4n + 2$) Hückel rule using Eqs (4)–(6) (Table 7-3). Unlike $C_{10}H_{10}$,[62] these larger rings ($n = 18$ and 30) only have relatively small strain effects.

[6]annulene: $C_8H_{10}\ (C_{2h}) - C_2H_4\ (D_{2h}) \rightarrow C_6H_6\ (D_{6h})$ (4)

[18]annulene: $3[C_8H_{10}\ (C_{2v}) - C_2H_4\ (D_{2h})] \rightarrow C_{18}H_{18}\ (D_{6h})$ (5)

[30]annulene: $5[C_8H_{10}(C_{2v}) - C_2H_4\ (D_{2h})] \rightarrow C_{30}H_{30}\ (D_{3h})$ (6)

TABLE 7-3 Comparison of $4n + 2$ Hückel Annulene Strain-Corrected Aromatic Stabilization Energies (ASE, kcal/mol)$^{a-d}$

Compound	Symmetry	BECKE3LYP/6-31G*	ASE	ASE/n
[6]annulene	D_{6h}	−232.24870	19.1^a	3.2
[10]annulene	D_{10h}	—	18^b	1.8
[18]annulene	D_{6h}	−696.64078	13.7^c	0.76
[30]annulene	D_{3h}	−1161.0880	15.5^d	0.52
C_8H_{10}	C_{2v}	−310.80013		
C_8H_{10}	C_{2h}	−310.80568		
C_2H_4	D_{2h}	−78.58745		

^{a} ASE from [$C_8H_{10}\ (C_{2h}) - C_2H_4\ (D_{2h}) \rightarrow C_6H_6(D_{6h})$].
^{b} Strain-corrected estimate for the D_{10h} from Ref. 62.
^{c} From $3[C_8H_{10}\ (C_{2v}) - C_2H_4\ (D_{2h})] \rightarrow C_{18}H_{18}\ (D_{6h})$; strain-corrected for the 1.9 Å approach of the six inner H values of D_{6h} [18]annulene.
^{d} From $5[C_8H_{10}\ (C_{2v}) - C_2H_4\ (D_{2h})] \rightarrow C_{30}H_{30}(D_{6h})$.

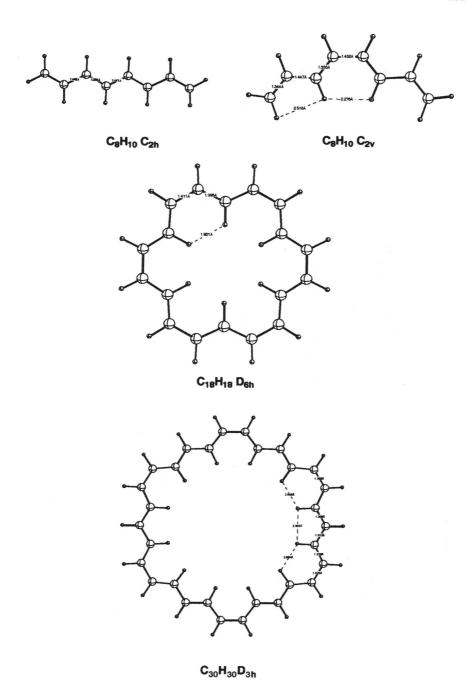

C_8H_{10} C_{2h}

C_8H_{10} C_{2v}

$C_{18}H_{18}$ D_{6h}

$C_{30}H_{30}D_{3h}$

Figure 7-3 The B3LYP/6-31G* optimized geometries of the Hückel annulenes: $C_{18}H_{18}$ (D_{6h}) and $C_{30}H_{30}$ (D_{3h}), as well as the C_8H_{10} (C_{2v}) and C_8H_{10} (C_{2h}) reference tetraenes.

Evaluated with appropriate acyclic polyene reference compounds (Figure 7-3) (chosen to balance the strain contributions), the ASEs of aromatic [n]annulenes vary surprisingly little with increasing value of n (Table 7-3). Consequently, the ASE per CH group (ASE/n) also shows *a strongly decreasing trend.* Thus, in agreement with earlier conclusions,[63] the stability of [n]annulenes, which are aromatic due to the extra stabilization provided by the cyclic delocalization, diminishes with increasing value of n. For example, the striking exothermicity (ΔE) of Eq. (7) demonstrates [18]annulene to be much less stable than three benzene molecules.

$$C_{18}H_{18} \text{ ([18]annulene)} \rightarrow 3C_6H_6 \text{(benzene)} \qquad \Delta E = -66 \text{ kcal/mol} \qquad (7)$$

The second comparison is provided by the polybenzenoid aromatics (Figure 7-4). Employing Cohen and Benson's aromatic group increments [64] [C_B–(H) = 3.30, C_B – (C_B) = 4.96], the computed ΔH_f° values for the polycyclic systems shown in Table 7-4 agree closely with the experimental values (from Pedley et al.[65]). However, the heats of formation of these aromatic compounds, estimated, instead, using conjugated olefin group increments $\Delta H_{f(con)}^\circ$[64] [(C_d–(C_d)(H) = 6.78, C_d–(C_d)(C) = 8.76)], are much larger than the experimental values. The differences reflect the additional resonance stabilization, i.e., the aromaticity. The aromatic stabilization energies (ASE) of polybenzenoid hydrocarbons can be based on the difference between ΔH_f° and the computed $\Delta H_{f(con)}^\circ$ (Table 7-4). While the total ASE increases upon each benzene annelation, the ASE per carbon (ASE/n) is remarkably constant.

TABLE 7-4 Polybenzenoid Hydrocarbon Heats of Formation[a,b,c,e] (kcal/mol) Aromatic Stabilization (ASE),[d] and Resonance Energies (RE)[f]

	Benzene	Naphthalene	Anthracene	Tetracene
$\Delta H_f^{\circ\,a}$	19.7 ± .2	35.9 ± .4	55.2 ± .6	69.6 ± 2.2
$\Delta H_{f(ar)}^{\circ\,b}$	19.8	36.3	52.8	69.4
$\Delta H_{f(con)}^{\circ\,c}$	40.7	72.0	103.3	134.6
ASE[d]	21.0	36.1	48.1	65.0
ASE/n	3.50	3.61	3.43	3.60
$\Delta H_{f(ncon)}^{\circ\,e}$	51.5	89.4	127.3	165.1
RE[f]	31.8	53.5	72.1	95.5
RE/n	5.30	5.35	5.15	5.30

[a] Experimental values from Pedley et al.[65]

[b] Based on Benson's group increments for aromatics [C_B – (H) = 3.30, C_B – (C_B) = 4.96] from Ref. 64.

[c] Based on Benson's increments for conjugated olefins [(C_d – (C_d)(H) = 6.78, C_d – (C_d)(C) = 8.76)] from Ref. 64.

[d] ASE = $\Delta H_{f(con)}^\circ$ – ΔH_f°.

[e] Based on Benson's increments for nonconjugated olefins [(C_d – (C)(H) = 8.59, C_d – (C)$_2$ = 10.34)] from Ref. 64.

[f] RE = $\Delta H_{f(ncon)}^\circ$ – ΔH_f°.

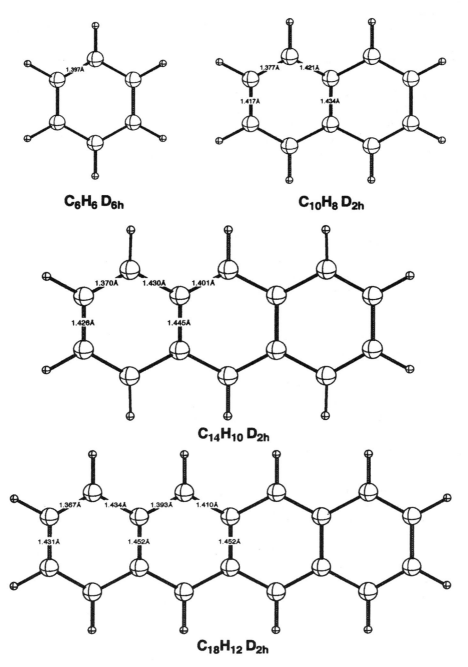

$C_6H_6 \ D_{6h}$

$C_{10}H_8 \ D_{2h}$

$C_{14}H_{10} \ D_{2h}$

$C_{18}H_{12} \ D_{2h}$

Figure 7-4 The B3LYP/6-31G* optimized geometries for illustrative polybenzenoid hydrocarbons.

Another set of resonance energies (RE) are obtained by using nonconjugated olefinic increments, $\Delta H^{\circ}_{f(nco)}$ [$(C_d - (C)(H) = 8.59$, $C_d - (C)_2 = 10.34$)] and taking the differences to the experimental ΔH°_f values. Note that such computed REs include both conjugation and aromaticity effects and are larger than the ASEs, but follow the same trends. The resonance contribution per carbon center RE (RE/n, Table 7-4) is also strikingly constant. This is also shown by the near-ther-moneutrality of the disproportionation reactions [eqs (8) and (9)]. These results show that *the aromatic delocalization and resonance energies of polybenzenoid aromatics are additive and depend primarily on the number of carbon centers.* Neither ASE/n nor RE/n become larger as the size of the system increases. There are energy variations among polybenzenoid hydrocarbon isomers (e.g., anthracene and phenanthrene),[65] but these are relatively small.

$$2 \text{ naphthalene} \rightarrow \text{benzene} + \text{anthracene} \qquad \Delta H = 2.8 \qquad (8)$$

$$\text{tetracene} + \text{benzene} \rightarrow \text{naphthalene} + \text{anthracene} \quad \Delta H = -1.8 \qquad (9)$$

Such additive stabilization, due to annelation, contrasts not only with the decreasing aromaticity exhibited by the monocyclic [n]annulenes (Table 7-3), but also with the *increasingly enhanced stabilization* of the *closo*-borane dian-ions with cluster size (Table 7-2). The exceptional behavior of the $B_nH_n^{2-}$ family is clear evidence for three-dimensional aromaticity.

7.3.3   Three-Dimensional Aromaticity of *Closo*-boranes $B_nH_n^{2-}$

Can the direct quantitative relationships among the geometric, energetic, and magnetic criteria of aromaticity,[8] which have been noted for related sets of organic molecules,[13] be extended to the three-dimensionally delocalized *closo*-boranes $B_nH_n^{2-}$ ($n = 5$–12)?

7.3.3.1 Geometric Criteria   Bond length equalization characterizes aromatic compounds, whereas large bond length alternations are found in antiaromatic compounds.[8,15] In *closo*-boranes, the difference between the longest and shortest skeletal bonds (ΔR, Tables 7-1 and 7-2, Figure 7-5) is the simplest geometric criterion. Small bond length alternations and high symmetry are associated with more efficient deltahedral skeletal bonding in the *closo*-borane dianions. Thus, all bond lengths are the same (i.e., $\Delta r = 0.0$ Å) in the most symmetrical boranes, $B_{12}H_{12}^{2-}$ (I_h) and $B_6H_6^{2-}$ (O_h) which have "perfect" delta-hedral bonding (Figure 7-5). Compound $B_{10}H_{10}^{2-}$ has a relatively small bond length alternation, $\Delta r = 0.139$ Å, followed by $B_5H_5^{2-}$ (0.149 Å) and $B_7H_7^{2-}$ (0.174 Å). The Δr of the remaining *closo*-boranes, $B_{11}H_{11}^{2-}$ (0.359 Å), $B_9H_9^{2-}$ (0.281 Å), and $B_8H_8^{2-}$ (0.293 Å) are significantly larger. The order of bond length alternation,

$$B_{12}H_{12}^{2-} \approx B_6H_6^{2-} > B_{10}H_{10}^{2-} > B_5H_5^{2-} > B_7H_7^{2-} > B_9H_9^{2-} > B_8H_8^{2-} > B_{11}H_{11}^{2-}$$

is plotted in Figure 7-5, and is discussed below.

7.3.3.2 Nucleus-Independent Chemical Shifts of Closo-borane Dianions

Nucleus-independent Chemical Shifts (NICS)[38] were introduced recently as an additional magnetic criterion of aromaticity.[8,17] The NICS is defined as the negative of the absolute magnetic shielding computed at the geometrical center of the cage. The effectiveness of NICS as a simple and efficient aromaticity probe has been shown for wide-ranging sets of planar and three-dimensional aromatic compounds.[38] In addition, the quantitative correspondence of NICS with ASE and geometric criteria has been demonstrated for five-membered aromatic and antiaromatic heterocycles.[38] Significantly negative NICS values imply diatropic ring currents: e.g., the 1,3-dehydro-5,7-adamantanediyl dication with only two interstitial electrons, and $B_{12}H_{12}^{2-}$, the borane dianion prototype, have, respectively, NICS values of –50.1 and –34.4 ppm.[38] These are much larger than that of benzene (–9.7 ppm). Positive values are associated with "antiaromaticity," e.g., +27.6 ppm for cyclobutadiene.[38]

As shown in Table 7-1 and Figure 7-5, the large negative NICS value at the cage centers of polyhedral boranes indicate the pronounced three-dimensional

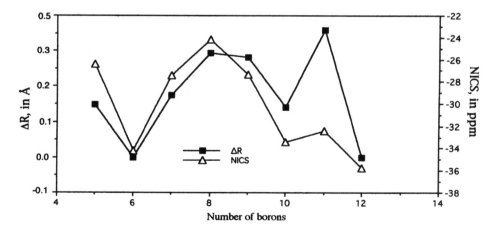

Figure 7-5 NICS values (in ppm) and the largest bond length difference (ΔR, in Å, see Figure 7-1) vs the number of boron atoms in the *closo*-borane dianions. Note the similar behavior patterns. Compare these trends with the deviations from the line defined in Figure 7-2.

delocalization in these molecules. The most symmetric $B_{12}H_{12}^{2-}$ (I_h), $B_6H_6^{2-}$ (O_h), and $B_{10}H_{10}^{2-}$ are the most "aromatic" among the *closo*-boranes. The NICS values suggest $B_9H_9^{2-}$, $B_7H_7^{2-}$, and $B_8H_8^{2-}$ to have nearly the same aromatic delocalization.

Compound $B_5H_5^{2-}$ has the smallest NICS value, but its aromaticity is still appreciable, in contrast to Aihara's conclusion[1] that this species is "nonaromatic." Recently, we demonstrated the "three-dimensional aromaticity" for $B_5H_5^{2-}$ and a related series of five-vertex 1,5-diheteroboranes,[33] e.g., by the ASE, the magnetic susceptibility exaltation, and the NICS criteria.[33,34,38] The three-dimensional delocalization of $B_5H_5^{2-}$ was demonstrated by the large ASE (34.8 kcal/mol), large magnetic susceptibility exaltation (–46.1 ppm cgs) and NICS (–28.1 ppm) values.[33] All these aromatic properties for $B_5H_5^{2-}$ contradict the classic description [66] and Aihara's "nonaromatic" Designation (RE = 0.0β, Table 7-2) for $B_5H_5^{2-}$.[1] Our results also question other interpretations, based on REs from graph-theoretical approaches, applied to the three-dimensional delocalization of polyhedral *closo*-borane dianions.[1]

The trends in the NICS values and the bond length alternation (Δr), compared in Figure 7-5, are remarkably similar, and demonstrate the relationship between geometric and magnetic criteria of aromaticity in *closo*-boranes. Note that both these trends also agree qualitatively with the energy trends (compare the deviations from the line shown in Figure 7-2 with Figure 7-5).

7.4   CONCLUSIONS

As has been noted previously,[27–31] not only do the stabilization energies of the *closo*-borane dianions tend to increase, but also the average stability per vertex tend to increase with increased cluster size along the $B_5H_5^{2-}$ to $B_{12}H_{12}^{2-}$ series (with variations). This behavior is quite different from that of the Hückel [n]annulenes and the polybenzenoid hydrocarbons. The strain-corrected aromatic stabilization energies per CH group in the [n]annulenes (ASE/n, Table 7-3) decreases with increasing ring size, e.g., the energy and the ASE of $C_{18}H_{18}$ is much less than that of three benzenes [Eq. (7)]. In contrast, the ASEs of benzenoid hydrocarbons, e.g., benzene, naphthalene, anthracene, and tetracene, are additive; the total ASEs increase with size, but the average energy per carbon (ASE/n and RE/n; Table 7-4) is quite constant.

The exceptional energetic behavior of the $B_nH_n^{2-}$ family is a direct indication of three-dimensional aromaticity. Rather than exhibiting additivity, the stability becomes proportionately greater with increasing cluster size.

We have taken this general increase into account by using $B_6H_6^{2-}$ and $B_{12}H_{12}^{2-}$ (the most symmetrical *closo*-borane dianions) as the comparison standards. Variations in the average stabilization energies of the other members of the family, evaluated on this basis, show $B_7H_7^{2-}$ and $B_{10}H_{10}^{2-}$ to be the next most stable systems; they also are rigid with high fluxional barriers. In contrast, $B_8H_8^{2-}$ and $B_{11}H_{11}^{2-}$ are quite fluxional on the NMR time scale, and are less stable thermodynamically. However, $B_5H_5^{2-}$ (which is not known experimentally) has the smallest average vertex energy, but its fluxional barrier is computed to be intermediate. The bonding in $B_5H_5^{2-}$ is delocalized, like that of the other *closo*-$B_nH_n^{2-}$ dianions.[33]

Instead of using "rigidity" as a structural criterion of aromaticity, we propose simply to employ Δr, the difference between the longest and the shortest bonds in the *closo*-borane dianions. This Δr criterion assumes that the deviations from perfectly symmetrical deltahedral bonding, imposed by the restrictions of less symmetrical cages, should be reflected in higher energies. This is just what we find. Figure 7-1 shows the closely parallel behavior or the Δr energetic evaluations.

Magnetic properties, as criteria of aromaticity, are most closely related to the degree of electron delocalization. Ring-current effects result in exalted diamagnetic susceptibilities and in abnormal magnetic shieldings (magnetic susceptibility anisotropies obviously are not applicable in the nearly spherical three-dimensional systems). While quite large magnetic susceptibility exaltations are found for the *closo*-borane dianions, the dependence is inherently complex since such exaltations are known in monocyclic systems to be dependent on the square of the ring area.[8b]

In contrast, NICS, which are based on the magnetic shieldings computed in the geometric centers of the cage systems,[38] provide a direct measure the ring-current effects (aromaticity, in this sense). There is remarkably similar behavior among the NICS values of the *closo*-borane dianions, the Δr geometric criterion, and the average energy (compare Figs 7-2 and 7-5). All this evidence supports Aihara's early proposal [1] that these *closo*-systems can be considered as three-dimensional aromatics, and it implies that *nido-*, etc., polyhedral electron-delocalized systems also exhibit the special characteristics associated with aromaticity.

ACKNOWLEDGMENTS

We thank Dr. Govindan Subramanian and Dr. Alexander M. Mebel for discussions and their calculations that contributed to this project. This work was supported by the Deutsche Akademische Austauschendienst (DAAD doctoral fellowship to K.N.), the Deutsche Forschungsgemeinschaft, and the Fonds der Chemischen Industrie.

REFERENCES

1. Aihara, J., *J. Am. Chem. Soc.*, 1978, **100**, 3339.
2. (a) Jemmis, E. D. and Schleyer, P. v. R., *J. Am. Chem. Soc.*, 1982, **104**, 4781; (b) Jemmis, E. D., *J. Am. Chem. Soc.*, 1982, **104**, 7071.
3. (a) Williams, R. E., *Inorg. Chem.*, 1971, **10**, 210; (b) Williams, R. E., *Adv. Inorg. Chem. Radiochem.*, 1976, **18**, 66; (c) Williams, R. E., *Chem. Rev.*, 1992, **92**, 177.

4. Muetterties, E. L. (ed.), *Boron Hydride Chemistry*, Academic Press, New York, 1975.

5. (a) Grimes, R. N., *Carboranes*, Academic Press, New York, 1970; (b) Liebman, J. F., Greenberg, A., and Williams, R. E., *Advances in Boron and the Boranes*, VCH Verlagsgesellschaft, New York, 1988; (c) Williams, R. E., in *Advances in Organometallic Chemistry* (Stone, F. G. A. and West, R., eds), Academic Press, New York, 1994.

6. Erlenmeyer, E., *Ann.* 1866, **137**, 327, and 344.

7. Faraday, M., *Phil. Trans. Roy. Soc. London*, 1825, 440.

8. (a) Garratt, P. J., *Aromaticity*, Wiley, New York, 1986; (b) Minkin, V. J., Glukhovtsev, M. N., and Simkin, B. Y., *Aromaticity and Antiaromaticity*, Wiley, New York, 1994.

9. Armit, J. W. and Robinson, R., *J. Chem. Soc.*, 1925, **127**, 1604.

10. Hückel, E., *Z. Physik*, 1931, **70**, 204; 1931, **72**, 310.

11. (a) Evans, M. G. and Polanyi, M., 1938, **34**, 11; Evans, M. G. and Warhust, E., *Trans. Faraday Soc.*, 1938, **34**, 614; (b) Evans, M. G. 1939, **35**, 824.

12. Woodward, B. R. and Hoffmann, R., *The Conservation of Orbital Symmetry*, Verlag Chemie, Weinheim, 1970.

13. Schleyer, P. v. R., Freeman, P., Jiao, H., and Goldfuss, B., *Angew. Chem. Int. Ed. Engl.*, 1995, **34**, 337.

14. Bernardi, F., Bottoni, A., and Venturini, A., *J. Mol. Struct. (Theochem.)*, 1988, **163**, 173, and references therein (more than 60 theoretical investigations are cited).

15. Schleyer, P. v. R. and Jiao, H., *Pure Appl. Chem.*, 1996, **68**, 209, and references therein.

16. Pascal, P., *Ann. Chim. Phys.*, 1910, **19**, 5.

17. (a) Dauben, H. J., Jr, Wilson, J. D., and Laity, J. L., *J. Am. Chem. Soc.*, 1968, **90**, 811 and 1969, **91**, 1991; (b) Dauben, H. J., Jr, Wilson, J. D., and Laity, J. L., in *Non-Bezenoid Aromatics* Vol. 2, (Snyder ed.), Academic Press, New York, 1971, and references therein; see also Ref. 16.

18. Pople, J. A., *J. Chem. Phys.*, 1956, **24**, 1111.

19. Pauling, L., *J. Chem. Phys.*, 1936, **4**, 673.

20. London, F. J., *Phys. Radium*, 1933, **8**, 397.

21. Kutzelnigg, W., Fleischer, U., and Schindler, M. *N. M. R., Basic Principles and Progress*, Springer Verlag, Berlin, 1990, p. 165.

22. (a) Wolinksi, K., Hinton, J. F., and Pulay, P., *J. Am. Chem. Soc.*, 1990, **112**, 8251; (b) Häser, M., Ahlrichs, R., Baron, H. P., Weis, P., and Horn, H., *Theor. Chim. Acta.*, 1992, **83**, 455, and references cited therein.

23. Minkin, V. I., Minyaev, R. M., and Zhdanov, Yu. A., *Nonclassical Structures of Organic Compounds*, Mir Publishers, Moscow, 1987.

24. (a) Wade, K., *Electron Deficient Compounds*, Nelson. London, 1971; (b) Olah, G. A., Wade, K., and Williams, R. E., *Electron Deficient Boron and Carbon Clusters*, John Wiley and Sons, New York, 1991.

25. Lipscomb, W. N., *Boron Hydrides*, W. A. Benjamin, New York, 1963, p. 43.

26. (a) Wade, K., *Adv. Inorg. Chem. Radiochem.*, 1976, **18**, 1; (b) Wade, K., *J. Chem. Commun.*, 1971, 792; (c) O'Neill, M. E.; Wade, K., in *Comprehensive*

Organometallic Chemistry (Wilkinson, G., Stone, F. G. A., and Abel, E., eds), Pergamon Press, New York, 1982.

27. King, R. B., *Inorg. Chimica Acta*, 1981, **49**, 237. According to topological and group-theoretical consideration, a diamond-square-diamond (DSD) rearrangement involves the switching of the edges shared by two deltahedral faces. When the new deltahedron (after DSD rearrangement) is identical to the original deltahedron, then the switching edge is degenerate. The 5-, 8-, 9-, and 11-vertex deltahedra, which have degenerate switching edges, are fluxional. Similarly, deltahedra without degenerate edges, as in the 6-, 10-, and 12-vertex cages, are rigid. Apparent qualitative relationships between the fluxional characteristic of polyhedron $B_nH_n^{2-}$ and their reactivities have been noted, along with some exceptions.

28. Dixon, D. A., Kleier, D. A., Halgern, T. A., Hall, J. H., and Lipscomb, W. N., *J. Am. Chem. Soc.*, 1977, **99**, 6226.

29. Porterfield, W. W., Jones, M. E., Gill, W. R., and Wade, K., *J. Inorg. Chem.*, 1990, **29**, 2914.

30. (a) Zhao, M. and Gimarc, B. M., *Inorg. Chem.*, 1993, **32**, 4700; (b) Ott, J. J. and Gimarc, B. M., *J. Comp. Chem.*, 1986, **7**, 673.

31. Zhao, M. and Gimarc, B. M., *Polyhedron*, 1995, **14**, 1315, and references therein.

32. King, R. B. and Rouvray, D. H., *J. Am. Chem. Soc.*, 1977, **99**, 7834.

33. Schleyer, P. v. R., Subramanian, G., and Dransfeld, A., *J. Am. Chem. Soc.*, 1996, **118**, 9988.

34. Subramanian, G., Schleyer, P. v. R., and Jiao, H., *Angew. Chem. Int. Ed. Engl.*, 1996, **35**, 2638.

35. Elvidge, J. A. and Jackman, L. M., *J. Chem. Soc.*, 1961, 859.

36. (a) Fleischer, U., Kutzelnigg, W., Lazzeretti, P., and Mühlenkamp, V., *J. Am. Chem. Soc.*, 1994, **116**, 5298, and references therein; (b) Benson, R. C. and Flygare, W. H., *J. Am. Chem. Soc.*, 1970, **92**, 7523.

37. (a) Bühl, M., Thiel, W., Jiao, H., Schleyer, P. v. R., Saunders, M., and Anet, F. A. L., *J. Am. Chem. Soc.*, 1994, **116**, 7429, and references therein; (b) Bühl, M. and van Wüllen, C., *Chem. Phys. Lett.*, 1995, **247**, 63.

38. Schleyer, P. v. R., Maerker, C., Dransfeld, A., Jiao, H., and Hommes, N. J. v. E., *J. Am. Chem. Soc.,* 1996, **118**, 6317.

39. (a) Frisch, M. J., Trucks, G. W., Schlegel, H. B., Gill, P. M. W., Johnson, B. G., Robb, M. A., Cheeseman, J. R., Keith, T., Petersson, G. A., Montgomery, J. A., Raghavachari, K., Al-Laham, M. A., Zakrzewski, V. G., Ortiz, J. V., Foresman, J. B., Cioslowski, J., Stefanov, B. B., Nanayakkara, A., Challacombe, M., Peng, C. Y., Ayala, P. Y., Chen, W., Wong, M. W., Andres, J. L., Replogle, E. S., Gomperts, R., Martin, R. L., Fox, D. J., Binkley, J. S., Defrees, D. J., Baker, J., Stewart, J. P., Head-Gordon, M., Gonzalez, C. and Pople, J. A., Gaussian 94 (Revision C.3), Gaussian Inc., Pittsburgh, PA, 1995; (b) Hehre, W., Radom, L., Schleyer, P. v. R., and Pople, J. A., *Ab initio Molecular Orbital Theory*, Wiley, New York, 1986; (c) Foresman, J. B. and Frisch. A. E., *Exploring Chemistry With Electronic Structure Methods*, 2nd Edition, Gaussian, Inc., Pittsburgh, PA 1996.

40. Scott, A. P. and Radom, L. *J., Phys. Chem.*, 1996, **100**, 16502, and references therein.

41. Housecroft, C. E. and Wade, K., *Inorg. Chem.*, 1983, **21**, 1391.

42. Lipscomb, W. N., *Science* (Washington, DC), 1966, **153**, 373–378.

43. McKee, M. L., *J. Am. Chem. Soc.*, 1992, **114**, 879.

44. Wales, D. J., *J. Am. Chem. Soc.*, 1993, **115**, 1557.

45. (a) Gimarc, B. N., Warren, D. S., Ott, J. J. and Brown, C., *Inorg. Chem.*, 1991, **30**, 1598.

46. Schaeffer, R., Johnson, O., and Smith, G. S., *Inorg. Chem.*, 1965, **4**, 917.

47. (a) Mebel, A. M., Charkin, O. P., Bühl, M., and Schleyer, P. v. R. J., *Inorg. Chem.*, 1993, **32**, 463; (b) Mebel, A. M., Charkin, O. P., and Schleyer, P. v. R. J., *Inorg. Chem.*, 1993, **32**, 469; (c) Bühl, M., Mebel, A. M., Charkin, O. P., and Schleyer, P. v. R. J., *Inorg. Chem.*, 1992, **31**, 3769.

48. (a) Knoth, W. H., Hertler, W. R., and Muetterties, E. L., *Inorg. Chem.*, 1965, **280**, 4; (b) Hertler, W. R., Knoth, W. H., and Muetterties, E. L., *J. Am. Chem. Soc.*, 1964, **86**, 5434.

49. Dewar, M. J. S. and McKee, M. L., *Inorg. Chem.*, 1978, **17**, 1569.

50. (a) Muetterties, E. L., *Tetrahedron* 1974, **30**, 1595; (b) Muetterties, E. L., Wiersama, R. J. and Hawthorne, M. F., *J. Am. Chem. Soc.*, 1973, **95**, 7520; (c) Klanberg, F., Eaton, D. R., Guggenberger, L. J., and Muetterties, E. L., *Inorg. Chem.*, 1967, **6**, 1271.

51. Ott, J. J., Brown, C. A., and Gimarc, B. M., *Inorg. Chem.*, 1989, **28**, 4269.

52. Guggenberger, L. J., *Inorg. Chem.,* 1969, **8**, 2771.

53. Klier, D. A. and Lipscomb. W. N., *Inorg. Chem.*, 1979, **18**, 1312; (b) Halgern, T. A. and Lipscomb, W. N., *J. Chem. Phys.*, 1973, **58**, 1569.

54. (a) Berry, T. E., Tebbe, F. N., and Hawthorne, M. F., *Tetrahedron Lett.*, 1965, **715**; (b) Toplin, E. I. and Lipscomb, W. N., *J. Am. Chem. Soc.*, 1973, **95**, 2384; (c) Klier, D. A. and Lipscomb., W. N., *Inorg. Chem.*, 1978, **17**, 166; (d) Muetterties, E. L., Hoel, E. L., Salentine, C. G., and Hawthorne, M. F. J., *Inorg. Chem.*, 1975, **14**, 950.

55. Kleier, D. A., Dixon, D. A., and Lipscomb, W. N., *Inorg. Chem.*, 1978, **17**, 166.

56. (a) Muetterties, E. L., Hoel, E. L., Salentine, C. G., and Hawthorne, M. F., *Inorg. Chem.,* 1975, **14**, 950; (b) Klanberg, F. and Muetterties, E. L., *Inorg. Chem.,* 1966, **5**, 1955.

57. Mebel, A. M., Charkin, O. P., Schleyer, P. v. R., and Najafian, K., *Inorg. Chem.*, in press.

58. Gimarc, B. M. and Ott, J. J., *Inorg. Chem.*, 1986, **25**, 2708.

59. Gimarc, B. M. and Ott, J. J., *Inorg. Chem.*, 1986, **25**, 83.

60. Wales, D. J. and Stone, A. J., *Inorg. Chem.*, 1987, **26**, 3845.

61. (a) Hofmann, M., Fox, M. A., Greatrex, R., Schleyer, P. v. R., Bausch, J. W., and Williams, R. E., *Inorg. Chem.*, 1996, **35**, 6170; (b) McKee, M. L., *J. Mol. Struct. (THEOCHEM)*, 1988, **168**, 191.

62. Sulzbach, H. M., Schleyer, P. v. R., Jiao, H., Xie, Y., and Schaefer, H. F., *J. Am. Chem. Soc.*, 1995, **117**, 1369.

63. Dewar, M. J. S. and Gleicher, G. J., *J. Am. Chem. Soc.*, 1965, **87**, 685.

64. Cohen, N. and Benson, S. W., *Chem. Rev.*, 1993, **93**, 2419.

65. Pedley, J. B., Naylor, R. D., and Kirby, S. P., *Thermochemical Data of Organic Compounds*, Chapman and Hall, London, 1986.

66. Jemmis, E. D. and Subramanian, G., *J. Phys. Chem.,* 1994, **98**, 9222.

8

ELECTRON-DEFICIENT COMPOUNDS AND THEORY-DEFICIENT PEOPLE*

ROBERT W. PARRY

Department of Chemistry, University of Utah, Salt Lake City, UT 84112-1194

Conventional bond pictures, which are familiar to every student completing a quality course in elementary chemistry, include the triple bond, the double bond, the single bond, and the metallic bond. These four names have long provided boxes that are useful in sorting out bonding patterns for a large number of the compounds of modern chemistry. When properly used, the bonding groups are useful in helping to define chemical and physical properties of substances under study.

Armed with the knowledge that carbon has four valence electrons, that hydrogen has one valence electron, that a single bond is formed when two atoms each contribute one electron to form an electron pair, and that electron pair bonds will assume that geometry that maximizes the distance between electron pairs, the geometry of CH_4, C_2H_6, C_3H_8, C_4H_{10}, BF_3, BCl_3, and even Al_2Cl_6, can be easily defined.

It would have been most reassuring if these generalizations could be effectively applied to the boron hydrides as well as to the boron halides. Compounds

*This chapter is, and the symposium was, respectfully dedicated to Dr. Robert E. Williams, the first person (with colleagues) to synthesize and recognize carboranes, and to propose correct structures for the simplest three carboranes. His early structures, *later verified completely*, challenged many sacred structural concepts of chemistry. In subsequent years, he has been highly successful in systematizing structural information on boranes, carboranes, and even their metal derivatives. His contributions are monumental. It is most appropriate that these symposium proceedings pay tribute to his contributions. The reader will find that Williams' work is the basis for many of the concepts considered here.

such as those shown below would be predicted:

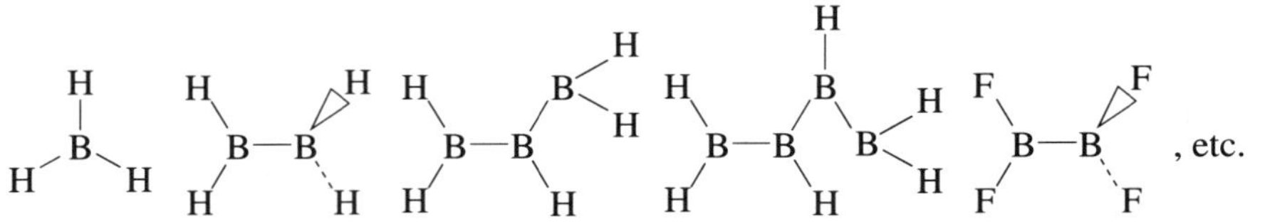

Although B_2F_4 is a known species, none of the above boron hydrides have ever been identified as stable molecules. The meticulous and exciting work done by Alfred Stock in Germany from 1912 to the late 1930s established clearly that boron hydrides, as opposed to boron halides, follow a different set of formula generalizations. Stock experimentally characterized B_2H_6 and B_4H_{10}.[2] The formal analogy between B_2H_6 and C_2H_6, and between B_4H_{10} and C_4H_{10}, was not lost on physical and organic chemists of the 1930s and 1940s. Several papers were published in that era, reporting that B_2H_6 had the same structure as C_2H_6 and that B_4H_{10} had the same structure as C_4H_{10}. These structural claims set off many vigorous debates over the bonding patterns for the boron hydrides. Since boron has only three valence electrons, not four as in carbon, it is not possible to construct a satisfactory conventional single-bond model for B_2H_6. While C_2H_6 has 14 electrons, which can give the seven single bonds needed for

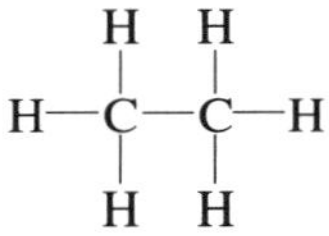

B_2H_6 has only 12 electrons, which can give only six electron pair bonds, not seven as in C_2H_6. One could write

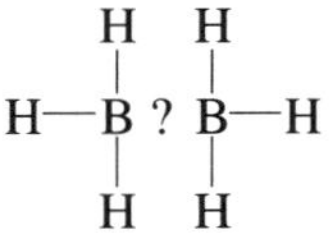

where ? expressed commonly held frustrations over why BH_3 is a dimer. Similar problems arose when one compared C_4H_{10} and B_4H_{10}. The C_4H_{10} molecule has 26 electrons to make the 13 electron pairs needed for the 13 single bonds of butane:

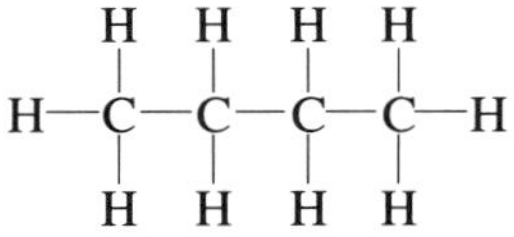

On the other hand, B_4H_{10} has only 22 electrons and could support only 11 conventional single bonds. An often-used practice in science was called upon: "If you don't understand a phenomenon, give it a name." As the name is accepted, the scientific mystery seems less formidable. The name "*electron-deficient compounds*" for the boranes and a number of other compounds, such as $Al_2(CH_3)_6$, became a part of the language of chemistry.

In 1946, when I was a young assistant professor at the University of Michigan, I used the term "electron-deficient molecule" in front of Professor Kasimir Fajans, a world-famous physical chemist on the University of Michigan faculty. Professor Fajans' scornful comment was, "There is no such a thing as an electron-deficient molecule; there is only a theory-deficient Parry." I could only take comfort, at the time, in knowing that I was not alone in my "theory deficiency." The structures and bonding patterns of the boranes posed very tough problems for all chemists in the late 1930s and early 1940s. Since those days, some 50 years ago, advances in structure determination and in bonding theory have provided a basis for a reasonable understanding of the boranes. However, the term "electron-deficient compounds" lives on. The term is harmless, as a part of our scientific legacy, but it may mask important lessons that lie in the Periodic Table. Let us explore this possibility.

Today, the four simple bonding patterns of the 1940s have been expanded. They are summarized in Table 8-1. *Quadruple bonds*, in which two metal atoms are held together by four pairs of electrons, have been identified and characterized by F. A. Cotton and his collaborators.[2] Farther down in Table 8-1, *three-center bonds*, in which *three* atoms are held together by a *single pair* of electrons, appear. These were used to rationalize the bonding in the simplest boron hydride, B_2H_6. The very early theoretical treatments of this bonding mode are associated with names such as Eberhardt, Crawford, and Lipscomb [3] and Longuet-Higgins and Roberts,[4] as well as others.

Lipscomb, using structural studies as a foundation for his theoretical work, expanded the idea of the three-centered bonds to include four-centered, five-centered, etc., bonds.[3] Molecular-orbital techniques [3] and ab initio methods [5] became useful as efforts to quantify bonding patterns developed. The metal bond is seen to be the final example of the multicentered bond where an infinite number of atoms are held together by pairs of electrons. One passes from highly localized bonds at the top of Table 8-1 to highly delocalized or metallic bonds at the bottom. As one would expect, boron, positioned to the left of carbon in the Periodic Table, would be expected to show some delocalization of bonding, but it is a far cry from the completely metallic bonds on the far left of the table or of the localized carbon bonds in the middle.

What options are available for boron bonding? It is somewhat more delocalized than carbon in its bonding patterns, but it is certainly NOT a metal. In considering this question it is useful to consider the changes in bonds seen in the *elements* as one goes across the Periodic Table from lithium to nitrogen. To facilitate this analysis, it is useful to record the number of valence electrons (or bonding electrons) available and the number of closest-neighbor atoms around a

TABLE 8-1   Bonding Patterns

Common Name	Number of Atoms Bonded	Number of Electron *Pairs* Used	Name of Structure and Occurrence
Quadruple bond	2	*4 and even* higher	*Quadruple* and possibly even higher order bonds between two atoms, such as Mo—Mo, W—W, Re—Re, etc.[2]
Triple bond	2	3	Classic *triple bonds*, particularly as seen in carbon chemistry and in related systems, such as N_2, CN^-, CO, etc.
Double bond	2	2	Classic *double bond*, particularly as seen in carbon chemistry and analogous systems (i.e., C_2H_4, etc.)
Single bond	2	1	Classic *single* bond seen in *many* areas of chemistry (organic and inorganic)
Three-centered bond	3	1	Three-centered bond, as seen in $[Al(CH_3)_3]_2$, $[Be(CH_3)_2]_n$, boranes, carboranes, etc.[3,5]
Multicentered bond	4,5,6	1	Four-, five-, six-center bonds as in some boranes, carboranes, etc.
Metal bond	∞	1	Metal bond

given atom (i.e., its coordination number or the number of "bonds" that must be formed). From these two quantities, *one can determine the number of bonds in which each valence electron must participate. This is one measure of the degree of delocalization of the bonding in the structures of the elements.* The data are summarized in Table 8-2. For example, a lithium atom has only one valence electron. Metallic lithium has a relatively open body-centered cubic structure with eight nearest lithium neighbors and six more lithiums at a distance only 15% longer than the distance of the eight nearest neighbors. This means that each valence electron of the lithium atom must be involved in making *at least eight* "bonds" and, more realistically, probably somewhere between 10 and 14, if one considers the very close group of six atoms at a distance only 15% larger than that of the nearest neighbors. Comparable analysis using the number of "valence" or bonding electrons and coordination numbers from known structures of the *elements* give the so-called *delocalization parameter* or the "number of bonds" in which each valence electron participates. This ranges from a value well above 10 for lithium to 0.33 for nitrogen. Boron, with a value of 3–4, lies midway between beryllium (6) and carbon (0.8–1). It cannot be exactly like

TABLE 8-2   Electron Delocalization in the Elements — The Delocalization Parameter

Atomic Number	Element	Number of "Valence" or Bonding Electrons	Number of Other Atoms around Each Atom (Coordination Number)	Lattice Type	average number of Atoms Bound by Each Bonding Electron[d] — *Measure of Electron Delocalization*
3	Lithium	1	8 plus part of $6 \approx 10$	Body centered cubic	About 10 to 12
4	Beryllium	2	12	Hexagonal close packed	6
5	Boron	3	6 in icosahedron; 2–6 outside icosahedron; use 9–12	packing of icosahedra of borons	3–4
6	Carbon	4	4	Diamond	1
		3.8[a]	3	Graphite	0.80
		3	1	Acetylene[b]	0.33
			1	Acetylide[c]	0.33
7	Nitrogen	3	1	Gaseous N_2^c	0.33
8	Oxygen	2	1	Gaseous O_2	0.5

[a]There is a question relative to the role played by the "double- bond electrons" in binding the layers together. Value 3.8 is an arbitrary estimate for number of electrons used in the plane.

[b]Consider only C—C linkage; only three electrons involved from reference carbon.

[c]Free electron pairs (or bonding–antibonding pairs in molecular- orbital model) are not counted.

[d]Delocalization parameter for "*electron-deficient*" systems is larger than 1. For single bond it is 1; for multiple bonds it is less than 1.

either one. A reasonable compromise would appear to be the linking together of delocalized units. These can be visualized as metalloid clusters with *three-dimensional*, delocalization, as contrasted with the two-dimensional delocalization in aromatic rings. What do these delocalized boron clusters look like? Elemental boron chooses an icosahedral shape. The 12-boron structure is shown in Figure 8-1.[6] The icosahedron may be visualized as two pentagonal boron pyramids joined base to base in a staggered position. This is shown in Figure 8-1 by representing the top pentagonal pyramid boron atoms as open circles (○) and the lower pentagonal pyramid as shaded circles (◉). Hydrogen atoms of $B_{12}H_{12}^{2-}$ are shown in Figure 8-1 to illustrate external hydrogen positions in this anion. A pertinent question is, Why are borons in an icosahedral pattern? Why doesn't

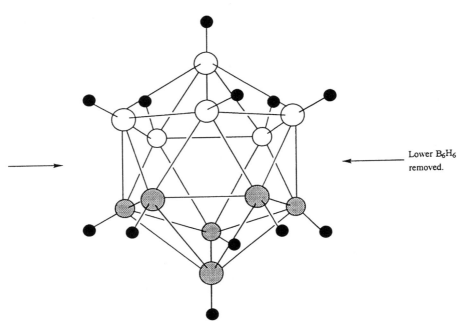

Lower B_6H_6
removed.

Figure 8-1   Modified picture of $B_{12}H_{12}^{2-}$ from the work of Wunderlich and Lipscomb.[6] Boron atoms are in an icosahedral array. The borons of the bottom pentagonal pyramid are shaded to differentiate them from the borons of the top pentagonal pyramid. Actually, all boron atoms are equivalent. Hydrogens (●) are shown attached to each boron.

one see either hexagonal or cubic close-packed arrays as in metals? Perhaps it is the nature of the icosahedron. If one arranges *12* spheres in an icosahedral array *with all spheres touching*, the volume of the *icosahedral unit* is *less* than that of any other 12-atom packing pattern.[7] The volume occupied by the spheres in the icosahedron is small and conditions for delocalization are excellent. The icosahedral units are found in all forms of elemental boron. It is the *arrangement* of the icosahedra that differentiates the different forms of boron. In a few cases, extra boron atoms, not in the icosahedron, are present to link the icosahedra together.[8] In other cases, the icosahedra appear as units in a lattice that are themselves linked by multicentered bonds. For example, in αB_{12} boron, all boron atoms belong to *icosahedral groups that are arranged in an approximately cubic close-packed array* (Figure 8-2).[8] Overall cubic and hexagonal packing of spheres gives a more dense structure than an *extended* icosahedral array. *Icosahedra do NOT pack well.* On the other hand, the 12 borons *in the icosahedron* are in a more dense arrangement than one can find in any other general close-packed structure. For boron, one finds delocalized units held together in some cases by multi-center bonds and in other cases by more localized linkages.

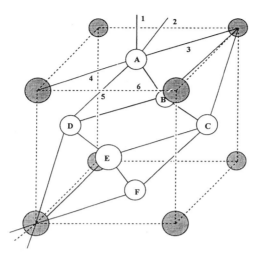

Figure 8-2 Packing of icosahedra of borons in αB_{12}: one of the forms of elemental boron.[8]

The carbon analog is seen in cases like naphthalene, or diphenyl, where benzene rings are combined in different ways.

Recognizing this pattern it is easier to understand the structures of the hydrides of both carbon and boron. If one cuts any carbon unit out of a diamond lattice (all C—C single bonds) and adds a hydrogen atom at the site of each C—C bond scission, one gets an alkane—an alkyl "carbon hydride." If one takes the graphite form of carbon and cuts out a carbon backbone, then adds a hydrogen wherever a C—C bond was cut, an aromatic hydrocarbon or "carbon hydride" results. *The dominant structural patterns found in alkyl and aromatic hydrocarbons are reflections of the structures of elemental carbon. In a similar way the structures of the boron hydrides are reflections of the icosahedral units found in elemental boron.*[10] Cutting out the B_{12} unit and adding hydrogens to each external boron gives the very stable $B_{12}H_{12}^{2-}$ anion. Molecular Orbital Theory indicated that the energy levels for the electrons holding the icosahedron together have the form shown in Fig. 8-3. One sees *thirteen* bonding orbitals in the bonding levels. Twenty-six cluster-bonding electrons are needed to give the stable structure. Each boron in $B_{12}H_{12}^{2-}$ can donate 2 electrons to the icosahedral cluster and one electron of each boron goes to give the observed B–H external bond. By this count we have 24 electrons for the icosahedral cage structure and 26 are needed to give the most stable "closed shell." The two extra electrons needed to complete the "closed shell of the icosahedron" can come from an external source, such as 2 sodium atoms to give $Na_2B_{12}H_{12}$, a very stable species.[5] This was an early triumph of theory.[5] An analogy to the formation of the Cl⁻ ion in atomic structure patterns is easily seen.

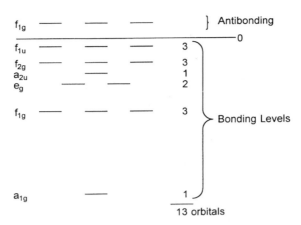

Figure 8-3 The $B_{12}H_{12}^{2-}$ molecular-orbital levels.[10] Thirteen orbitals require 26 electrons to give the closed electronic structure.

If one breaks up aromatic rings when cutting out an aromatic hydrocarbon from graphite, the addition of hydrogens is a little more complicated depending on how the aromatic systems are broken. Similarly, if the icosahedron is broken, some three-center bonded B⁀H⁀B linkages are required (along with one B—H of each BH_2 group) to assist in stabilizing the "semicage" boron structure.[12] For example, B_5H_{11} can be represented by the modified structure diagram (Figure 8-4) patterned after a paper by Robert E. Williams.[11] How is this structure tied to the icosahedral structure of $B_{12}H_{12}^{2-}$? Figure 8-1 and Figure 8-4 are helpful in this exercise. First, imagine a plane that cuts off the lower B_6H_6 unit in Figure 8-1 (see arrows cutting off the shaded boron unit). The remaining top B_6H_6 unit now has a hydrogen attached to each boron. Three B⁀H⁀B bonds are inserted at the base to partially [12] stabilize the structure after removal of the lower B_6H_6 unit (see Figure 8-4). Now, one B—H is removed from the *pentagonal base* of the above B_6H_6. To stabilize this structure, *after removal of the B—H*, an extra hydrogen is added to each of the three remaining adjacent B—H groups. Two hydrogens are shown as ▢ (endo) and one as the ◯ (semibridge), which is attached to the top boron seen in Figure 8-5. All data, physical, chemical, and theoretical, suggest that the hydrogen represented by ◯ is a bridging hydrogen, but energy differences between the bridging and the endo-BH_2 configuration, ▢, are small.[13]

Since electrons of the bridge hydrogens and of the one B—H linkage in each of the two BH_2 groups seem to be useful in maintaining the core boron structure, the so-called "core electrons" can be counted. The *total* number of electrons in B_5H_{11} is:

$$5 \text{ borons} \times \frac{3 \text{ electrons}}{\text{boron}} + 11 \text{ hydrogens} \times \frac{1 \text{ electrons}}{\text{hydrogen}} = 26 \text{ electrons} \atop \text{total}$$

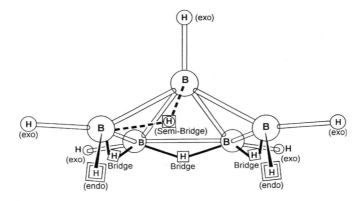

B) Boron atom - Gives 2 electrons to "core."

H) <u>Exo</u> Hydrogen atom - Electrons for bond **NOT** counted in "core."

H) Unique Hydrogen atom - "semi-bridging" gives 2 electrons to "core."

H] Bridging Hydrogen atom - Gives 2 electrons to "core."

H]] <u>Endo</u> Hydrogen atom - Gives 2 electrons to "core."

Figure 8-4   Modified structure diagram for B_5H_{11}.

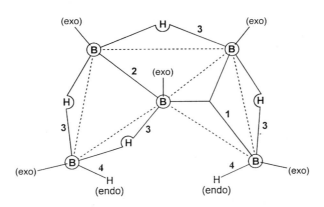

Figure 8-5   Bonding in B_5H_{11} using two-center and three-center bonds. The five exo-hydrogens, one to each boron, point outward and are shown simply as a line from each boron, B^-. A three-center $B\!-\!B\!-\!B$ bond is shown as $\overset{B}{\underset{B\quad B}{\diagup\!\!\diagdown}}$ (1). One $B\!-\!B$ two-center boron bond is shown as $B\!-\!B$ (2). The $B\!-\!H\!-\!B$ bridge bonds are shown as $B\!-\!H\!-\!B$ (3). The (endo) or internal $B\!-\!H$ bonds are shown as $B\!-\!H$ (4). Dashed lines represent geometry *only*. Total electron pair bonds for core: one three-center BBB link, one two-center $B\!-\!B$, four $B\!-\!H\!-\!B$ bridges, two $B\!-\!H$ (endo) bonds = 8 total.

There are five exo-hydrogens, each one bound to a boron by a two-center two electron bond; thus, 10 electrons are required for the *exo*-hydrogens. Then, 26 (total) electrons minus 10 electrons for exo-bonds gives 16 electrons for the "core electrons." Sixteen electrons require eight orbitals, as seen earlier in Figure 8-5. The number of core orbitals is a useful piece of information for clas-

TABLE 8-3   Types of Boranes and Carboranes Based on Electron Counts

Class Name	Geometry	Core or Framework Electrons	Source of Name
Closo	A *closed* polyhedron, tetrahedron, octahedron, icosahedron, etc.	$(2n + 2)$ electrons where n = number of framework nuclei	Boron or carbon framework atoms at corners of a regular *closed* polyhedron
Examples:			
$B_{12}H_{12}^{2-}$	Regular icosahedron (Figure 8-1)[a]	$(2 \times 12) + 2 = 26$ "core" or framework electrons	Closed polyhedron
$C_2B_{10}H_{12}$	Regular icosahedron (Figure 8-7)[b]	$(2 \times 12) + 2* = 26$ *The two extra electrons needed come from two carbon atoms	Closed polyhedron
$B_6H_6^{2-}$	Regular octahedron[c]	$(2 \times 6) + 2 = 14$ "core" electrons	Closed polyhedron
Nido	Visualized as a *closo*-polyhedron in which one B—H unit has been removed along with electrons of ion; then H atoms are added to give necessary electron count	$(2n + 4)$ electrons in "core" where n = number of framework atoms	Like a nest:
Examples:			
,B_5H_9	Remove one B—H unit and two electrons from $B_6H_6^{2-}$. Add 4 H atoms to give necessary electron count[d]	Boranes + 4 B—H—B bridges = $(2 \times 5) + 4$ = 14 "core" electrons	Nido—B_5H_9
B_6H_{10}	Remove one B—H (apex) from pentagonal bipyramid. Add H to provide necessary electron count. See also Table 8-5[e]	$(2 \times 6) + 4 = 16$ "core" electrons	Nido—B_6H_{10}

TABLE 8-3   Types of Boranes and Carborane Based on Electron Counts

Class Name	Geometry	Core or Framework Electrons	Source of Name
Arachno	A closo-polyhedron in which two adjacent B—H units are removed	$(2n + 6)$ = electrons in core where n = number of framework atoms	Like a spider:
Example: B_4H_{10}	A $B_6H_6^{2-}$ in which two adjacent units are removed along with the ion charge[f]	$(2 \times 4) + 6 = 14$ "core" electrons	

Arachno-B_4H_{10}

Planar projection

4 Bridge hydrogens
4 Exo-hydrogens
2 Endo-hydrogens $\Longleftarrow$

○ Boron atom
⊶⊚ B-H unit removed from octahedron
○ Exo-hydrogen
• Endo-hydrogen
◎ Bridge hydrogen

| Hypho | | $(2n + 8)$ electrons in core | |
| Klado | | $(2n + 10)$ electrons in core | |

More about the electron count:

[a] For $B_{12}H_{12}^{2-}$: *total* electrons = boron + hydrogen + charge = $(3 \times 12) + (1 \times 12) + 2 = 50$ total. Electrons for *exo*-hydrogen bonds = $(2 \times 12) = 24$. *"Core" electrons = total − exo = 50 − 24 = 26.*

[b] For $C_2B_{10}H_{12}$: *total* electrons = carbon + boron + hydrogen = $(4 \times 2) + (3 \times 10) + (1 \times 12) = 50$ total. Electrons for *exo*-hydrogen bonds = $(2 \times 12) = 24$. *"Core" electrons = total − exo = 50 − 24 = 26.*

[c] For $B_6H_6^{2-}$: *total* electrons = boron + hydrogen + charge = $(3 \times 6) + (1 \times 6) + (2) = 26$ total. Electrons for exo hydrogen bonds = $(2 \times 6) = 12$. *"Core" electrons = total − exo = 26 − 12 = 14.*

[d] For B_5H_9: *total* electrons = boron + hydrogen = $(3 \times 5) + (1 \times 9) = 24$ total. Electrons for *exo*-hydrogen bonds = $(2 \times 5] = 10$. *"Core" electrons = total − exo = 24 − 10 = 14.*

[e] For B_6H_{10}: *total* electrons = boron + hydrogen = $[3 \times 6) + (1 \times 10) = 28$ total. Electrons for *exo* B—H bonds = $(2 \times 6) = 12$. *"Core" electrons = total − exo = 28 − 12 = 16.*

[f] For B_4H_{10}: *total* electrons = boron + hydrogen = $(3 \times 4) + (1 \times 10) = 22$ total. Electrons for *exo* B—H bonds = $(2 \times 4) = 8$. *"Core" electrons = total − exo = 22 − 8 = 14.*

sifying boron hydrides, carboranes, and related molecules. The five classifications *are based* on "core electron count", these are called *closo-, nido-, arachno-, hypho-,* and *klado*-classes. Some examples are shown in Table 8-3.

Recognizing the differences noted earlier between the geometry of boron hydrides and the geometry of carbon hydrides, one can ask the question: Can we mix CH and BH in the same hydride structure? If so, what happens to the resulting geometry? Experiments by Landesman [14] in the laboratories of the Olin Matheson Corporation, were done as part of the High Energy Fuels Program using B_2H_6 and C_2H_2 as reactants. *Very small* total yields (less than 1%) of the compounds $C_2B_3H_x$, $C_2B_4H_y$, and $C_2B_5H_z$ were identified on the basis of mass

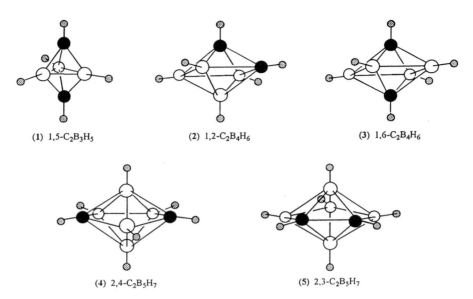

    (1) 1,5-$C_2B_3H_5$ (2) 1,2-$C_2B_4H_6$ (3) 1,6-$C_2B_4H_6$

             (4) 2,4-$C_2B_5H_7$ (5) 2,3-$C_2B_5H_7$

Figure 8-6   Structures first proposed by R. E. Williams and collaborators for the compounds $C_2B_3H_5$ (**1**), $C_2B_4H_6$, (**2** and **3**), and $C_2B_5H_7$ (**4**). A second isomer for $C_2B_5H_7$ (**5**) was found as expected.

spectral data, but results were not reported in the open literature.[15c] Because hydrogen could be easily lost, x, y, and z were not established. From these data, it appeared that one could mix boron and carbon in special hydrides, but the structural questions or even the formulas, were not resolved.[15] It remained for Williams, Good and coworkers [15e] to refine the new reaction between B_5H_9 and acetylene enough to permit accumulation of a small quantity of the "hydride mixture," and then to separate the mixture into three components: $B_3C_2H_x$, $B_4C_2H_y$, and $B_5C_2H_z$. They then used carefully obtained mass spectral and infrared data, plus NMR data obtained from very primitive early instruments, plus amazing insight, to suggest structures for these three compounds. The structures, which were later verified by X-ray, are shown in Figure 8-6. Needless to say, the appearance of carbons bound to five and six other atoms raised serious questions. Opposition to the structures was fierce. The Williams models [15] suggested that boron was even more of a "rogue element" than people had previously thought. Boron could corrupt the erstwhile proper geometry of carbon as displayed in its classic bonding models. Further, only triangular B_3 groups paid any homage to the icosahedrons of borons in elemental boron. The vigor of the structural debate and even its effect on publishability and patentability of the excellent experimental work have been described by Williams and coworkers elsewhere.[15b] A quotation from Willams' 1994 paper is pertinent. I quote, "Disputes concerning classical versus non-classical carbons in electron deficient clusters seem to incite outrage more akin to religious zealotry than science."[15d] In short, discussion was vigorous.

Far to the east, in the main laboratories of Olin Matheson, and in the laboratories of the Reaction Motors Division of Thiokol Corporation, work on high-energy fuels was underway. Because of military secrecy, there was little communication between the groups. It was not until about 1960 that the secrecy blanket began to disappear. Finally, the laboratories could submit their work for publication. Two groups [16] simultaneously reported the following reactions:

$$B_{10}H_{14} + 2 \text{ Lewis base ligands} \rightarrow B_2H_{12} \cdot L_2 + H_2$$

Pictorially, this is seen as:

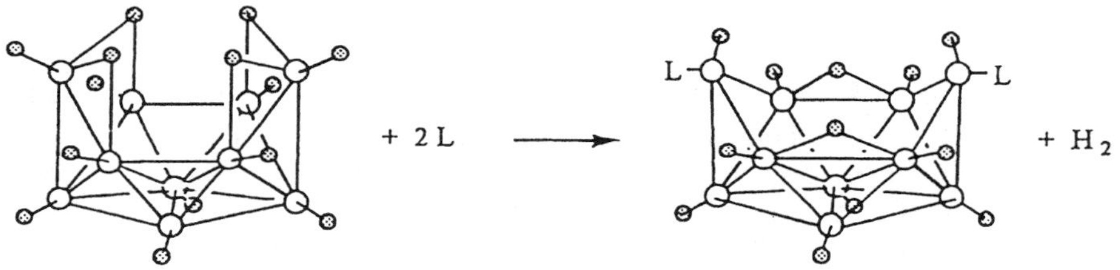

$$L = -NCCH_3, -S(CH_3)_2$$

Over 75 bases are known to give the reaction shown above. The second very interesting reaction involved the mixing of $B_{10}H_{12}L_2$ with acetylene. If L was either $(CH_3)_2S$ or H_3CCN, and proper conditions were established, a very interesting HB and CH hydride (carborane) appeared as a product. A pictorial representation of the process is shown here:

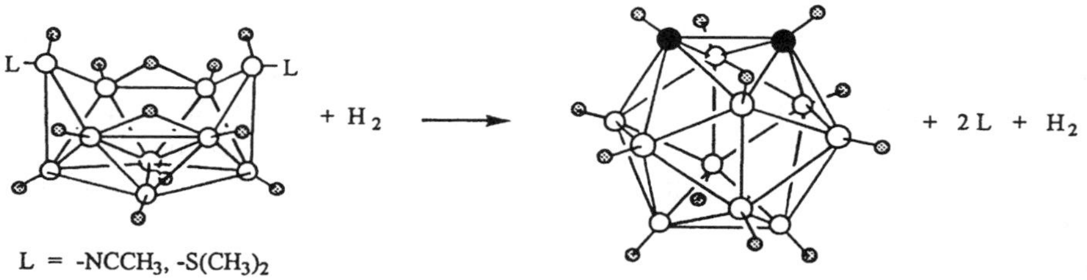

$$L = -NCCH_3, -S(CH_3)_2$$

O-carborane or $1,2\text{-}C_2B_{10}H_{12}$

The equation is

$$B_{10}H_{12}L_2 + C_2H_2 \rightarrow C_2B_{10}H_{12} + 2L + H_2$$

The remarkably stable product, $C_2B_2H_{12}$, is of major importance in the question of the geometry of "mixed BH and CH hydrides." This carborane has two

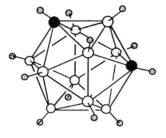

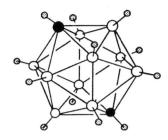

1,7-$C_2B_{10}H_{12}$ or m-carborane

1,12-$C_2B_{10}H_{12}$ or p-carborane

Figure 8-7 Two additional isomers of $C_2B_{10}H_{12}$.[15d]

carbon *atoms* inserted into the $B_{10}H_{12}$ structure. A closed icosahedron results. Recall that for $B_{12}H_{12}$, two extra electrons were needed to complete the stable "closed-shell" pattern. Thus, the stable structure was the ion $B_{12}H_{12}^{2-}$. Recall also that each boron provided two electrons to the icosahedral framework structure to give a total of 24 electrons. Twenty-six electrons were needed to give the stable closed-shell pattern. These two extra electrons usually came from metal atoms, such as two sodium atoms or one calcium atom to give $Na_2B_{12}H_{12}$ or $CaB_{12}H_{12}$, respectively. If one now inserts two carbon atoms into the B_{10} framework, the electron count is $10 \times 2 = 20$ electrons from boron and $2 \times 3 = 6$ electrons from carbon,[17] giving a total of 26 framework (or core) electrons. One has a very stable *neutral* icosahedral structure. This icosahedral unit is remarkably nonreactive. If the *ortho*-carborane shown above is heated, carbons in the icosahedron can move to give two other isomers of $C_2B_{10}H_{12}$. These are shown in Figure 8-7. Their structures are now well known.

The icosahedral carboranes established a very important point. *Carbon can contribute three electrons to the icosahedral framework to give the very stable closed-shell icosahedral structure* comparable to $B_{12}H_{12}^{2-}$. The theory was developed by Hoffmann and Lipscomb.[18] In this case, one might conclude that the geometry of the carborane structure is determined by the boron parent.

A new question is now obvious: What is the progenitor of the simpler carboranes described first by Williams et al.? The Williams structures are not related to icosahedra in any way. The geometry of $C_2B_4H_6$ is based on an octahedron (see Figure 8-6, structure **2** and **3**). It was noted in the earlier discussion of carbon hydrides that the triple-bonded acetylene can be related to the acetylides such as calcium acetylide (CaC_2 or calcium carbide).[9] It is not strange that a boron hydride derivative and a carborane should relate to a metal boride, even calcium boride, CaB_6. The unit cell of CaB_6 is cubic and contains one formula weight of CaB_6. The metal atom is at the center of the cube and borons are present in *octahedral units*, one at each corner of the cubic cell (Figure 8-8).[19, 20] The molecular-orbital-derived energy level diagram for six atoms in an

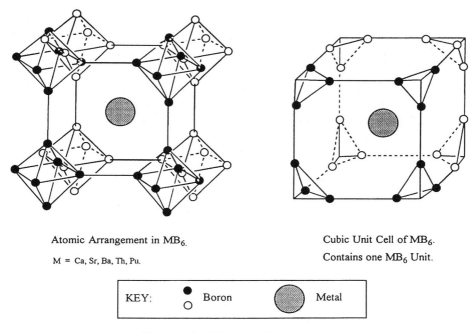

Atomic Arrangement in MB_6.

M = Ca, Sr, Ba, Th, Pu.

Cubic Unit Cell of MB_6.

Contains one MB_6 Unit.

KEY: ● / ○ Boron ⬤ Metal

Figure 8-8 Metal borides, MB_6.[19]

octahedral array[10b] is shown below:

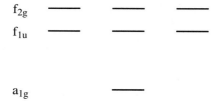

f_{2g} —— —— ——

f_{1u} —— —— ——

a_{1g} ——

There are seven orbitals in the aray. If each boron has one electron that is used in external bonding to hydrogen atom (or other species), each boron can donate two electrons to the "core." This gives a total of $6 \times 2 = 12$ electrons for binding the six borons. Two more electrons are needed. A metal can supply these electrons to give CaB_6. This was also predicted.[4] Remembering that the two necessary extra bonding electrons can be supplied by replacing two borons with two carbon atoms in the octahedron; it is quite logical that a molecule $C_2B_4H_6$ should exist. If we consider an *idealized* system with "core electrons holding together an array of BH^{2+} and CH^{3+} *pseudo* ions, it would seem logical that the two CH^{3+} units would get as far away from each other as possible because of the higher positive charge on each. This was formally stated by Williams many years ago when he noted (without reference to ions) that the structure with the lowest energy ground state should have carbons as far apart as possible.[11b]

Up till now, we have considered carboranes in which two carbon atoms have replaced two boron atoms in a stable boron polyhedron. Since two carbons were inserted into a known stable boron structure, it is not too surprising that boron geometry prevailed. The geometry of carbon was unconventional in these carboranes, to say the least, but carbon was clearly a minority guest.

A new question now arises: Do we have to have a majority of boron atoms in a structure containing carbons in order to stabilize these nontraditional structures for carbon? This question will be probed by considering replacement of borons with carbons in molecules such as B_5H_9 and B_6H_{10}. Since the B_6H_{10} structure ($C_xB_{6-x}H_{10-x}$) series has more available data, it will be considered first. The first member of the series, B_6H_{10}, was isolated and the molecular formula was established by Stock.[1b] The energy levels for a C_xB_{6-x} "core" of pentagonal-pyramidal geometry (C_{5v} symmetry) is given below:(10b)

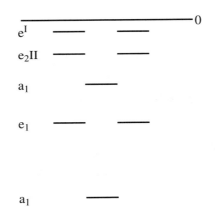

Eight core orbitals are seen. A closed-shell configuration will require 16 electrons. In B_6H_{10}, where $x = 0$, we have a total of $3 \times 6 = 18$ electrons from the boron, plus 10 electrons from the hydrogen, for a *total* of 28 electrons. There are six (exo) hydrogens, so the "core" electron count for B_6H_{10} is: $28 - 12 = 16$. The existence of B_6H_{10} and its electron count support the theoretical model. If one puts a C—H in place of one B—H of B_6H_{10}, the process is equivalent to adding an extra proton to a core boron nucleus and an *extra electron to the valence shell*. In short, $(C—H) = (B—H) + H = (B—H) + H^+ + e^- = (C—H)^+ + e$. As noted above, the "core" structure of the BC-framework of (C_{5v}) symmetry needs 16 electrons to achieve stability. In the CB_5 "core," the borons will contribute $2 \times 5 = 10$ electrons and carbon will contribute 3 for a total of 13. One then needs only *three* electrons from *three* hydrogen atoms placed in *three-center two-electron bridge bonds* to complete the stable electron configuration. The structures of B_6H_{10}, CB_5H_9, and other members of the series are shown in Table 8-4. Note that replacing one B—H by a C—H permits removal of one hydrogen atom from a bridge bond position. Formulas of other members in the series, obtained by this procedure, are shown in Table 8-4. When we get to $C_4B_2H_6$ (Table 8-4) all of the bridge hydrogen atoms have been removed and $C_4B_2H_6$ is a neutral, nonbridged, fairly stable molecule. When the next boron is removed to give a C_5B core, the electron count is $(3 \times 5) + 2 = 17$. We have *one too many core electrons*.[21] The stable structure (16 core) can be achieved by removing

TABLE 8-4 **Carboranes Based on the $B_7H_7^{2-}$ Polyhedron (The B_6H_{10} Structure)**

Formula	References	Geometry
B_6H_{10}	(a) Stock, A., *Hydrides of Boron and Silicon*, Cornell University Press: Ithaca, NY, 1957, pp. 74–79 (b) Hirshfeld, F. L.; Eriks, K.; Dickerson, R. E.; Lippert, E. L., Jr, and Lipscomb, W. N., *J. Chem. Phys.*, 1958, **28**, 56	B_6H_{10}
$2\text{-}CB_5H_9$	(a) Onak, T. P., Dunks, G. B., Spielman, J. R., Gerhart, F .J., and Williams, R. E., *J. Am. Chem. Soc.*, 1966, **88**, 2061 (b) Dunks, G. B. and Hawthorne, M. F., *Inorg Chem.*, 1969, **12**, 2667 (c) Dunks, G. B. and Hawthorne, M. F., *J. Am. Chem. Soc.*, 1968, **90**, 7355. (d) Cheung, C. C. S and Beaudet, R. A., *Inorg Chem.*, 1971, **10**, 1144.	$2\text{–}CB_5H_9$
$2,3\text{-}C_2B_4H_8$	(a) Onak, T .P., Williams, R. E., and Weiss, H. G., *J. Am. Chem. Soc.*, 1962, **84**, 2830. (b) Streib, W. E., Boer, F. P., and Lipscomb, W. N., *J. Am. Chem. Soc.*, 1963, **85**, 2331. (c) Boer, F. P., Streib, W. E., and Lipscomb, W. N., *Inorg. Chem.*, 1964, **3**, 1666: (d) Brown, C. L., Gross, K. P., and Onak, T. P., *J. Am. Chem. Soc.*, 1972, **94**, 8055. (e) Schleyer, P., *Chem. Commun.*, 1993, 1776. (f) Schleyer, P., *J. Am. Chem. Soc.*, 1993, **115**, 12385.	$2,3\text{–}C_2B_4H_8$
$2,3,4\text{-}C_3B_3H_7$	(a) Franz, D. A. and Grimes, R. N., *J. Am. Chem. Soc.*, 1970, **92**, 1438. (b) Franz, D. A., Miller, V. R., and Grimes, R. N., *J. Am. Chem. Soc.*, 1972, **94**, 412. (c) See also: Grimes, R. N., *Carboranes*, Academic Press, New York, NY, 1970, p. 23.	$2,3,4\text{–}C_3B_3H_7$

(Contd.)

TABLE 8-4   Carboranes Based on the $B_7H_7^{2-}$ Polyhedron (The B_6H_{10} Structure) (contd)

Formula	References	Geometry
2,3,4,5-$C_4B_2H_6$	(a) Binger, P., *Tet. Letters,* 1966, 2675. (b) Onak, T. P. and Wong, G. F., *J. Am. Chem. Soc.,* 1970, **92**, 5226. (c) Miller, V. R. and Grimes, R. N., *Inorg. Chem.,* 1972, **11**, 862. (d) Beudet, R., *J. Chem. Phys.,* 1974, **61**, 683. (e) Wrackmeyer, B., *Chem. Commun.,* 1996, 1219.	 $C_4B_2H_6$
$[C_5BH_6]^+$	(a) Jutzi, P., *Angew. Chemie.,* 1977, **89**, 339 (b) Jutzi, P., *J. Org. Met. Chem.,* 1978, **61**, C-5 (iodinated and phenyl cation). (c) Jutzi, P., *Chem. Ber.,* 1979, **112**, 2488. (d) Jutzi, P., *J. Org. Met. Chem.,* 1995, **487**, 127.	 $[C_5BH_6]^+$
$[C_6H_6]^{2+}$	(a) Hogaween, H. and Kwant, P. W., *J. Am. Chem. Soc.,* 1974, **96**, 2208 (permethyl derivative). (b) Williams, R. E. and Field, L. D., *Boron Chemistry,* Vol. 4, (Pergamon Press, New York, 1979, p. 131.	 $[C_6H_6]^{2+}$

an electron from the core to get a positively charged ion. The ion $[C_5BH_6]^+$ is shown in Table 8-4. For $[C_6H_6]$ we have $3 \times 6 = 18$ core electrons. To gain stability, two electrons must be removed from the core to give $[C_6H_6]^{2+}$. This whole series is a very interesting one. We pass from a structure dictated in large measure by the boron geometry of the well-known B_6H_{10} to the well established nonclassic carbon cation, $[C_6H_6]^{2+}$. *Interestingly enough, boron is not essential to generate a carbon ion with a core structure completely analogous to its boron hydride analog.* In an important paper presented in 1979, Williams and Field [22] showed that (and I quote), "Certain 'non-classical' carbocations are isoelectronic and isostructural carbon copies of polyboranes." This paper provided strong support for the thesis that the cores of B_6H_{10} and $C_6H_6^{2+}$ are isostructural. The paper is based on a correlation between the chemical shifts of boron-11 in B_6H_{10} and those of carbon-13 in the fully methylated $[C_6H_6]^{2+}$. The background relationship had been established by Spielvogel and Purser [23] in their 1971 paper, where an equation relating the chemical shift of boron-11 (^{11}B) to that of carbon-13 (^{13}C) was developed for the case where the two atoms, C and B, are measured in a comparable electronic environment. This relationship was confirmed. with slight modification, by Nöth and Wrackmeyer,[24] and was refined somewhat by Spielvogel, Nutt, and Izydore.[25] The relationship, as used by Williams and Field,[22] gave convincing proof that the $[C_6H_6]^{2+}$ ion and B_6H_{10} are isostructural.

An analogous boron hydride, B_5H_9, has also been examined. The ion $B_6H_6^{2-}$, based on the parent B_6^{2-} ion from metal borides, is the parent *closo*-polyhedron for B_5H_9. It has an octahedral geometry. If one removes both a BH unit from the $B_6H_6^{2-}$ and the two extra electrons that give the ionic charge, the remaining species is a B_5H_5 fragment. Energy levels from molecular-orbital (MO) calculations for a B_5H_5 fragment in the form of a square pyramid are shown below:[10b]

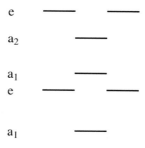

There are seven core bonding orbitals to be filled. For the five $B—H_{(exo)}$ bonds, we have $2 \times 5 = 10$ electrons. These are NOT part of the core. *Total* electrons are $15 + 5 = 20$. The number of "core electrons" in B_5H_5 is $20 - 10 = 10$. Ten electrons require five orbitals. We need four more core electrons to stabilize the B_5 "core." These can be provided by four hydrogen atoms placed as $B—H—B$ bridges on the square face of the B_5H_5 pyramid. This gives the 14 electrons needed to stabilize B_5H_9. Since hydrogen *atoms* were added to the neutral B_5H_5, the product is electronically neutral B_5H_9.

If we now systematically replace the $B—H$ of B_5H_9 with $C—H$ units and use the procedures already described for the $B_6H_{10} \rightarrow [C_6H_6]^{2+}$ series, a comparable series of compounds would be expected. As far as I am aware, only three members of this series are known. These are B_5H_9, $B_3C_2H_7$, and $[C_5H_5]^+$. The currently known members are clearly shown in Table 8-5. The ion $[C_5H_5]^+$ is of particular interest. In a paper published in *Inorganic Chemistry* in 1971 (received, June 1970) Williams [26a] *predicted* that the ion $[C_5H_5]^+$ would exist and would have the same core carbon arrangement as the borons in B_5H_9. This was a very bold move since the ion $[C_5H_5]^+$ had been sought by organic chemists and was expected to have geometry more closely related to that of the well-known cyclopentadienyl anion.[26b] Resonance stabilization was invoked in support of the stability of $[C_5H_5]^+$. In 1992, Masamune [27] reported the synthesis of $[C_5H_5]^+$ and used NMR data to support the structure shown in Table 8-5. The structure was confirmed unequivocally by the work of Williams and Field,[22] which was discussed on p. 136–137.

TABLE 8-5   Carboranes Based on the $B_7H_7^{2-}$ Polyhedron (The B_6H_{10} Structure)

Formula	References	Structure
B_5H_9 (Known)	(a) Stock, A., *Hydrides of Boron and Silicon*, Cornell University Press, Ithaca, NY, 1957, p. 72. (b) Dulmage, J. W. and Lipscomb, W. N., *J. Am. Chem. Soc.*, 1951, **73**, 3539. (c) Hrostowski, H. J. Meyers, R. J., and Pimentel, G. C., *J. Chem. Phys.*, 1952, **20**, 518. (d) Hedberg, K., Jones, M. E., and Schomaker, V., *J. Am. Chem. Soc.*, 1951, **73**, 3538.	*Nido*–B_5H_9
$B_3C_2H_7$ (Known)	Franz, D. A., Miller, V. R., and Grimes, R. N., *J. Am. Chem. Soc.*, 1972, **94**, 412 (low stability).	$B_3C_2H_7$
$[C_5H_5]^+$ (Known)	(a) William, R. E., *Inorg. Chem.*, 1971, **10**, 210 (formula and structure suggested). (b) Masumune, S., *J. Am. Chem. Soc.*, 1972, **94**, 8955 (synthesized). (c) Williams, R. E. and Field, L. D., *Boron-4*, IME Boron-IV, Snowbird, UT, 1979; *Boron Chemistry*, Vol. 4, Pergamon Press, New York, 1979.	$[C_5H_5]^+$

Probable structures for still *unreported* members of the $B_5H_9 \rightarrow C_5H_5^+$ series:

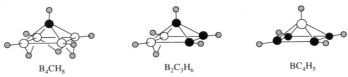

B_4CH_8 $B_2C_3H_6$ BC_4H_5

We now come to the "64-dollar question:" What factors carry carbon geometry beyond the single-, double-, or triple-bond models and into the realm of multicentered bonds? This is a very tricky question. The answers will undoubtedly invoke as much "religious zealotry" (as described by Williams [15c]) as science, but here is one opinion.

There are two *ions* of empirical formula C_5H_5. One is the very-well-known cyclopentadienyl anion, $C_5H_5^-$. The other is the less well-known and, at least in earlier times, more controversial cation, $C_5H_5^+$. The $C_5H_5^-$ ion with carbons arranged in a symmetrical planar ring is obtained easily and in good yield from C_5H_6: it is easily represented by conventional, traditional carbon bonds. It has MO energy levels as shown below *for the six π electrons of the $C_5H_5^-$ ion* (cyclopentadienyl ion):[28]

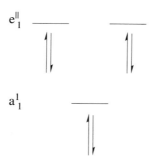

The six electrons complete this electron pattern, which is consistent with Huckel's rule for aromaticity ($4n + 2$ electrons, where $n = 1$). The cyclopentadienyl anion is a stable structure. If one removes two electrons from the $C_5H_5^-$ anion to get $C_5H_5^+$ *in a planar five-membered ring*, the MO pattern is NOT complete (only four electrons, not six, can be placed in the three orbitals):

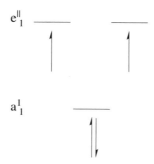

and aromaticity as given by Huckel's rule is not observed. The cation $C_5H_5^+$ has NOT been observed as a stable species in a planar form.[26] On the other hand, if the core carbons are arranged in a square pyramid, the seven bonding molecular-orbital levels are as shown earlier on p. 209. Fourteen electrons are required. We have 15 core electrons in C_5H_5 ($3 \times 5 = 15$), so one electron is lost to give the required 14. The ion $[C_5H_5]^+$ can be observed under special, rather uncommon conditions. Carbon can respond to changes in circumstances; the product will not always have the lowest possible energy (that is why we can have *ortho*, *mera*-, and *para*- isomers of benzene and $C_2B_{10}H_{12}$), but a depression in the potential energy curve defines the higher energy structure, which can, under appropriate conditions, be detected. Carbon is resourceful. In recent years, its allegiance to diamond and graphite geometry (alkanes) and to carbide geometry, CaC_2, has been stretched a bit by Buckeyball geometry. It is not surprising that it can "borrow geometry" from its close friend and neighbor, the "electron-deficient" boron, to result in "nonclassic carbon structures."

In all of the foregoing odyssey, Bob Williams has been leading the way. His many colleagues join me in thanking him for his experimental skills, his scientific insights, his ability to correlate reams of data, for his courage in facing very vigorous opposition, and for being a "good guy." He is a chemist's chemist.

REFERENCES

1. (a) Stock, A., *Hydrides of Boron and Silicon*, Cornell University Press, Ithaca, NY, 1932, pp. 51, 60, (b) p. 74.

2. Cotton, F. A. and Walton, R. A., *Metal Atoms*, John Wiley and Sons, New York, 1982, and references therein.

3. A few of the many references are listed below:
   (a) Eberhardt, W. H., Crawford, B.L., and Lipscomb, W. N., *J. Chem. Phys.*, 1954, **22**, 989;
   (b) Dickerson, R.E. and Lipscomb, W. N., *J. Chem. Phys.*, 1957, **27**, 212;
   (c) Lipscomb, W. N., *Adv. Inorg. Chem. Radiochem.*, 1959, **1**, 117;
   (d) Lipscomb, W. N., *Boron Hydrides*, W. A. Benjamin, New York, 1963, p. 89;
   (e) Lipscomb, W. N., *Boron Hydride Chemistry*, (E. L. Muetterties, ed.), Academic Press, New York, 1975, Chap. 2, p. 39.

4. Longuet-Higgins, H. C. and Roberts, M. DeV., *Proc. R. Soc.*, 1955, **A230**, 110.

5. Buehl, M. and Schleyer, P. von R., *Electron Deficient Boron and Carbon Clusters* (G. Olah, K. Wade, R.E. Williams, eds), John Wiley and Sons, New York, 1991, p. 113, and references therein.

6. (a) Wunderlich, J. A. and Lipscomb, W. N., *J. Am. Chem. Soc.*, 1960, **82**, 4427;
   (b) Hoard, J. L., *Boron Synthesis, Structure, and Properties* (J. A. Kohn, W. F. Nye, and G. K. Gaule, eds)., Plenum Press, New York, 1960, p. 1–6;
   (c) Hughes, R. E., Kennard, C. H. L., Sullenger, D. B., Weakliem, D. B., Sands, D. B., and Hoard, J. L., *J. Am. Chem. Soc.*, 1963, **85**, 361;
   (d) Decker, B. F., and Kasper, J. S., *Acta Cryst.*, 1959, **12**, 503.

7. This is true only if there is no sphere in the center. In this case (only 12 borons), all spheres are in contact and there is insufficient room for the central sphere.

8. Wells, A. F., *Structural Inorganic Chemistry*, Clarendon Press, Oxford, U.K., 1984, p. 1049.

9. For the alkyne hydrocarbon structures, one must start with triple-bonded units, such as $Ca(C\equiv C)$ (i.e., calcium carbide or calcium acetylide). Some of the boron hydrides can be visualized as derivatives of the B_6^{2-} units found in CaB_6, etc., or in other borides. The $B_6H_6^{2-}$ ion is a stable species, but not as stable as $B_{12}H_{12}^{2-}$.

10. (a) Lipscomb, W. N., *Boron Hydrides*, W.A. Benjamin, New York, 1963, p. 89.
    (b) Muetterties, E. L. and Knoth, W. H., *Polyhedral Boranes*, Marcel Decker, New York, 1968, p. 40.

11. (a) Williams, R. E., *Chem. Rev.*, 1992, **92**, 198;.(b) ibid., 1992, **92**, 204.

12. Compound B_6H_6 with C_{5v} symmetry has the molecular-orbital pattern shown in the text. Assuming that this diagram is reasonable,10b we can conclude that 16 "core" or structural electrons will be needed to stabilize the B_6H_6 pentagonal pyramid. After removal of the lower pentagonal pyramid, we have $6 \times 2 = 12$ core electrons.

Four more are needed to give the required 16. Three $B \longrightarrow \underset{H}{\smile} \longrightarrow B$ are specified in our description. We are still short one electron so the three bridge linkages *partially* stabilize the structure (four bridges would give the stable B_6H_{10}). At this point additional changes in the pentagonal pyramid are indicated.

13. (a) The X-ray data,[13b] calculations,[13c,d] and electron diffraction data [13e] indicate that the second hydrogen attached to the top boron of B_5H_{11} is distorted toward a bridging position between the top boron and one of the borons in the base of the structure. This would justify marking the semibridge hydrogen as ⊡, but energy differences are small.

(b) Huffman, J. C., Ph.D. Thesis, Indiana University, 1974;

(c) McKee, M. L., and Lipscomb, W. N., *Inorg. Chem.*, 1981, **20**, 4442;

(d) Buehl, M. and von R. Schleyer, P., *Electron Deficient Boron and Carbon Clusters* (G. Olah, K. Wade, and R.E. Williams, eds), John Wiley and Sons, New York, 1991, p. 127;

(e) Greatrex, R., Greenwood, N. N., Rankin, D. W., and Robertson, H.E., *Polyhedron,* 1987, **6**, 1849.

14. (a) Landesman, H., *Production of Boranes* (R. T. Holtzman, R. L. Hughes, I. C., Smith, and E. W. Lawless, eds), Academic Press, New York, 1967, pp. 163–184;

(b) Williams, R.E., *Progress in Boron Chemistry*, Vol. 2, (R. J. Brotherton and H. Steinberg, eds), Chap. 2, Pergamon Press, Oxford, U.K., 1970, p. 37.

15. (a) Keilin and coworkers [15b] observed these same compounds as trace products of the reaction between gaseous B_5H_9 and C_2H_2 in a silent electric discharge.

(b) Shapiro, I., Keilin, B., Good, C. D., and Williams, R. E., *J. Am. Chem. Soc.*, 1963, **85**, 3167.[15c]

(c) Publication of these early findings (1950s) was delayed by military security until the early 1960s.

(d) Williams, R. E. *Advances in Organometallic Chemistry,* 1994, **36**, 1–55;

(e) Shapiro, I., Good, C. D, and Williams, R. E. *J. Am. Chem. Soc.*, 1962, **84**, 3837;

(f) Shapiro, I., Keilin, B., Williams, R. E., and Good, C. D., *J. Am. Chem. Soc.*, 1963, **85**, 3167.

16. (a) One group was at the Eastern Laboratories of Olin Matheson and was led by Ted Heying.[16b] The other was at the Reaction Motors Division of Thiokol Corporation and was led by Murray Cohen.[16c,d,e] (b) Heying, T. L., Ager, J. W., Jr, Clark, S. L., Mangold, D. J., Goldstein, H. L., Hillman, M., Polak, R. J., Szymanski, J. W., Schroeder, H., Reiner, J. R., Alexander, R. P., Papelli, S., and Trotz, S. I., *Inorg. Chem.*, 1963, **2**, 1089–1110 (five papers from members of the groups of authors above); (c) Cohen, M., Fein, M. M., Bobinski, J., Mayes, N., Schwartz, N., Grafstein, D., Paustian, J. E., Lichstein, M., Dvorak, J., Smith, H., Karlan, S., and Vogel, C., *Inorg. Chem.*, 1963, **2**, 1111–1133 (five papers with authors from the above list); (d) Bobinski, J. J., *J. Chem. Educ.*, 1964, **41**, 500; (e) Rose, C. D., U. S. Patent 3,028,432, 1962.

17. One electron from each carbon is used in bonding the exo-hydrogen, so each carbon atom can contribute only three electrons to the structure.

18. Hoffmann, R. and Lipscomb, W. N., *J. Chem. Phys.*, 1962, **36**, 3489–93.

19. Post, B., *Boron, Metalloboron Compounds, and Borides* (R. Adams, ed.), Interscience, New York, 1964, pp. 322–326.

20. In CaB_6, octahedra use exo-electrons to link other octahedra.

21. Calculating this in an alternative fashion, we get: total electrons = $(4 \times 5) + (1 \times 3) + (1 \times 6) = 29$ electrons. Then, $29 - 12 = 17$ core electrons. Remove one for a stable shell.

22. Williams, R. E. and Field, L. D., *Boron Chemistry-4*, Lectures of the Fourth International Meeting on Boron Chemistry, Snowbird, UT, 1979; R. W. Parry and G. Kodama (eds), Pergamon Press, Oxford, U.K., 1980.

23. Spielvogel, B. F. and Purser, J. M., *J. Am. Chem. Soc.*, 1971, **93**, 4418–4426.

24. Nöth, H. and Wrackmeyer, B., *Chem. Ber.*, 1974, **107**, 3089–3103.

25. Speilvogel, B. F., Nutt, W. R., and Izydore, R. A., *J. Am. Chem. Soc.*, 1975, **97**, 1609.

26. (a) Williams, R. E., *Inorg. Chem.*, 1971, **10**, 210–214;
    (b) Roberts, J. D., and Caserio, M. C., *Basic Principles of Organic Chemistry*, W. A. Benjamin, New York, 1964, pp. 241–242.
    (c) In Williams' 1971 paper,[26a] it was suggested that closed polyhedra should be used as the starting points for the systematics of carborane and borane geometries, particularly the *closo-*, *nido-*, and *arachno-* classifications. Williams properly pointed out that this has merit as an educational tool. On the other hand, I feel that the tie between the structures of boron hydrides and the carboranes with the icosahedron of elemental boron and the octahedron of a number of borides should not be lost. In some cases, the geometry of the fragment can be seriously distorted, even to the point of giving new basic polyhedra.

27. Masamune, S., *J. Am. Chem. Soc.*, 1972, **94**, 8956.

28. Carey, F. A. and Sundberg, R. J., *Advanced Organic Chemistry*, 3rd Edition, Part A, Plenum Press, New York, 1993, p. 45.

PART III

UNTANGLING MOLECULAR STRUCTURES

9

COMPUTATIONAL STUDIES OF *NIDO*-$C_4B_7H_{11}$ CARBORANES AND *NIDO*-8-VERTEX BORANES AND CARBORANES

JOSEPH W. BAUSCH and ANDREW. J. TEBBEN

Department of Chemistry, Villanova University, Villanova, Pennsylvania, 19085

9.1 INTRODUCTION

In the first part of this chapter, we demonstrate that the ab initio/IGLO/NMR method can be used to establish the correct structures of the three known *nido*-$C_4B_7H_{11}$ carboranes.[1] Furthermore, the relative energies of these and other possible cage structures are compared with predictions using empirical patterns governing carbon placement. In the second part of this chapter, an extensive computational study of *nido*-8-vertex boranes and carboranes is reported.[2] The preference for a five- or six-membered open face and the relative importance of hydrogen (bridge and endo) and carbon placement in the clusters are evaluated.

Computational chemistry has made a significant impact on the understanding of the structure and bonding in boron hydrides and related electron-deficient clusters.[3] With the power of modern computers and commercially available program packages, this area of chemistry is no longer inhabited solely by theoretical chemists. Synthetic and structural chemists are rapidly appreciating how quickly and easily even high-level ab initio calculations can be carried out on large molecules. One of the most exploited areas involves calculating structural isomers for a given molecular formula. From a simple study of this type, much information is gained, including the relative energies and detailed structural data (i.e., bond lengths, angles, etc.). Many of the empirically derived trends and rules concerning the structure of electron-deficient clusters can be evaluated from this type of computational study.

In the early 1980s, the ability to solve structural ambiguities, which are not uncommon in polyboranes and related systems, was improved significantly when Kutzelnigg, Schindler, and coworkers developed an ab-initio-based method for the accurate calculation of NMR chemical shifts based upon the *I*ndividual *G*auge for *L*ocalized *O*rbitals (IGLO).[4] The Schleyer group recognized the potential of the IGLO method to distinguish between structural alternatives for a variety of polyboranes and carboranes, provided that ab initio optimized structures are employed as input for the IGLO calculations.[5] They called this technique the ab initio/IGLO/NMR method; this method has since been applied to a variety of other systems.[6]

9.2 COMPUTATIONAL METHODS

The structures in this study were fully optimized within the specified symmetry constraints employing either the GAUSSIAN 92 or GAUSSIAN 94 programs [7] using the standard Pople basis sets included. A vibrational frequency analysis was carried out on each optimized geometry at HF/6-31G* to determine the nature of the stationary point. The NMR chemical shifts were calculated using the IGLO method.[4] The primary reference for the ^{11}B NMR chemical shifts is B_2H_6, and the δ values were converted to the experimental $BF_3 \cdot O(C_2H_5)_2$ scale using the experimental value of $\delta(B_2H_6) = 16.6$ ppm.[8]

9.3 AB INITIO/IGLO/NMR STUDY OF *NIDO*-$C_4B_7H_{11}$ CARBORANES

9.3.1 Background

Three compounds (**I**,[9] **II**,[10] and **III** [9]) of molecular formula *nido*-$C_4B_7H_{11}$ have been synthesized by the following reactions:

$$\textit{arachno-}\mu\text{-}6,9\text{-}CH_2\text{-}5,6,9\text{-}C_3B_7H_{11} + NaH \xrightarrow{\Delta} \textit{nido-}C_4B_7H_{11} \qquad (1)$$
$$\textbf{I}$$

$$\textit{closo-}1,5\text{-}C_2B_3H_5 \xrightarrow[\text{hot/cold reactor}]{400°C/0°C} \textit{nido-}C_4B_7H_{11} \qquad (2)$$
$$\textbf{II}$$

$$\textit{arachno-}4,5\text{-}C_2B_7H_{13} + HC\equiv CH \xrightarrow[\text{hexane}]{120°C} \textit{nido-}C_4B_7H_{11} \qquad (3)$$
$$\textbf{III}$$

Based upon skeletal electron-counting rules,[11] all three compounds were proposed to have an open-cage structure. The three initially proposed structures, **Ia**, **IIa**, and **IIIa** for compounds **I**, **II**, and **III**, respectively, are shown in Figure 9-1, along with a recently proposed [12] alternative structure, **IIb**, for compound **II**.

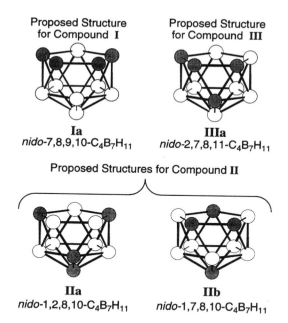

Figure 9-1 Four previously proposed structures for the three known isomers of the *nido*-C₄B₇H₁₁ carboranes: **Ia**, **IIa** and **IIb**, and **IIIa** (borons are white and carbons are shaded circles; hydrogens have been omitted for clarity).

9.3.2 Computational Details

Each C_1 symmetry isomer was optimized up to the HF/6-31G* level of theory, and each C_s symmetry isomer was optimized up to the electron-correlated level MP2(FULL)/6-31G*. For the relative energy determination, a single-point calculation at MP2(FULL)/6-31G* using the HF/6-31G* optimized geometry was carried out for each structure (Table 9-1). The zero-point energies at HF/6-31G* have been scaled by 0.89.[13] For the IGLO calculations, most of the C₄B₇H₁₁ isomers were calculated employing a DZ basis set, except for isomers **2a**, **2b**, and **2d**, where a basis set II (as denoted by the Bochum group [4]) was employed.

9.3.3 Results

9.3.3.1 Isomer I Compound **I** was proposed to have the structure *nido*-7,8,9,10-C₄B₇H₁₁ (**Ia** Figure 9-1), on the basis of 1D ¹¹B and 2D COSY ¹¹B-¹¹B NMR experiments. Structure **Ia** is also that originally predicted by geometrical systematics [11a] to have the most stable arrangements of carbons. Of the 18 arrangements of four carbons over the *nido*-11-vertex icosahedral fragment with C_s symmetry, only six of these would give the 2:2:1:1:1 ratios of boron

TABLE 9.1 **Absolute (–a.u.) and Relative (kcal/mol) Energies of** $nido\text{-}C_4B_7H_{11}$ **structures.**[a]

Isomer	6-31G*//6-31G*	ZPE[b]	E_{rel}[c]	MP2/6-31G* //6-31G*	E_{rel}[d]
7,8,9,10 (**1a**)	330.67890	99.32	0.8	331.92467	0.0
1,7,8,10 (**2b**)	330.67927	98.75	0.0	331.91905	3.0
2,7,9,10 (**3b**)	330.66312	98.65	10.0	331.90826	9.7
2,7,8,10 (**3c**)	330.66339	98.53	9.7	331.90432	12.0
2,8,9,10 (**2c**)	330.65407	98.75	15.8	331.90117	14.2
1,7,8,9 (**2d**)	330.64183	98.62	23.4	331.88766	22.5
2,7,8,9 (**3d**)	330.62898	98.52	31.3	331.87933	27.7
1,2,8,10 (**2a**)	330.62927	97.70	30.3	331.86574	35.4
2,7,8,11 (**3a**)	330.59742	97.61	50.2	331.84489	48.4

[a] Absolute (–a.u.) energies of C_s symmetry structures optimized at MP2(FULL)/6-31G* are the following: 331.92649 (**1a**), 331.87126 (**2a**), 331.92092 (**2b**), 331.90520 (**2c**), and 331.88966 (**2d**).

[b] Zero-point energy at the HF/6-31G* level.

[c] Relative energy at the 6-31G*//6-31G* + ZPE (6-31G*) level.

[d] Relative energy at the MP2(FULL)/6-31G*//6-31G* + ZPE (6-31G*) level.

resonances observed experimentally. Based upon this fact combined with the two-dimensional NMR spectra, the method of synthesis, and the geometrical preferences of carbon, the proposed structure, **Ia**, appeared to be secure.

Structure **Ia** was subjected to the ab initio/IGLO/NMR method. The ab initio optimized geometry for $nido\text{-}7,8,9,10\text{-}C_4B_7H_{11}$, structure **1a**, is given in Figure 9-2 along with the calculated ^{11}B NMR chemical shifts, which are compared with the experimental values for compound **I**. The agreement between the calculated and experimental values is excellent, thus the proposed isomer, structure **Ia**, is reconfirmed as correct by the ab initio/IGLO/NMR method.

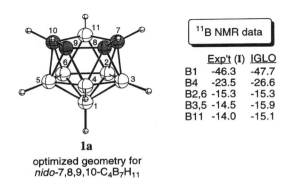

^{11}B NMR data		
	Exp't (I)	IGLO
B1	-46.3	-47.7
B4	-23.5	-26.6
B2,6	-15.3	-15.3
B3,5	-14.5	-15.9
B11	-14.0	-15.1

1a
optimized geometry for
$nido\text{-}7,8,9,10\text{-}C_4B_7H_{11}$

Figure 9-2 Ab initio optimized geometry and IGLO calculated ^{11}B NMR chemical shifts for $nido\text{-}7,8,9,10\text{-}C_4B_7H_{11}$, structure **1a**.

9.3.3.2 Isomer II Compound **II** was first proposed [10] to be *nido*-1,2,8,10-$C_4B_7H_{11}$, structure **IIa**, and, later,[12] to have the structure *nido*-1,7,8,10-$C_4B_7H_{11}$, structure **IIb** (Figure 9-1). There are 12 arrangements of four carbons over the 11-vertex icosahedral fragment with C_s symmetry that would be compatible with the multiplicities observed in the ^{11}B NMR spectrum (2:2:2:1 ratios). Both proposed structures (**IIa** and **IIb**) contain a mirror plane of symmetry and were selected on the basis of two conflicting empirical viewpoints. Structure **IIa** was originally proposed [10] primarily on empirical line-width patterns in its 1D ^{11}B NMR spectrum, its 1D ^{13}C NMR spectrum (2:1:1 ratios), and the absence of a boron atom in the "apex" position. Additional support came from the fact that this structure maximizes boron–carbon interactions in the cage, but it was acknowledged, at the time, that alternative structures, such as the 1,7,8,10 isomer (structure **IIb**), could not be ruled out. Subsequently, it was proposed [12] that structure **IIb** should, in fact, be the favored geometry, based upon carbon's preference for occupying low-coordination sites in carboranes.

In order to assign, conclusively, the structure of compound **II**, both candidate structures, *nido*-1,2,8,10-$C_4B_7H_{11}$ (**IIa**), and *nido*-1,7,8,10-$C_4B_7H_{11}$, (**IIb**), (Figure 9-1), were subjected to the ab initio/IGLO/NMR method. The ab initio optimized structures, **2a** and **2b**, and the calculated ^{11}B NMR chemical shifts, together with the experimental data for compound **II**, are included in Figure 9-3.

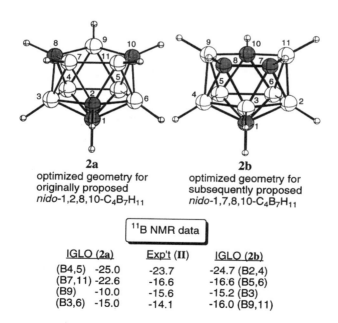

2a
optimized geometry for originally proposed *nido*-1,2,8,10-$C_4B_7H_{11}$

2b
optimized geometry for subsequently proposed *nido*-1,7,8,10-$C_4B_7H_{11}$

^{11}B NMR data

IGLO (2a)		Exp't (II)	IGLO (2b)	
(B4,5)	-25.0	-23.7	-24.7	(B2,4)
(B7,11)	-22.6	-16.6	-16.6	(B5,6)
(B9)	-10.0	-15.6	-15.2	(B3)
(B3,6)	-15.0	-14.1	-16.0	(B9,11)

Figure 9-3 Ab initio optimized geometries and IGLO calculated ^{11}B NMR chemical shifts for the proposed *nido*-1,2,8,10-$C_4B_7H_{11}$, structure **2a**, and the alternative isomer *nido*-1,7,8,10-$C_4B_7H_{11}$, structure **2b**.

The correlation between the experimental data for compound **II** and the calculated ^{11}B NMR chemical shifts for structure **2b** is significantly better than with structure **2a**. For example, the calculated resonance (in ppm) for the unique boron in structure **2b** (B3, −15.2) agrees quite well with the experimental value (−15.6), while the calculated value for the unique boron in structure **2a** (B9, −10.0) is at over 5 ppm lower field from the experimental value (−15.6). Thus, the IGLO ^{11}B NMR chemical shift calculations favor structure **2b** over **2a** for the geometry of compound **II**; the relative energies also favor structure **2b** by 32.4 kcal/mol (see Table 9-1).

Two other candidate configurations with C_s symmetry (structures **2c** and **2d** in Figure 9-4) that are predicted to be energetically less favorable than structure **2b**, but more favorable than structure **2a** based upon carbon location preferences,[11a] were also subjected to the ab initio/IGLO/NMR method. The 2,8,9,10 isomer, structure **2c**, gives good agreement of the calculated shifts with the experimental data for three of the ^{11}B resonances, but the calculated shift for B1 (−50.7) is more than 35 ppm to lower field than the experimental value (−15.6). Thus, structure **2c** is ruled out as the geometry for compound **II**. The 1,7,8,9 isomer, structure **2d**, gives fairly good correlation of the IGLO-calculated ^{11}B NMR chemical shifts with the experimental data for compound **II**, with only B6 (−10.8) differing appreciably from the experimental value (−15.6). However, the agreement between the calculated and experimental ^{11}B NMR

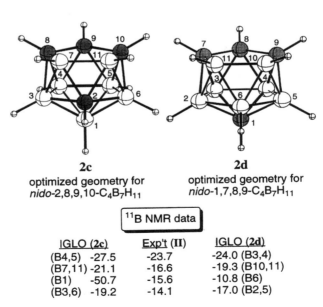

2c
optimized geometry for
nido-2,8,9,10-$C_4B_7H_{11}$

2d
optimized geometry for
nido-1,7,8,9-$C_4B_7H_{11}$

^{11}B NMR data		
IGLO (2c)	Exp't (II)	IGLO (2d)
(B4,5) −27.5	−23.7	−24.0 (B3,4)
(B7,11) −21.1	−16.6	−19.3 (B10,11)
(B1) −50.7	−15.6	−10.8 (B6)
(B3,6) −19.2	−14.1	−17.0 (B2,5)

Figure 9-4 Ab initio optimized geometries and IGLO calculated ^{11}B NMR chemical shifts for *nido*–2,8,9,10-$C_4B_7H_{11}$, structure **2c**, and *nido*-1,7,8,9-$C_4B_7H_{11}$, structure **2d**.

data for structure **2b** is better than for structure **2d**. In addition, isomer **2b** is calculated to be 19.5 kcal/mol more stable than isomer **2d** (Table 9-1).

9.3.3.3 Isomer III

Compound **III**, with C_1 symmetry, was originally assigned the 2,7,8,11-structure, (**IIIa**, Figure 9-1), on the basis of 1D ^{11}B and 2D COSY ^{11}B–^{11}B NMR studies.[9] A number of the resonances in the ^{11}B spectrum are closely spaced, however, making an unambiguous assignment difficult. Empirical carbon-location preferences [11a] do not favor the 2,7,8,11 isomer, structure **IIIa**, as it contains far too many adjacent carbon–carbon connections, an unprecedented arrangement, and would not be expected to be stable. There are a number of alternative C_1 symmetry isomers possible for compound **III**; thus, the ab initio/IGLO/NMR method appeared ideally suited for solving this structural dilemma.

An effort to determine the ab initio optimized geometry for the proposed 2,7,8,11 isomer of C$_4$B$_7$H$_{11}$, structure **IIIa**, (Figure 9-1) gave an unexpected structure, **3a** (Figure 9-5). Structure **3a** is unusual in that it contains a four-membered square face (C2–C7–C11–B6) in addition to the anticipated five-membered open face, with several bond distances not within normal B—B and B—C

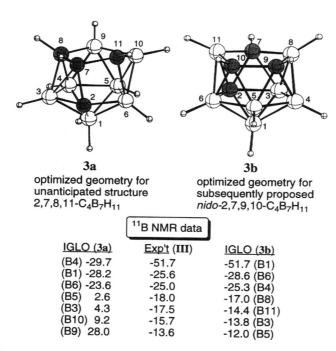

3a
optimized geometry for
unanticipated structure
2,7,8,11-C$_4$B$_7$H$_{11}$

3b
optimized geometry for
subsequently proposed
nido-2,7,9,10-C$_4$B$_7$H$_{11}$

^{11}B NMR data		
IGLO (3a)	Exp't (III)	IGLO (3b)
(B4) -29.7	-51.7	-51.7 (B1)
(B1) -28.2	-25.6	-28.6 (B6)
(B6) -23.6	-25.0	-25.3 (B4)
(B5) 2.6	-18.0	-17.0 (B8)
(B3) 4.3	-17.5	-14.4 (B11)
(B10) 9.2	-15.7	-13.8 (B3)
(B9) 28.0	-13.6	-12.0 (B5)

Figure 9-5 Ab initio optimized geometries and IGLO calculated ^{11}B NMR chemical shifts for the proposed *nido*-2,7,8,11-C$_4$B$_7$H$_{11}$, structure **3a**, and the favored alternative isomer *nido*-2,7,9,10-C$_4$B$_7$H$_{11}$, structure **3b**.

bond ranges. Carbon-preference systematics predicts [11a] that incorporating a triangular facet of carbon atoms is disfavored. The calculated ^{11}B NMR chemical shifts of structure **3a** correlate poorly with the experimental data (Figure 9-5); thus, structure **IIIa** may be ruled out as the structure of compound **III**.

An alternative structure, $nido$-2,7,9,10-$C_4B_7H_{11}$, appeared to be the most logical candidate for compound **III** and the ab initio optimized geometry, structure **3b**, is given in Figure 9-5. Carbon-location preferences predict configuration **3b** to be the most stable C_1 symmetry $nido$-$C_4B_7H_{11}$ structure. Structure **3b** contains three of the four carbon atoms on the open face in an "as-nonadjacent-as-possible" configuration to enhance charge smoothing. There are only two other locations available for the fourth carbon, and the 2,7,9,10 configuration, structure **3b**, also appears to be the most logical based upon the synthesis from two "carbons-adjacent" precursors: $arachno$-4,5-$C_2B_7H_{13}$ and acetylene. The calculated ^{11}B NMR chemical shifts of structure **3b** correlate well with the experimental data for compound **III** (Figure 9-5); thus, $nido$-2,7,9,10-$C_4B_7H_{11}$, structure **3b**, is favored as the revised structure for compound **III**.

Two other possible candidate configurations of $nido$-$C_4B_7H_{11}$ with C_1 symmetry (structures **3c** and **3d**, Figure 9-6) that would be predicted by empirical carbon-location preferences to be energetically less favorable than structure **3b**, but more favorable than structure **3a**, were also subjected to the ab

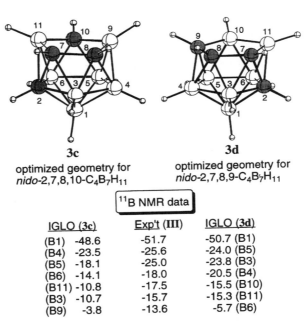

	3c		**3d**	
	optimized geometry for $nido$-2,7,8,10-$C_4B_7H_{11}$		optimized geometry for $nido$-2,7,8,9-$C_4B_7H_{11}$	

^{11}B NMR data		
IGLO (3c)	Exp't (III)	IGLO (3d)
(B1) −48.6	−51.7	−50.7 (B1)
(B4) −23.5	−25.6	−24.0 (B5)
(B5) −18.1	−25.0	−23.8 (B3)
(B6) −14.1	−18.0	−20.5 (B4)
(B11) −10.8	−17.5	−15.5 (B10)
(B3) −10.7	−15.7	−15.3 (B11)
(B9) −3.8	−13.6	−5.7 (B6)

Figure 9-6 Ab initio optimized geometries and IGLO calculated ^{11}B NMR chemical shifts for $nido$-2,7,8,10-$C_4B_7H_{11}$, structure **3c**, and $nido$-2,7,8,9-$C_4B_7H_{11}$, structure **3d**.

initio/IGLO/NMR method. The 2,7,8,10 isomer, structure **3c**, gives poor correlation of the calculated ^{11}B NMR chemical shifts with the experimental data for compound **III**. For example, the calculated chemical shifts (in ppm) for B11 (−10.8) and B5 (−18.1) are at around 7 ppm lower field from the experimental values (−17.5 and −25.0, respectively).

The 2,7,8,9 isomer, structure **3d**, gives fairly good correlation between the calculated ^{11}B NMR chemical shifts and the experimental values for compound **III**, but the 2,7,9,10 isomer, structure **3b**, gives better correlation. In particular, the calculated shift (in ppm) for B6 (−5.7) in structure **3d** is located at about 8 ppm lower field than the experimental value (−13.6). Thus the ab initio/IGLO/NMR method selects structure **3b** in favor of structure **3c** or **3d** as the correct geometry for the asymmetrical C$_4$B$_7$H$_{11}$ carborane. There is also a substantial difference in the relative energies of structure **3b** and **3d**, with structure **3b** being 18.0 kcal/mol more stable than structure **3d** (Table 9-1). Note also that the 2D ^{11}B NMR spectrum obtained for compound **III** could be compatible with the 2,7,9,10 isomer, structure **3b**, because of the closely spaced resonances.

9.3.4 Discussion

9.3.4.1 General Comments about Cage Carbon Site Preferences
The relative energies of all the *nido*-C$_4$B$_7$H$_{11}$ isomers considered in this study are included in Table 9-1. It is satisfying to find that the three isomers (structures **1a**, **2b**, and **3b**) favored by the NMR chemical shift calculations are also the three most stable of those examined computationally. These results agree completely with the previously existing empirical carbon-location preferences.[11a] It should be noted that the *nido*-11-vertex cage is ideally suited for the testing of these empirical rules as it contains only three kinds of vertices, and no complications are expected due to the lack of structural vertex homogeneity. It is also satisfying to find that the 2,7,8,11 isomer, structure **3a**, is calculated to be much higher in energy than all of the other isomers calculated in this study. This result is in agreement with the prediction from carbon-location preferences that indicates any proposed structure (compound **IIIa**, Figure 9-1) containing a triangular facet of three carbons should be energetically disfavored. The fact that the optimized geometry of this structure (Figure 9-5) contains a square open face supports this proposition.

A comparison of the relative energies of the *nido*-C$_4$B$_7$H$_{11}$ carboranes calculated in this study also enables us to address a residual question concerning the relative importance of the previously mentioned carbon-location preferences for (1) low-coordinated sites versus (2) nonadjacent locations when both are available. The first seven isomers of *nido*-C$_4$B$_7$H$_{11}$, in terms of estimated relative stability based upon empirical geometrical systematics, are the first seven structures illustrated in Figure 9-7b, drawn in a fashion (as illustrated in Figure 9-7a) to facilitate easy determination of vertex connectivity (the vertices on the open face are four-coordinate and the rest are five-coordinate) and number of carbon–carbon contacts. The order chosen assumes that the preference for

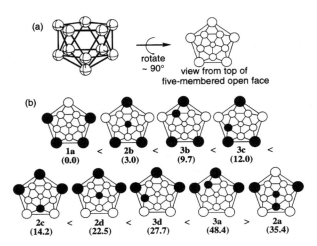

Figure 9-7 (a) Illustration of two views of a *nido*-11-vertex cluster: "side" view and "top" view. (b) Stability order among the *nido*-C$_4$B$_7$H$_{11}$ isomers.

carbons to occupy low-coordination sites is significantly more important than the preference for carbons to occupy nonadjacent locations. The ab initio determined relative energies (from Table 9-1) are given in parentheses. That the top seven configurations displayed in Figure 9-7b match the top entries in Table 9-1 confirms this assumption. Prior to this study, some residual uncertainty remained regarding low-coordination versus adjacent-carbon preferences. For example, structure **1a**, which is 3.0 kcal/mol more stable than structure **2b** (see Table 9-1), incorporates *three* C—C carbons-adjacent situations (which is associated with instability), while structure **2b** incorporates only *one* C—C carbons-adjacent situation but does contain *one* carbon in a higher coordinated location (also indicative of instability). Apparently, as previously proposed based upon empirical observations, the occupation of one additional low-coordinated site (a positive effect) more than offsets the destabilizing effects of adding two additional carbons-adjacent situations.

Further examination of the structures in Figure 9-7b confirms these trends. Decreasing stability results as the single carbon not on the open face in structure **2b** moves adjacent to the lone peripheral carbon in structure **3b**, then adjacent to one of the pair of peripheral carbons in structure **3c**. An analogous trend is found upon examination of structures **2c**, **2d**, **3d**, and **3a**, each having three adjacent carbons on the open-face. As the lone carbon not on the open face in structure **2c** migrates closer to the open-face carbons in structures **2d** and **3d**, the result is decreased stability. When this carbon makes contact with two peripheral carbons, a highly unstable situation occurs and the cage, instead, adopts the *arachno*-geometry structure **3a**.

In 1983, it was originally noted [10] that compound **II**, shown in this study to be structure **2b**, undergoes isomerization via thermolysis at 330°C to give an unisolated material with ^{11}B NMR resonances of −13.0, −14.0, −22.9, and −45.5 ppm. These chemical shifts are similar to those reported for compound **I** (isolated in 1988), shown in this study to be structure **1a**, and reexamination of the original (1983) ^{11}B NMR spectrum of the material from the isomerization confirms this assignment. This isomerization is consistent with the relative energies of structure **1a** and **2b**, with structure **2b** being only 3 kcal/mol higher in energy (see Table 9-1).

9.3.5 Conclusions

The ab initio/IGLO/NMR method has been successfully applied to establish the structures of the three isomers of *nido*-$C_4B_7H_{11}$. The method confirms compound **I** to be *nido*-7,8,9,10-$C_4B_7H_{11}$, structure **1a**, and compound **II** to be *nido*-1,7,8,10-$C_4B_7H_{11}$, structure **2b** (rather than structures **2a**, **2c**, or **2d**) and determines compound **III** to be *nido*-2,7,9,10-$C_4B_7H_{11}$, structure **3b** (rather than structures **3a**, **3c**, or **3d**). The relative energies obtained from the ab initio optimized geometries indicate the thermodynamically most stable arrangement of carbon atoms in these 11-vertex *nido*-clusters has all four carbon atoms in adjacent, low-coordinate positions about the open face (structure **1a**). Thus, in complete agreement with the empirically determined trends,[11a] carbon occupation of low-coordinated sites appears to be a more important factor than avoiding carbon–carbon connections in the cage. However, moving adjacent carbons from the open face to nonadjacent higher coordination vertices remains of secondary importance.

9.4 COMPUTATIONAL STUDY OF *NIDO*-8-VERTEX BORANES AND CARBORANES

9.4.1 Background

The second part of this chapter revisits the question of the preferred structure for the *nido*-8-vertex cluster class. On the basis of Williams' original geometrical systematics,[11a] the anticipated gross structure of a *nido*-8-vertex electron-count cluster would be generated (Figure 9.8a) by removing a high-coordination vertex from a nine-vertex polyhedron (tricapped trigonal prism), giving an eight-vertex five-membered open-face geometry ("ni-8⟨V⟩" in the Williams' nomenclature [11e]). However, the first eight-vertex polyborane cluster to be structurally characterized via X-ray crystallography, B_8H_{12},[14] was shown to have a *nido*-6-membered open-face geometry ("ni-8⟨VI⟩"), differing from the "expected" structure simply by the absence of one edge connection (Figure 9-8b). This ni-8⟨VI⟩ geometry is the same as that predicted by geometrical systematics for *arachno*-8-vertex clusters, i.e., derived from a 10-vertex polyhedron by removal of two high-coordinated vertices (Figure 9-8c). Based, in part, on these results, the original geometrical systematics were modified.[11e] The

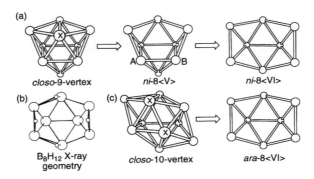

Figure 9-8 (a) Derivation of ni-8⟨V⟩ and ni-8⟨VI⟩ frameworks from a nine-vertex polyhedron. (b) X-ray-determined geometry for B_8H_{12} (minus terminal hydrogens). (c) Derivation of ara-8⟨VI⟩ framework from a 10-vertex polyhedron.

removal of a high-coordinated vertex from a *closo*-polyhedron was not considered the sole method of generating a *nido*-structure. Another approach is to remove either a low- or high-coordinated vertex, and, subsequently, high-coordinated edge connectivities are sequentially broken to see how many different *nido*-fragments can be reasonably generated. Thus, the ni-8⟨VI⟩ framework can be thought of as being generated from a nine-vertex polyhedron, either by removing a $4k$ vertex followed by two high-coordinated edge connectivities (this route not shown), or by removal of a $5k$ vertex followed by breaking one high-coordinated edge connection ("A–B") as shown in Figure 9-8a.

A limited number of other *nido*-8-vertex electron-count clusters have also been structurally characterized via X-ray crystallography. The following compounds have been shown to have the same ni-8⟨VI⟩ "*arachno*-type" configuration: *nido*-$(\eta^6\text{-}C_6Me_6)Fe(Me)_4C_4B_3H_3$,[15] *nido*-$(\eta^5\text{-}C_5H_5)Co(Ph)_4C_4B_3H_3$,[16] and *nido*-$4,5\text{-}C_2B_6H_9^-$.[17] The only X-ray crystallographically characterized *nido*-8-vertex cluster with a ni-8⟨V⟩ configuration is *nido*-$(\eta^5\text{-}C_5H_5)_2Co_2SB_5H_7$.[18] Three other *nido*-8-vertex electron-count clusters have recently been structurally characterized computationally employing the ab initio/IGLO/NMR method and shown to have ni-8⟨VI⟩ configurations. These include the 3,6- [19] and 4,5-isomers [17] of *nido*-$C_2B_6H_{10}$, and *nido*-$3,5\text{-}C_2B_6H_9^-$.[20]

Thus, we were curious to see if any *nido*-8-vertex electron-count polyboranes or carboranes with ni-8⟨V⟩ configurations could be found computationally. A reasonably comprehensive search of clusters has been carried out ranging from $B_8H_8^{4-}$ to $C_4B_4H_8$. Ab-initio-based IGLO basis set II NMR chemical shift calculations were also carried out and should serve as a useful guide for structural confirmation if any new *nido*-8-vertex boranes or carboranes are synthesized.[21]

9.4.2   Computational Details

All structures were optimized at the MP2(FULL)/6-31G* level of theory, except where indicated. The relative energies were determined at MP2(FULL)/6-31G* + ZPE (6-31G*) and are given in the figures, while the absolute and zero-point energies can be found in the Appendix Table 9A.1. The zero-point energies at HF/6-31G* have been scaled by 0.89.[13] The NMR chemical shifts for the *nido*-8-vertex clusters were calculated using IGLO basis set II. It should be pointed out that input structures for the carboranes always had the carbons on the open face, and, if non-exo-terminal hydrogens were present, they were usually inputted as either bridge and/or *endo*-BHs. It should also be pointed out that the input frameworks used were usually a ni-8⟨VI⟩, but sometimes a ni-8⟨V⟩ was used. From the results, it appears as though the size of the inputted open face did not necessarily determine whether a structure optimized to a ni-8⟨VI⟩ or ni-8⟨V⟩ configuration. For clarity, those clusters that are experimentally known have been indicated by italicizing their number designation.

9.4.3   Results

9.4.3.1 $B_8H_8^{4-}$ The only minimum found for $B_8H_8^{4-}$ (structure **4**) displays a ni-8⟨VI⟩ configuration with overall C_{2v} symmetry (Figure 9-9a). All attempts failed to optimize a $B_8H_8^{4-}$ structure containing a ni-8⟨V⟩ configuration. The first geometry employed was based on the most spherical nine-vertex polyhedron with B–B and B–H bond lengths of 1.7 and 1.1 Å, respectively, minus a 5*k* vertex (see ni-8⟨V⟩ in Figure 9-8a). This input geometry, starting either at HF/STO-3G, HF/3-21G, or HF/6-31G*, optimized to nido-8⟨VI⟩ structure **4**. A second input geometry employed was based upon the ni-8⟨V⟩ structure **5** for $B_8H_9^{3-}$ (see

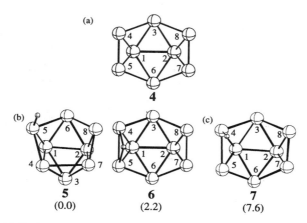

Figure 9-9   (a) Optimized geometry for $B_8H_8^{4-}$ (structure **4**). (b) Optimized geometries for $B_8H_9^{3-}$ boranes (structures **5** and **6**). (c) Transition state (structure **7**) connecting structures **5** and **6** (relative energies, in kcal/mol, in parentheses).

above) with the endo-hydrogen removed. At HF/6-31G*, this input geometry also optimized to **4**.

Lee [22] has predicted, using second-moment scaled Hückel theory, that the ni-8⟨VI⟩ geometry for $B_8H_8^{4-}$ is preferred energetically over the ni-8⟨V⟩ by ~60 kcal/mol. An estimation of the difference in energy between the ni-8⟨VI⟩ and ni-8⟨V⟩ $B_8H_8^{4-}$ at an ab initio level was made by comparing the energy of the optimized geometry **4** and the energy of the input geometry based upon the $B_8H_9^{3-}$ structure **5** minus the endo-hydrogen. At HF/6-31G*, the energy difference is 53.8 kcal/mol.

9.4.3.2 $B_8H_9^{3-}$ Two local minima were found for the $B_8H_9^{3-}$ system (Figure 9-9b), and the lowest energy structure (**5**) has a ni-8⟨V⟩ configuration with an *endo*-hydrogen on a boron atom that is connected to three other cage atoms (a "3*k*" vertex [11e]). Isomer **5** is calculated to be 2.2 kcal/mol more stable than the ni-8⟨VI⟩ structure **6**, which has a bridge hydrogen spanning five-coordinated boron vertices ("55-bridge" hydrogen). All the B–B bond lengths around the open face in structure **5** are within 2.00 Å, supporting the designation of this structure as a true ni-8⟨V⟩ cluster. Using the QST2 and IRC methods within Gaussian 94, the transition state (structure **7**, Figure 9-9c) connecting structure **5** and **6** was located. The barrier for the isomerization of structure **6** to structure **5** is predicted to be 5.4 kcal/mol. A simple explanation for why structure **5** is more stable than structure **6** is not readily apparent, thus, a detailed molecular-orbital analysis is likely needed. No isomers of $B_8H_9^{3-}$ isomers are known experimentally.

9.4.3.3 $B_8H_{10}^{2-}$ Four local minima were found for the $B_8H_{10}^{2-}$ system (Figure 9-10a). The lowest energy structure (**8**) has a ni-8⟨VI⟩ configuration with two

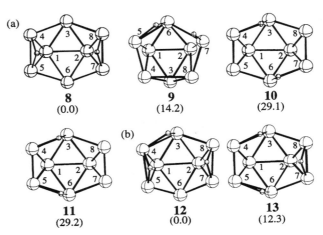

Figure 9-10   Optimized geometries for boranes (a) $B_8H_{10}^{2-}$ and (b) $B_8H_{11}^{-}$.

55-bridge hydrogens. Isomer **9**, 14.2 kcal/mol higher in energy than isomer **8**, has a ni-8⟨V⟩ configuration. The two remaining structures (**10** and **11**) both have a ni-8⟨VI⟩ configuration and two *65*-bridge hydrogens. The high relative energies of these systems compared with that of structure **8** can be rationalized as being due to the less favorable locations of bridge hydrogens. No isomers of $B_8H_{10}^{2-}$ are known experimentally.

9.4.3.4 $B_8H_{11}^-$

A variety of input geometries were employed for the $B_8H_{11}^-$ system, but only two local minima were found (structures **12** and **13**), with each displaying a ni-8⟨VI⟩ configuration (Figure 9-10b). Isomer **12** has *55*-, *65*-, and *66*-bridge hydrogens, while structure **13** has a *55*- and two *65*-bridge hydrogens. The C_1 symmetry structure (**12**) is over 12 kcal/mol lower in energy than the C_s symmetry structure (**13**), which seems surprising as structure **13** would appear to have the more favorable bridge-hydrogen locations. No isomers of $B_8H_{11}^-$ are known experimentally.

9.4.3.5 B_8H_{12}

As mentioned previously, the polyborane B_8H_{12} has been structurally characterized via X-ray crystallography [14] and shown to display a ni-8⟨VI⟩ configuration with overall molecular C_s symmetry (Figure 9-8b). However, in solution, B_8H_{12} is fluxional and on the NMR time scale shows C_{2v} symmetry (three resonances in the ^{11}B NMR spectrum in 4:2:2 ratios). This fluxional behavior in solution was considered to involve two equivalent C_s symmetry structures. McKee [23] has shown that the calculated geometry (using HF/3-21G theory) for the C_s symmetry B_8H_{12} reproduces the solid-state structure. He also predicted a low barrier for the fluxional process as the energy for a C_{2v} symmetry B_8H_{12} (containing two *endo*-BH groups) was only slightly higher than that for the C_s isomer. Schleyer and coworkers have examined the B_8H_{12} C_s and C_{2v} isomers at higher levels of theory and computed their ^{11}B NMR chemical shifts using the IGLO method.[5] Their results also show the C_s isomer to be more stable than the C_{2v} form (3.0 kcal/mol at the MP2/6-31G* level of theory), and the IGLO-calculated ^{11}B NMR shifts for the C_s isomer (in ppm) [−16.8 (av. B1,2), −22.3 (B4,7), 9.4 (av. B3,5,6,8)] gave satisfactory agreement with the experimental data [−22.0 (B1,2), −19.4 (B4,7), 7.5 (B3,5,6,8)], and even better agreement is obtained if the assignments for the B1,2 and B4,7 experimental resonances [24] are reversed.

As part of our extensive calculations on *nido*-8-vertex electron-class clusters, the B_8H_{12} system was also investigated. Three local minima were found for B_8H_{12} (Figure 9-11a), one of which (structure *14*) is the C_s symmetry isomer, as mentioned previously. Another isomer is the C_2 symmetry structure **15**, which is only 1.7 kcal/mol higher in energy (at MP2/6-31G* level of theory) than structure *14*. A third structure (**16**) of much higher energy was also found that has two unfavorable *76*-bridge hydrogens.

If the X-ray geometry for B_8H_{12} or ab initio calculated energies were not available, one would predict, based upon empirically derived rules, that structure **15** would be the most stable isomer if bridge-hydrogen placement was the

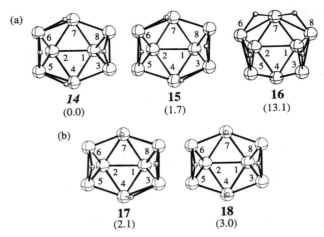

Figure 9-11 Optimized geometries for (a) B_8H_{12} boranes and (b) higher order B_8H_{12} boranes.

determining factor for stability. Structure **15** contains two 65- and two 66-bridge hydrogens, while structure **14** contains one 65- and three 66-bridge hydrogens. A detailed molecular-orbital explanation is probably needed to explain why structure **14** is the most stable geometry for B_8H_{12}.

At first glance, the C_2 symmetry B_8H_{12} (structure **15**) is appealing as it has three boron environments in 4:2:2 ratios, the same ratio as shown by the ^{11}B NMR spectrum. To check whether the C_2 symmetry isomer **15** is the preferred solution-phase structure for B_8H_{12}, ab initio chemical shift calculations were carried out. The averaged IGLO-calculated shifts for structure **15** (in ppm) [−14.9 (B1,2), −34.9 (B4,7), 12.8 (av. B3,5,6,8)] are not in satisfactory agreement with the experimental values [−22.0 (B1,2), −19.4 (B4,7), 7.5 (B3,5,6,8)].

Although the C_2 isomer of B_8H_{12}, structure **15**, is probably not a significant contributor to the overall composition of B_8H_{12} based upon the above IGLO calculations, it is likely the structure involved in the fluxional process that converts one C_s symmetry isomer (**14**) to its mirror-image C_s isomer. Using the LST and IRC methods within GAUSSIAN 92, a C_1 symmetry transition state (structure **17**, Figure 9-11b) was located for the isomerization of structure **15** to structure **14**, the barrier for which is 0.4 kcal/mol (at MP2/6-31G* level of theory). This low barrier agrees with the experimental observation [23] of fluxional behavior for B_8H_{12}. The C_{2v} symmetry B_8H_{12} (structure **18**) optimizes to a second-order saddle point (two imaginary frequencies); thus, it is not a true transition state.

9.4.3.6 B_8H_{10}·L   Another *nido*-8-vertex electron-count cluster known experimentally [25] is *nido*-B_8H_{10}·L (where L = NEt₃), which was characterized by NMR spectroscopy. The originally proposed static structure (**19**, Figure 9-12a) has a ni-8⟨V⟩ configuration with the Lewis base at the unique boron of the open

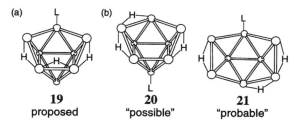

Figure 9-12   (a) Originally proposed static structure for *nido*-B_8H_{10}·L and (b) possible and probable alternative dynamic structures.

face. This proposed structure was suspect because it contains one unacceptable *77-* and two dubious *75*-bridge hydrogens. Two recently proposed [11e] alternative structures for *nido*-B_8H_{10}·L (Figure 9-12b) include the "possible" structure (**20**) with the Lewis base located on a boron, not on the ni-8⟨V⟩ open face, and the more favorable *65*- and *66*-bridge hydrogens, while the "probable" structure (**21**) has a similar structure but with a ni-8⟨VI⟩ configuration.

For either the "possible" structure (**20**) or the "probable" structure (**21**) to be correct, a fluxional "windshield-wiper" process involving the *66*-bridge hydrogen would have to be invoked to rationalize the observed 2:2:2:1:1 ratio of signals in the ^{11}B NMR spectrum for *nido*-B_8H_{10}·NEt_3. Because of the variety of structures possible, the ab initio/IGLO/NMR method seemed ideally suited for solving this structural quandary.

All attempts to optimize to a framework corresponding to the proposed ni-8⟨V⟩ structure (**19**) failed: input geometries of C_s symmetry similar to structure **19** (where L = NH_3 or NMe_3) always gave the ni-8⟨VI⟩ framework of structure **22** with an *endo*-BH (Figure 9-13a). A vibrational frequency analysis of these optimized geometries (**22-NH_3** and **22-NMe_3**) indicated one imaginary frequency. Reoptimization of these structures under no symmetry constraints gave a C_1 symmetry framework like structure *21* (Figure 9-13b), where the *endo*-BH in structure **22** becomes a *66*-bridge hydrogen in structure **21**. All attempts to optimize to a framework corresponding to the "possible" ni-8⟨V⟩ structure (**20**) also failed: input geometries of C_1 symmetry similar to structure **20** (where L = NH_3 or NMe_3) always optimized to structure **21**.

To allow the most favorable comparison with the experimental data, II//HF/6-31G* calculations were carried out on *21*-NEt_3. The ^{11}B NMR chemical shifts for *21*-NEt_3 (see Figure 9-13b) are in satisfactory agreement with the experimental values. Note that the "static" shifts are averaged, as shown, to give the "dynamic" values, to allow comparison with the experimental data. The proposed fluxional process requires migration of the *66*-bridge hydrogen at B4–B5 in **21-L** to the B5–B6 position, generating a mirror image, **21-L** (Figure 9-13c). The transitional structure is **22-L**, which the calculations (at MP2/6-31G*) show is only 1.0 kcal/mol higher in energy than **21-L** (when L = NMe_3).

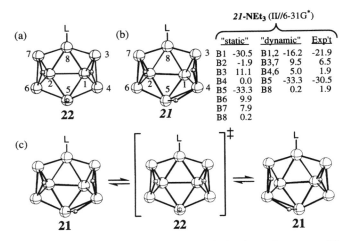

Figure 9-13 (a) Structure (**22**) resulting from optimization of structure **19**. (b) Calculated structure and IGLO ^{11}B NMR data for *nido*-B$_8$H$_{10}$·NEt$_3$ (**21**-NEt$_3$). (c) Proposed fluxional process between enantiomeric structures **21**.

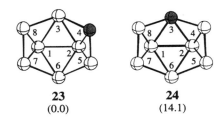

Figure 9-14 Optimized geometries for CB$_7$H$_8^{3-}$ systems.

9.4.3.7 CB$_7$H$_8^{3-}$ Only two local minima were found for the CB$_7$H$_8^{3-}$ system, each having a ni-8⟨VI⟩ configuration (Figure 9-14). Isomer **23** is of lowest energy, with the cage carbon located in a 3*k* position, while isomer **24**, 14.1 kcal/mol higher in energy, has the cage carbon in a 4*k* location on the open face. No isomers of CB$_7$H$_8^{3-}$ are known experimentally.

9.4.3.8 CB$_7$H$_9^{2-}$ Six local minima were found for the CB$_7$H$_9^{2-}$ system (Figure 9-15). The lowest energy structure (**25**) has a ni-8⟨VI⟩ configuration with the cage carbon in a 3*k* site and a *55*-bridge hydrogen. Isomer **26**, only 2.1 kcal/mol higher in energy than isomer **25**, has a 3*k* cage carbon and an *endo*-BH on a ni-8⟨V⟩ framework. Isomers **25** and **26** are similar to structures **5**, ni-8⟨VI⟩, and **6**, ni-8⟨V⟩, for B$_8$H$_9^{3-}$ in Figure 9-9b, except that the ni-8⟨V⟩ configuration, structure **5**, is more stable. Two other CB$_7$H$_9^{2-}$ structures (**28** and **30**) with ni-8⟨V⟩ configurations were also located, but these are considerably higher in energy

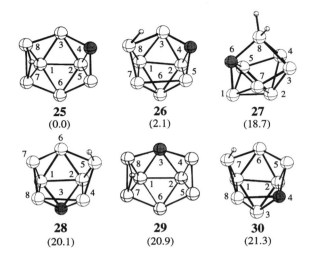

25
(0.0)

26
(2.1)

27
(18.7)

28
(20.1)

29
(20.9)

30
(21.3)

Figure 9-15 Optimized geometries for $CB_7H_9^{2-}$ carboranes.

than structures **25** and **26**. Structure **28** is interesting in that the input geometry had a ni-8⟨VI⟩ framework with the carbon located on the open face in a $4k$ vertex and a 65-bridge hydrogen, but, upon optimization, a ni-8⟨V⟩ structure with the carbon no longer on the open face and an *endo*-BH resulted. The $4k$-value in this case is the same as if it had remained on the open face. The more stable isomers, **25** and **26**, incorporate $3k$ carbons, while the less stable isomers, **21**, **22**, **23**, and **24** incorporate $4k$ carbons. Structures **28** and **29** are analogous to structures **25** and **26**, except that the former "pair" have $4k$ carbons and the ni-8⟨V⟩ configuration is slightly more stable. Another interesting $CB_7H_9^{2-}$ isomer is structure **27**, which looks like a typical *nido*-7-vertex system, but is capped with a BH_2 group. It can be considered to be a *nido*-8-vertex system with two four-membered open faces (ni-8⟨IV + IV⟩). The input geometry for structure **27** was a ni-8⟨VI⟩ structure with a $3k$ cage carbon adjacent to an *endo*-BH on a $3k$ cage boron atom. No isomers of $CB_7H_9^{2-}$ are known experimentally.

9.4.3.9 $CB_7H_{10}^-$ Five local minima were found for the $CB_7H_{10}^-$ system (Figure 9-16), all but one of which have a ni-8⟨VI⟩ configuration. The lowest energy structure (**31**) has a $4k$ cage carbon and two 55-bridge hydrogens. Isomers **32** and **33** have the cage carbon atom in a $3k$ location, with structure **32** having adjacent 65- and 66-bridge hydrogens, while structure **33** has nonadjacent 65- and 66-bridge hydrogens. It is somewhat surprising that structure **32** is more stable than structure **33** as the latter appears to have the more desirable locations for the bridge hydrogens. Isomer **34** has a $4k$ cage carbon and nonadjacent 55- and 65-bridge hydrogens. The highest energy isomer (**35**) has a ni-8⟨V⟩ framework with an *endo*-BH and a 65-bridge hydrogen, and the cage carbon is

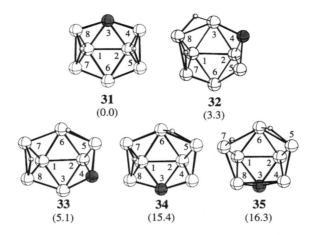

31
(0.0)

32
(3.3)

33
(5.1)

34
(15.4)

35
(16.3)

Figure 9-16   Optimized geometries for $CB_7H_{10}^-$ carboranes.

located in the $4k$ vertex not on the open face. No isomers of $CB_7H_{10}^-$ are known experimentally.

9.4.3.10 CB_7H_{11}   Two local minima were found for the CB_7H_{11} carborane system (Figure 9-17), both having a ni-8⟨VI⟩ configuration. The lowest energy isomer (**36**) has a $4k$ cage carbon and three bridge hydrogens (*55-*, *65-*, and *66*-types). The higher energy isomer (**37**) has a $3k$ cage carbon and three *66*-bridge hydrogens. No isomers of CB_7H_{11} are known experimentally. The unknown most stable isomer (**36**) in Figure 9-17 is isoelectronic and isostructural with the known most stable isomer for $B_8H_{10}\cdot L$, structure **21**, in Figure 9-13b, since a C–H group is isoelectronic with a B^-—L^+ group.

9.4.3.11 $C_2B_6H_8^{2-}$ Six local minima were found for the $C_2B_6H_8^{2-}$ carborane system [26] (Figure 9-18), all of which have a ni-8⟨VI⟩ configuration. The lowest energy structure (**38**) has nonadjacent $3k$ cage carbons in overall molecular C_s symmetry. No isomers of $C_2B_6H_8^{2-}$ are known experimentally.

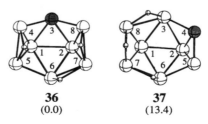

36
(0.0)

37
(13.4)

Figure 9-17   Optimized geometries for CB_7H_{11} carboranes.

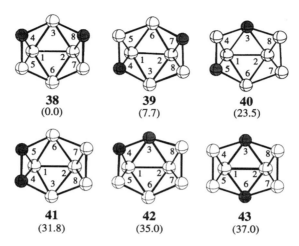

38
(0.0)

39
(7.7)

40
(23.5)

41
(31.8)

42
(35.0)

43
(37.0)

Figure 9-18   Optimized geometries for $C_2B_6H_8^{2-}$ carboranes.

9.4.3.12 $C_2B_6H_9^-$ Eight local minima were found for the $C_2B_6H_9^-$ carborane system (Figure 9-19), seven of which have a ni-8⟨VI⟩ framework; the other has a ni-8⟨V⟩ configuration. Two of the isomers are known experimentally (**44** and **47**) and ab initio/IGLO ^{11}B NMR chemical shift calculations have been reported previously [17,20] and agree satisfactorily with the experimental data. Isomer **44** is of the lowest energy and has nonadjacent $3k$ and $4k$ cage carbons together with a 55-bridge hydrogen. The other experimentally known isomer (**47**) is only 4.4 kcal/mol higher in energy than isomer **44** and has adjacent $3k$ cage carbons and a 55-bridge hydrogen. A ni-8⟨V⟩ configuration $C_2B_6H_9^-$ carborane (structure **48**) was also found and is 7.4 kcal/mol higher in energy than structure **44**. It has nonadjacent $3k$ and $4k$ carbons on the open face and an *endo*-BH.

9.4.3.13 $C_2B_6H_{10}$ Seven local minima were found for the $C_2B_6H_{10}$ carborane system (Figure 9-20), all of which have a ni-8⟨VI⟩ framework. Two isomers of $C_2B_6H_{10}$ are known experimentally (structures **52** and **55**) and each has been structurally   characterized   by   computationally   employing   the   ab initio/IGLO/NMR method.[17,19] The lowest energy isomer is the known [27] carborane (structure **52**), which has two $4k$ cage carbons and two 55-bridge hydrogens. According to the calculations, the next most stable isomer energetically is structure **53**, which could potentially be synthesized via protonation of the known $C_2B_6H_9^-$ carborane, structure **44**.[20] The other experimentally known $C_2B_6H_{10}$ carborane (structure **55**),[16] calculated to be 22.5 kcal/mol higher in energy than structure **52**, is fluxional, with likely rapid bridge proton rearrangements across the B3–B8, B7–B8, and B6–B7 edges. The symmetrical $C_2B_6H_{10}$ carborane structure **57** is much higher in energy than structure **55**; thus, it is not a contributing structure.

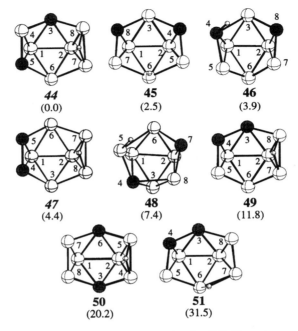

Figure 9-19 Optimized geometries for $C_2B_6H_9^-$ carboranes.

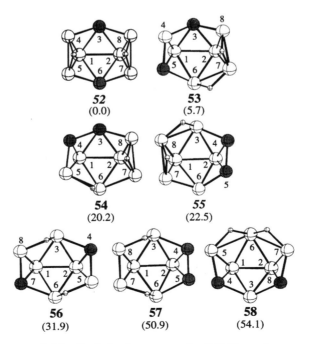

Figure 9-20 Optimized structures for $C_2B_6H_{10}$ carboranes.

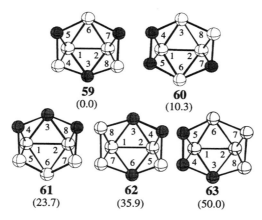

Figure 9-21   Optimized structures for $C_3B_5H_8^-$ carboranes.

9.4.3.14 $C_3B_5H_8^-$ Five local minima were found for the $C_3B_5H_8^-$ carborane system (Figure 9-21), all of which have a ni-8⟨VI⟩ framework. The lowest energy isomer (**59**) contains three nonadjacent $3k$ cage carbons. No isomers of $C_3B_5H_8^-$ are known experimentally.

9.4.3.15 $C_3B_5H_9$ Eight local minima were found for the $C_3B_5H_9$ carborane system (Figure 9-22), most of which have a ni-8⟨VI⟩ configuration. The excep-

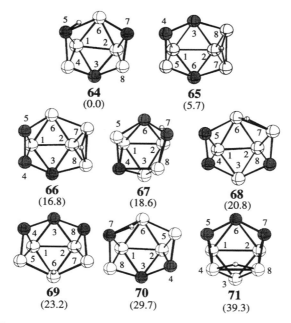

Figure 9-22   Optimized structures for $C_3B_5H_9$ carboranes.

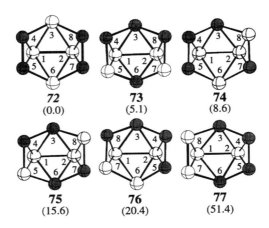

Figure 9-23   Optimized structures for $C_4B_4H_8$ carboranes.

tions are the highest energy isomer, **71** (ni-8⟨V⟩ framework with three adjacent cage carbons and a *55*-bridge hydrogen), and the fourth most stable isomer, **67** (ni-8⟨V⟩ framework with an endo-CH). The input geometry for structure **67** was a ni-8⟨VI⟩ framework containing two *4k* and one *3k* cage carbons and an *endo*-BH on the boron situated between the *3k* and one of the *4k* cage carbons. The lowest energy $C_3B_5H_9$ carborane (structure **64**) contains three nonadjacent *3k* cage carbons and an *endo*-CH group. No isomers of $C_3B_5H_9$ are known experimentally.

9.4.3.16 $C_4B_4H_8$ Six local minima were found for the $C_4B_4H_8$ carborane system (Figure 9-23), all of which have a ni-8⟨VI⟩ configuration. The lowest energy isomer (**72**) has all four cage carbons in *3k* vertices. Alkylated versions of isomer **72** are known experimentally [28] and are proposed to have this ni-8⟨VI⟩ geometry, based upon the spectroscopic data. Our ab initio/IGLO calculations for isomer **72** (in ppm) [−10.8 B(1,2), −15.3 (B3,6)] are in excellent agreement with the experimental data (−10.8 and −12.4, respectively) as reported by Fehlner [27b] for the C-alkylated *nido*-$(CH_3)_4C_4B_4H_4$ carborane; thus, confirming a ni-8⟨VI⟩ configuration in solution.

9.4.4   Discussion

9.4.4.1   *General Comments About Ni-8⟨VI⟩ Versus Ni-8⟨V⟩ Configurations*
From the above calculations, it seems clear that a ni-8⟨VI⟩ "*arachno*-type" framework is nearly always the preferred configuration for *nido*-8-vertex electron-count boranes and carboranes. In only one isomeric system, $B_8H_9^{3-}$, is a ni-8⟨V⟩ configuration the most stable structure. The likelihood of the synthesis of

this cluster to validate this computational finding is slim, although it is conceivable that $B_8H_9^{3-}$ (structure **5**) could be generated under highly basic, low-nucleophilic conditions from B_8H_{12}. In the $B_8H_{10}^{2-}$ system, the second most stable structure (**9**) has a ni-8$\langle V \rangle$ configuration, but if a $B_8H_{10}^{2-}$ cluster should ever be made, it is likely to isomerize into the more stable structure **8**. In the $CB_7H_9^{2-}$ carborane system, the ni-8$\langle V \rangle$ structure (**26**) is calculated to be only 2.1 kcal/mol higher in energy than the most stable structure (**25**), and it may be synthetically feasible. However, the barrier to isomerization of structure **26** to structure **25** is probably small. Two other $CB_7H_9^{2-}$ carboranes (structures **28** and **30**) contain ni-8$\langle V \rangle$ configurations, but both are much higher in energy than structure **25**. Although isomer **30** may easily isomerize to isomer **25**, isomer **28** could be kinetically stable. Isomer **71** of the $C_3B_5H_9$ system contains a ni-8$\langle V \rangle$ framework, but it is synthetically unlikely as isomerization to lower energy nido-8$\langle V \rangle$ structure **69** seems likely.

Perhaps the best candidate for a synthetically feasible *nido*-8$\langle V \rangle$ cluster is isomer **48** of formula $C_2B_6H_9^-$. Protonation of the unknown $C_2B_6H_8^{2-}$ cluster structure **39** may generate structure **48**. The $C_2B_6H_8^{2-}$ carborane anion **39** could possibly be made by two-electron reduction of the known carborane *closo*-1,7-$C_2B_6H_8$.

9.4.4.2 *General Comments about Cage Carbon and "Extra" Hydrogen Placements*

In the first part of this chapter, it was shown that the relative energies of *nido*-$C_4B_7H_{11}$ carboranes completely agreed with the previously existing empirical carbon location preferences,[11a] which indicates that carbons prefer to occupy low-coordinate sites on the cage and that this is more important than avoiding carbon–carbon connections. Attempts to test out these same empirical carbon–placement rules, and also bridge and *endo*-hydrogen location preferences, in the *nido*-8-vertex clusters were met with mixed results. In those carborane systems with no bridge or *endo*-hydrogens ($C_4B_4H_8$, $C_3B_5H_5^-$, $C_2B_6H_8^{2-}$, and $CB_7H_8^{3-}$), the most stable isomer does have the carbon(s) in the empirically preferred low-coordination site(s). However, the relative ordering within the rest of the isomers within a given system (i.e., $C_4B_4H_8$) does not always follow the empirically derived rules (e.g., isomer **73** has two 4*k* and two 3*k* carbons and is more stable than isomer **74**, which has one 4*k* and three 3*k* carbons). In the all boron-containing clusters with bridge and *endo*-hydrogens ($B_8H_9^{3-}, B_8H_{10}^{2-}, B_8H_{11}^-$, and B_8H_{12}), no definite trends are found regarding hydrogen placement. In the carborane systems containing bridge and/or *endo*-hydrogens, no definite trends can be found regarding competition for carbon placement versus hydrogen location. Thus, it appears that, unlike the *nido*-11-vertex framework (which has totally homogeneous 4*k* peripheral vertices and totally homogeneous 5*k* cage vertices), the two *nido*-8-vertex configurations, ni-8$\langle VI \rangle$ and ni-8$\langle V \rangle$ (which have inhomogeneous 3*k* and 4*k* peripheral vertices, and inhomogeneous 4*k* and 5*k* cage vertices in the latter), are less suited for the application of these empirically derived rules. This may be partially due to their being less "rigid" than the *nido*-11-vertex configuration.

9.4.5 Conclusions

Ab initio calculations were used in an extensive search of boranes and carboranes falling into the *nido*-8-vertex electron-count class. The results of this study indicate a ni-8⟨VI⟩ framework to be the usually preferred configuration energetically, although it is clear that there are some isomers that prefer a ni-8⟨V⟩ framework. Another isomer of B_8H_{12} (structure *15*), with C_2 symmetry, was also found, and is energetically only slightly higher than the C_s isomer (structure *14*). The transition state connecting the C_s and C_2 symmetry B_8H_{12} structures was also found (structure *17*) and this has C_1 symmetry. IGLO calculations were also employed to show that the experimentally known *nido*-B_8H_{10}·NEt_3 has the "probable structure" ni-8⟨VI⟩ projected [11e] configuration *21*, and the calculations confirmed that the known alkylated derivatives of the *nido*-$C_4B_4H_8$ carborane have the ni-8⟨VI⟩ configuration *72*.

ACKNOWLEDGMENTS

We thank the Bochum group for permission to use the IGLO program, and J.W.B. thanks R. C. Rizzo, L. G. Sneddon, and A. E. Wille for their contribution to the $C_4B_7H_{11}$ calculations, and R. E. Williams for his significant contributions to all of this work.

SUPPLEMENTARY MATERIAL

The Cartesian coordinates for the optimized geometries at the highest level of theory employed and the IGLO-calculated NMR chemical shifts for all the systems calculated in this study are available on the World Wide Web at: http://rs6chem.vill.edu/faculty/bausch.htm/suppinfo.

NOTES AND REFERENCES

1. This work has been recently published: Bausch, J. W., Rizzo, R. C., Sneddon, L. G., Wille, A. E., and Williams, R. E., *Inorg. Chem.*, **35**, 131, 1996.

2. A more comprehensive study, including nitrogen, sulfur, and oxygen containing *nido*-8-clusters, will be published elsewhere: Bausch, J. W., Tebben, A. J., Ji, G., and Williams, R. E., submitted to *Inorg. Chem.*, Dec. 1997.

3. Bühl, M. and Schleyer, P. v. R., in *Electron Deficient Boron and Carbon Clusters* (G. A. Olah, K. Wade, and R. E. Williams, eds), Wiley, New York, 1991, Chap. 4, pp. 113–142.

4. Kutzelnigg, W., Fleischer, U., and Schindler, M., in *NMR, Basic Principles and Progress* (P. Diehl, E. Fluck, H. Günther, R. Kosfeld, and J. Seelig, eds), Vol. 23, Springer-Verlag, Berlin, 1990, pp. 165–262, and references therein.

5. Bühl, M. and Schleyer, P. v. R., *J. Am. Chem. Soc.*, **114**, 477, 1992.

6. See footnote 17 in: Diaz, M., Jaballas, J., Tran, D., Lee, H., Arias, J., and Onak, T., *Inorg. Chem.*, **35**, 4536, 1996.

7. Frisch, M. J., Trucks, G. W., Head-Gordon, M., Gill, P. M. W., Wong, M. W., Foresman, J. B., Johnson, B. G., Schlegel, H. B., Robb, M. A., Replogle, E. S., Gomperts, R., Andres, J. L., Raghavachari, K., Binkley, J. S., Gonzalez, C., Martin, R. L., Fox, D. J., Defrees, D. J., Baker, J., Stewart, J. J. P., and Pople, J. A., Gaussian 92 (Revision E-2), Gaussian, Inc., Pittsburgh, PA, 1992; (b) Frisch, M. J., Trucks, G. W., Schlegel, H. B., Gill, P. M. W., Johnson, B. G., Robb, M. A., Cheeseman, J. R., Keith, T., Petersson, G. A., Montgomery, J. A., Raghavachari, K., Al-Laham, M. A., Zakrzewski, V. G., Ortiz, J. V., Foresman, J. B., Cioslowski, J., Stefanov, B. B., Nanayakkara, A., Challacombe, M., Peng, C. Y., Ayala, P. Y., Chen, W., Wong, M. W., Andres, J. L., Replogle, E. S., Gomperts, R., Martin, R. L., Fox, D. J., Binkley, J. S., Defrees, D. J., Baker, J., Stewart, J. P., Head-Gordon, M., Gonzalez, C., and Pople, J. A., Gaussian 94 (Revision B.1), Gaussian, Inc., Pittsburgh, PA, 1995.

8. Onak, T. P., Landesman, H. L., and Williams, R. E., *J. Phys. Chem.*, **21**, 1533, 1959.

9. Štíbr, B., Jelínek, T., Drdáková, E., Heřmánek, S., and Plešek, J., *Polyhedron*, **8**, 669, 1988.

10. Astheimer, R. J., and Sneddon, L. G., *Inorg. Chem.*, **22**, 1928, 1983.

11. (a) Williams, R. E., *Inorg. Chem.*, **10**, 210, 1971; (b) Wade, K., *Adv. Inorg. Chem. Radiochem.*, **18**, 1, 1976; (c) Williams, R. E., *Adv. Inorg. Chem. Radiochem.*, **18**, 67, 1976; (d) Rudolph, R. W., *Acc. Chem. Res.*, **9**, 446, 1976; (e) Williams, R. E., in *Electron Deficient Boron and Carbon Clusters* (G. A. Olah, K. Wade, and R. E. Williams, eds), Wiley, New York, 1991, Chap. 2, pp. 11–93; (f) Williams, R. E., *Chem. Rev.*, **92**, 177, 1992.

12. (a) Ref. 11(f); (b) Williams, R. E., in *Advances in Organometallic Chemistry* (F. G. A. Stone and R. West, eds), Vol. 36, Academic Press, New York, 1994, pp. 1–55.

13. Pople, J. A., Krishnan, R., Schlegel, H. B., DeFrees, D., Binkley, J. S., Frisch, M. J., Whiteside, R. F., Hout, R. F., and Hehre, W. J., *Int. J. Quantum Chem., Symp*, **15**, 269, 1981.

14. Enrione, R. E., Boer, F. P., and Lipscomb, W. N., *Inorg. Chem.*, **3**, 1659, 1964.

15. Micciche, R. P., Briguglio, J. J., and Sneddon, L. G., *Organometallics*, **3**, 1396, 1984.

16. Zimmerman, G. J. and Sneddon, L. G., *Inorg. Chem.*, **19**, 3650, 1980.

17. Kang, S. O., Bausch, J. W., Carroll, P. J., and Sneddon, L. G., *J. Am. Chem. Soc.*, **31**, 3763, 1992.

18. Zimmerman, G. J., and Sneddon, L. G., *J. Am. Chem. Soc.*, **103**, 1102, 1981.

19. Bausch, J. W., Prakash, G. K. S., Bühl, M., Schleyer, P. v. R., and Williams, R. E., *Inorg. Chem.*, **31**, 3060, 1992.

20. Onak, T., Tseng, J., Tran, D., Herrera, S., Chan, B., Arias, J., and Diaz, M., *Inorg. Chem.*, **31**, 3910, 1992.

21. The IGLO data is available in the Supplementary Material section and will also be reported elsewhere (see note 2).

22. Lee, S., *Inorg. Chem.*, **31**, 3063, 1992.

23. McKee, M. L., *J. Phys. Chem.*, **94**, 435, 1990.

24. Maruca, R., Odom, J. D., and Schaeffer, R., *Inorg. Chem.*, **7**, 412, 1968.

25. Briguglio, J. J., Carroll, P. J., Corcoran, E. W., Jr, and Sneddon, L. G., *Inorg. Chem.*, **25**, 4618, 1986.

26. The isomers of $C_2B_6H_8^{2-}$ have also been investigated computationally by T. Onak (personal communication to J.W.B.).

27. (a) Gotcher, A. J., Ditter, J. F., and Williams, R. E., *J. Am. Chem. Soc.*, **95**, 7514, 1973; (b) Reilly, T. and Burg, A. B., *Inorg. Chem.*, **13**, 1250, 1974.

28. (a) Fehlner, T. P., *J. Am. Chem. Soc.*, **99**, 8355, 1977; (b) Fehlner, T. P., *J. Am. Chem. Soc.*, **102**, 3424, 1980; (c) Siebert, W. and El-Essawi, M. E. M., *Chem. Ber.*, **112**, 1480, 1979.

APPENDIX

TABLE 9A-1   **Absolute (MP2/6-31G*) and Zero-Point Energies (in a.u.) for the** *Nido*-**8-Vertex Clusters Calculated in this Study; Amine Complexes of** B_8H_{10} **at HF/6-31 G***

Structure	ZPE	MP2/6-31G*	Structure	ZPE	MP2/6-31G*
	$B_8H_8^{4-}$			$C_2B_6H_8^{2-}$	
4	0.0958	−201.90475	**38**	0.1149	−229.21564
			39	0.1147	−229.20311
	$B_8H_9^{3-}$		**40**	0.1146	−229.17788
5	0.1119	−203.00828	**41**	0.1143	−229.16506
6	0.1130	−203.00568	**42**	0.1148	−229.15970
7	0.1107	−202.99513	**43**	0.1151	−229.15676
	$B_8H_{10}^{2-}$			$C_2B_6H_9^-$	
8	0.1294	−203.94211	*44*	0.1300	−229.94434
9	0.1288	−203.91892	45	0.1267	−229.93750
10	0.1270	−203.89354	*46*	0.1300	−229.93754
11	0.1274	−203.89371	*47*	0.1300	−229.93727
			48	0.1284	−229.93120
	$B_8H_{11}^-$		49	0.1300	−229.92567
12	0.1434	−204.64941	50	0.1299	−229.91207
13	0.1426	−204.63924	51	0.1284	−229.89273
	B_8H_{12}			$C_2B_6H_{10}$	
14	0.1562	−205.16537	52	0.1443	−230.48772
˙15	0.1551	−205.16260	53	0.1430	−230.47755
16	0.1556	−205.14453	54	0.1430	−230.45439
17	0.1537	−205.16200	55	0.1428	−230.45050
18	0.1520	−205.16063	56	0.1416	−230.43443
	$CB_7H_8^{3-}$		57	0.1407	−230.40348
23	0.1058	−215.63198	58	0.1396	−230.39735
24	0.1062	−215.60980			

(Contd.)

TABLE 9A-1 (*Continued*)

Structure	ZPE	MP2/6-31G*	Structure	ZPEMP2/6-31G*	
	$CB_7H_9^{2-}$			$C_3B_5H_8^-$	
25	0.1221	−216.57595	59	0.1224	−242.59368
26	0.1211	−216.57178	60	0.1223	−242.57722
27	0.1204	−216.54460	61	0.1227	−242.55616
28	0.1205	−216.54252	62	0.1223	−242.53638
29	0.1220	−216.54257	63	0.1218	−242.51348
30	0.1200	−216.54012		$C_3B_5H_9$	
	$CB_7H_{10}^-$		64	0.1361	−243.12427
31	0.1375	−217.30273	65	0.1366	−243.11564
32	0.1361	−217.29630	66	0.1365	−243.09971
33	0.1362	−217.29342	67	0.1367	−243.09491
34	0.1361	−217.27705	68	0.1351	−243.09010
35	0.1359	−217.27538	69	0.1335	−243.08505
	CB_7H_{11}		70	0.1344	−243.07557
36	0.1501	−217.82758	71	0.1342	−243.05989
37	0.1487	−217.80503		$C_4B_4H_8$	
	$B_8H_{10} \cdot L$		72	0.1289	−255.75999
$21 \cdot NH_3$	0.1795	−259.42970	73	0.1292	−255.75205
$21 \cdot NMe_3$	0.2711	−376.51197	74	0.1271	−255.74490
$21 \cdot NEt_3$	0.3633	−493.59311	75	0.1288	−255.73469
$22 \cdot NH_3$	0.1776	−259.42634	76	0.1288	−255.72734
$22 \cdot NMe_3$	0.2693	−376.50894	77	0.1280	−255.67731

10

APPLICATIONS OF IGLO/GIAO NMR COMPUTATIONS TO CARBORANE SYSTEMS: PRODUCTS OF TRIMETHYLAMINE/*CLOSO*-CARBORANE REACTIONS; CARBORANE ^{13}C CHEMICAL SHIFTS

THOMAS ONAK

Department of Chemistry, California State University, Los Angeles, CA 90032

10.1 AMINE REACTIONS

Selective, and progressive, cage opening of *closo*-carboranes have been observed in certain instances with the use of nucleophiles, such as hydride ion and fluoride ion.[1–3] Amines have also been effective in this regard, but primary and secondary amines have often resulted in complete, or nearly complete, breakdown of the carbon–boron cage. Tertiary amines, such as trimethyamine, have been more delicate, however, in effecting selective cage-opening. Thus, primary and secondary amines tear apart *closo*-1,6-$C_2B_4H_6$ into monoboron species whereas trimethylamine with this same carborane gives a quantitative yield of a B-Me$_3$N-*nido*-2,4-$C_2B_4H_6$ (Figure 10-1).

One of the smallest known carboranes, *closo*-1,5-$C_2B_3H_5$ was predicted[4] to be unreactive toward R$_3$N (R = H), but experimental observations [5,6] indicate that a mixture of the carborane and the amine (R = Me) forms a solid, the structure of which has proven very elusive. In an effort to shed more light on this, we have recently carried out ab initio calculations with 1,5-$C_2B_3H_5$ and NH$_3$, allowing the NH$_3$ to attach to a single boron of the trigonal bipyramidal *closo*-1,5-$C_2B_3H_5$ and allowing both the hydrogen on that boron, and the cage itself, to assume any configuration it wishes. At the 6-31G* level of theory, we find a

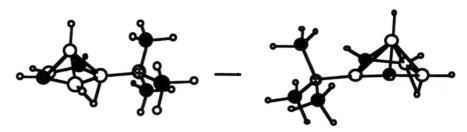

Figure 10-1   5-Me$_3$N-*nido*-2,4-C$_2$B$_4$H$_6$, produced quantitatively from Me$_3$N and *closo*-1,6-C$_2$B$_4$H$_6$, rearranges slowly (in THF solution) to 3-Me$_3$N-*nido*-2,4-C$_2$B$_4$H$_6$. Large white spheres = boron; small white spheres = hydrogen; black spheres = carbon; textured spheres = nitrogen.

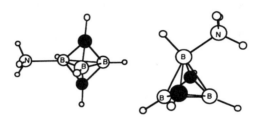

Figure 10-2   Two spatial representations of 2-NH$_3$-1,5-C$_2$B$_3$H$_5$ (geometry optimized at the 6-31G* level of theory; no imaginary frequencies were found from a frequency calculation).

stable adduct, Figure 10-2, which is about 10 kcal/mol more stable than the total energy of the dissociated compounds.[7]

Recent IGLO calculations by us indicates that ^{11}B shifts can be expected at ca. $\delta = -20$ ppm (B—N) and at + 30 to + 32 ppm for the other two borons of the adduct. Our NMR studies on the initially observed solid product between tertiary amines and 1,5-C$_2$B$_3$H$_5$ (dissolved in various solvents) show that there are boron resonances that are consistent with the IGLO predictions of these resonances (taking into account the expected upfield shift when H is substituted for an alkyl group of the amine)[8] along with other resonances, at the initial stages of the reaction, but in time disappear with concurrent intensification of still other resonances. Variable temperature work is obviously needed to clarify the nature of this reaction.

The parent *closo*-2,4-C$_2$B$_5$H$_7$ does not appear to react with Me$_3$N, whereas this same tertiary amine react under ambient conditions with halogenated derivatives of this same five-boron carborane to yield 1:1 adducts (Figure 10-3), which can be converted to [B-Me$_3$N-*closo*-2,4-C$_2$B$_5$H$_6$]$^+$ ions by way of halide extraction.[9,10]

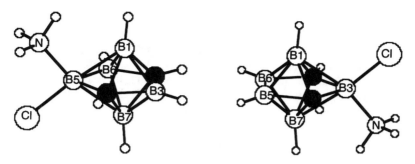

Figure 10-3 Presumed 5-R_3N-5-Cl-2,4-$C_2B_5H_6$ and 3-R_3N-3-Cl-2,4-$C_2B_5H_6$ adduct structures for products from reactions between Me$_3$N and B-Cl- derivatives of $C_2B_5H_7$. Stable adducts with the above structures have also been found from ab initio calculation attempts.[11]

The next higher *closo*-carborane, 1,7-$C_2B_6H_8$, initially forms a very weak adduct with Me$_3$N, but gives rise to substantial quantities of a somewhat stronger adduct upon allowing a mixture to stand at ambient temperature for a day or two (Figure 10-4).[8] However, in the two solvents tetrahydrofuran (THF) and benzene, this latter adduct is in slow equilibrium with the dissociated materials, 1,7-$C_2B_6H_8$ and Me$_3$N, with adduct formation favored to a degree of about 2.5:1 in THF (at ambient temperatures) and the dissociated materials favored to the extent of about 6:1 in benzene.

None of the following four *closo*-carboranes, 1,10-$C_2B_8H_{10}$ and the three *closo*-$C_2B_{10}H_{12}$ isomers (1,2-, 1,7-, 1,12-), appear to react with trimethylamine under reasonable thermal conditions.[1,12] The *closo*-2,3-$C_2B_9H_{11}$ carborane reacts with trimethylamine [13] to afford what is believed to be 3-Me$_3$N-*nido*-7,9-$C_2B_9H_{11}$.

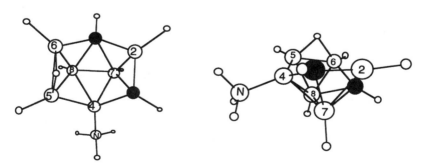

Figure 10-4 Two perspectives of the structure of 4-R_3N-1,3-$C_2B_6H_8$, formed from the interaction of Me$_3$N and *closo*-1,7-$C_2B_6H_8$.

Very recently, we have found that, although there appears to be little or no interaction between trimethylamine and *closo*-1,6-$C_2B_7H_9$ under ambient conditions, the ^{13}C and ^{11}B NMR of a mixture of the two show significant chemical shift changes at reduced temperatures.[14] When examining the *closo*-1,6-$C_2B_7H_9$, without the amine, but in a freon solvent, there is very little indication that low temperatures effect the NMR chemical shifts of the carborane itself. With the amine present, some dramatic chemical shift changes are observed in both the ^{11}B and ^{13}C NMR spectra. The carborane ^{13}C resonance shifts a little over 9 ppm upfield upon lowering the temperature from ambient to $-35°C$; and, although the position of the area-2 ^{11}B resonance remains not far from its original position, the area-4 ^{11}B peak shifts nearly 5 ppm to lower field; also, the area-1 ^{11}B NMR resonance shifts a little over 6 ppm upfield. These chemical shift changes are reversible, and are suggestive of a significant temperature-dependent interaction between the trimethylamine and the $C_2B_7H_9$ carborane.

We embarked on an ab initio calculational effort in order to determine if an amine adduct of the "open" $C_2B_7H_9$ structure could account for the experimental observations. The initial calculational study of such a species necessarily involved NH_3 as the Lewis base rather than NMe_3, owing to obvious computational resource limitations. After subjecting reasonable B—N bonded $NH_3·C_2B_7H_9$ structures to geometry optimizations at both the 3-21G and 6-31G* levels of theory, it was determined that the isomer 3 (Figure 10-5), with the nitrogen of NH_3 axially attached to the open-face boron between the two carbon atoms, is the most stable among structurally related isomers that differ in the position of the NH_3 attachment to any of the boron atoms along the "open" portion of the cage. The IGLO NMR calculations at the double-ζ level on all of the $NH_3·C_2B_7H_9$ adducts were then carried out, and only the most stable one of these adducts (structure 3, Figure 10-5) appears to give ^{11}B and ^{13}C chemical shift results that can satisfactorily account for the experimental NMR observations. At the lowest temperature ($-35°C$) at which the adduct essentially remains in solution, very favorable comparisons between the experimental NMR data and the calculational IGLO/NMR information are found upon assuming that about $25 \pm 1\%$ of the (calculationally) most stable $NH_3·C_2B_7H_9$ adduct (structure 3) is present in rapid equilibrium with the dissociated compounds, NR_3 (R = H, Me for the calculational portion of the study, and R = Me for the experimental portion of the study) and *closo*-$C_2B_7H_9$. Similarly, the results of GIAO/NMR calculations on the *closo*-$C_2B_7H_9$ species averaged in with those on the most stable $NR_3·C_2B_7H_9$ adduct give the best results when about $25 \pm 1\%$ of the adduct is considered in rapid equilibrium with the dissociated *closo*-carborane at $-35°C$. The correlation coefficient derived from a comparison of the experimental results with either the ab initio/IGLO or the ab initio/GIAO calculational results gave correlation coefficients, r^2, of between 0.99 and 1.00.

When the sum of the electronic energies for NH_3 and *closo*-$C_2B_7H_9$ are compared with the most stable $NH_3·C_2B_7H_9$ adduct, it is noticed that at the 3-21G level of theory the difference in energies favors the dissociated materials,

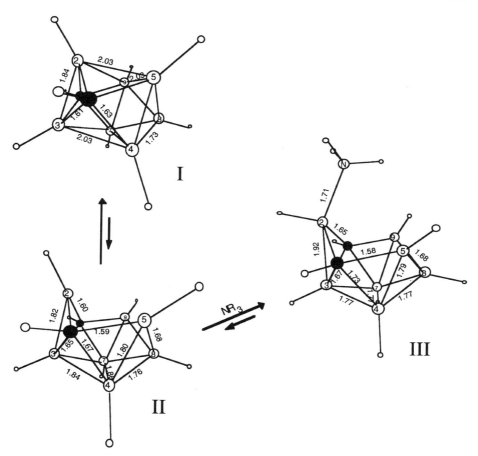

Figure 10-5 Depiction of 6-31G* optimized structures for the *closo*-structure of $C_2B_7H_9$ (structure **1**), the "open" structure of $C_2B_7H_9$ (structure **2**), and the calculationally most stable NR_3 (R = H) *B*-bonded adduct of $C_2B_7H_9$ (structure **3**).

whereas at the 6-31G* level the calculated energy difference is nearly the same (within a couple of kcal/mol). At the correlated levels, the energies begin slightly to favor the adduct. A similar calculational comparison using the open structure of $C_2B_7H_9$ for the starting material does not significantly change this. It is therefore concluded, from theoretical considerations alone, that adduct formation is possible but not necessarily exceptionally favored over dissociation. This is in good agreement with the experimental findings in that ca. 25% of the carborane is found as the adduct at −35°C. Frequency calculations carried out at the 6-31G* level of theory yield an entropy change of −43 cal/mol K for the reaction $NH_3 + closo\text{-}C_2B_7H_9 \rightarrow H_3NC_2B_7H_9$ (structure **3**). For the temperature change of ambient to −35°C, this is equivalent to about 2.6 kcal/mol in favor of the adduct. This is in the range expected (ca. 2+ kcal/mol in favor of the adduct) for a phenomenon in which a 75:25 ratio for the dissociated-carborane:adduct at the lower temperature ($-35°C$) proceeds to mostly dissociated materials (> 98%) when the temperature is raised to ambient.

Of the two freon/$C_2B_7H_9$ samples, the one with freon-12 initially seemed to us to be the more promising in producing a solvent-soluble $NMe_3 \cdot C_2B_7H_9$ adduct at low temperatures; but even then, after adding NMe_3 to the $C_2B_7H_9$/freon-12 mixture, the adduct precipitates out of solution at ca. $-50°C$, and at slightly above that temperature only small ^{11}B and ^{13}C chemical shift changes, favoring less than 5% of the $NMe_3 \cdot C_2B_7H_9$ adduct, are evident (ca. 2–3% from the ^{11}B NMR shift data and about 1–2% from the ^{13}C shift data, each with an estimated 1% error). It is concluded that the effect of the freon solvent permits the temperature to be lowered to ca. $-50°C$ before a significant amount of adduct precipitates out, but that the NMR observations on the dissolved materials are strongly dependent on a "dilution effect." The dilution of the reagents by the freon obviously results in a shift of the equilibrium quantities of the reaction $R_3N + C_2B_7H_9 = R_3N \cdot C_2B_7H_9$ toward the left, which obviously means that less association is taking place in comparison to when trimethylamine is used as the sole solvent.

10.2   CARBON-13 NMR SHIFTS OF CARBORANES; CALCULATIONAL VERSUS EXPERIMENTAL

10.2.1   Introduction

Accounts of ^{13}C NMR chemical shifts of carboranes, following two brief initial reports,[15,16] and one early review,[17] are sparse and have been principally confined to individual chemical shift determinations for a few specific compounds in the course of synthetic studies.[1] Theoretical calculations of carborane ^{13}C chemical shifts are meager, but an important IGLO/NMR study [18] that principally emphasized ^{11}B correlations between theoretical and experimental findings also resulted in ^{13}C chemical shift predictions for a few carboranes. The IGLO study,[18] included ^{13}C information for closo-1,5-$C_2B_3H_5$, closo-1,2-and 1,6-$C_2B_4H_6$, closo-CB_5H_7, nido-2-CB_5H_9, nido-2,3-$C_2B_4H_8$, and nido-2,3,4,5-$C_4B_2H_6$. In the instances of 1,2-$C_2B_4H_6$, CB_5H_7, 2-CB_5H_9, and 2,3,4,5-$C_4B_2H_6$, no previous experimental work was available; but for the remaining three carboranes in this earlier study, the match between theoretical and experimental ^{13}C chemical shifts was found to be within a few ppm at the DZ//MP2/6-31G* (IGLO//ab initio geometry optimization level) level of computation; i.e.; δ (in ppm) = 78.5 (expt), δ = 80.3 (theor.) for 1,6-$C_2B_4H_6$; δ = 122 (expt), δ = 123.9 (theor.) for 2,3-$C_2B_4H_8$; and δ = 103 (expt), δ = 96.9 (theor.) for 1,5-$C_2B_3H_5$.

We extend the ab initio/IGLO/NMR approach to include our recently collected ^{13}C NMR information on more than 40 carborane compounds available to us and to further evaluate the predictive value of the IGLO calculational tool. It was felt important to assess the confidence level for any linear relationship that could be found for ^{13}C chemical shifts—experimentally obtained versus calculated values. In this respect, it should be mentioned that previous studies [19] have shown that ^{13}C shifts have been usefully predicted by the ab initio IGLO/NMR technique for many organic systems.

In order to include several fairly large carboranes (i.e., those with eight or more nonhydrogen atoms), we undertook to correlate the experimental data with IGLO results obtained at the DZ level on geometries optimized at the 6-31G* level of theory. The application of this level of theory toward reasonable-size (up to 12 nonhydrogen atoms) carboranes is attainable in many laboratories without unreasonable computational resource demands. Should high confidence levels for carborane ^{13}C correlations be found at this level of theory, this, along with the already developed and reported ^{11}B NMR correlations,[18] would greatly add to the arsenal of structure proof procedures for polyhedral compounds of carbon and boron. This not only would be helpful in making overall structural assignment(s) to any new carboranes, but also would have the benefit of providing reasonably useful geometrical (bond-distance, bond-angle, and bond-dihedral) parameters as a result of the ab initio geometry optimization procedure that is a part of the overall ab initio//IGLO/NMR scheme.[18] Although this is necessarily "gas-phase"-derived information, it could, in some instances, supplement any diffraction information and/or microwave-derived data, or, in many other instances, provide the initial structural information on a compound.

10.2.2 Results and Discussion

The ab Initio/IGLO/NMR calculated ^{13}C NMR chemical shifts for a considerable number of carboranes, given in Table 10-1, are compared with the corre-

TABLE 10-1 The ^{13}C NMR Chemical Shifts—Experimentally Obtained and Those Obtained from IGLO and GIAO Calculations on ab initio Optimized Geometries (δ Given in ppm Relative to Tetramethylsilane)

Compound	Carbon Position	δ, expt	δ, IGLO DZ// 6-31G*[a]	δ, GIAO 6-31G* //6-31G*
2,3,4,5-$C_4B_2H_6$	C(2,5)	88.9	86.2[g]	77.3
2,3,4,5-$C_4B_2H_6$	C(3,4)	101.4	104.2[g]	92.5
6-Me-2,3,4,5-$C_4B_2H_5$	C_{cage}(2,5)	86.9	82.3	74.2
6-Me-2,3,4,5-$C_4B_2H_5$	C_{cage}(3,4)	99.9	103.6	92.0
6-Me-2,3,4,5-$C_4B_2H_5$	Me	−2.5	−3.4	-3.0
1,5-$C_2B_3H_5$	C(1,5)	103.3[b,c]	94.9[g]	88.0
1,2-$C_2B_4H_6$	C(1,2)	50.3	42.6[g]	37.5
1,6-$C_2B_4H_6$	C(1,6)	78.5[b,c]	76.3[g]	70.6
2-Cl-1,6-$C_2B_4H_5$	C(1,6)	81.1	78.0	69.8

(Contd.)

TABLE 10-1 *(Continued)*

Compound	Carbon Position	δ, expt	δ, IGLO DZ// 6-31G*[a]	δ, GIAO 6-31G* //6-31G*
2,4-Cl$_2$-1,6-C$_2$B$_4$H$_4$	C(1,6)	80.9	79.2	70.1
2,3-C$_2$B$_4$H$_8$	C(2,3)	123.7[b]	121.1[g]	114.4
2-Me-2,3-C$_2$B$_4$H$_7$	C$_{cage}$(2)	141.3	132.8	129.4
2-Me-2,3-C$_2$B$_4$H$_7$	C$_{cage}$(3)	125.7	121.8	114.5
2-Me-2,3-C$_2$B$_4$H$_7$	Me	26.2	21.4	22.8
2,3-Me$_2$-2,3-C$_2$B$_4$H$_6$	C$_{cage}$(2,3)	136.7	131.6	126.8
2,3-Me$_2$-2,3-C$_2$B$_4$H$_6$	Me	20.3	17.8	18.6
[2,4-C$_2$B$_4$H$_7$]$^-$	C(2,4)	75.2	68.3	58.0
3-Me$_3$N-2,4-C$_2$B$_4$H$_6$	C(2,4)	63.4	59.1	51.0
3-Me$_3$N-2,4-C$_2$B$_4$H$_6$	Me	54.1	46.1	45.4
1-Me-B$_5$H$_8$	Me	−11.2[d]	−10.3	−10.5
1-CB$_5$H$_7$	C(1)	58.7	45.5[g]	41.0
2-Me-1-CB$_5$H$_6$	C$_{cage}$(1)	58	46.9	42.5
2-Me-1-CB$_5$H$_6$	Me	-5.5	−5.3	−6.5
2-CB$_5$H$_9$	C(2)	107.6	98.9[g]	91.4
2-Me-2-CB$_5$H$_8$	C$_{cage}$(2)	127.4	111.7	108.6
2-Me-2-CB$_5$H$_8$	Me	24.2	20.3	21.3
3-Me-2-CB$_5$H$_8$	C$_{cage}$(2)	104.8	94.4	87.7
3-Me-2-CB$_5$H$_8$	Me	2.3	1.4	1.2
4-Me-2-CB$_5$H$_8$	C$_{cage}$(2)	106.4	97.4	90.0
4-Me-2-CB$_5$H$_8$	Me	−2.6	−2.8	−2.7
2,4-C$_2$B$_5$H$_7$	C(2,4)	80.0[b,c]	73.9[g]	70.5
1-Me-2,4-C$_2$B$_5$H$_6$	C$_{cage}$(2,4)	83.3	73.9	70.8
1-Me-2,4-C$_2$B$_5$H$_6$	Me	−3.5	−3.7	−4.2
2-Me-2,4-C$_2$B$_5$H$_6$	C$_{cage}$(2)	86.0	80.7	82.9
2-Me-2,4-C$_2$B$_5$H$_6$	C$_{cage}$(4)	82.9	73.1	69.1
2-Me-2,4-C$_2$B$_5$H$_6$	Me	20.6	16.7	17.6
3-Me-2,4-C$_2$B$_5$H$_6$	C$_{cage}$(2,4)	81.3	71.8	69.4
3-Me-2,4-C$_2$B$_5$H$_6$	Me	−2.5	−1.4	−2.5
5-Me-2,4-C$_2$B$_5$H$_6$	C$_{cage}$(2)	80.1	72.1	68.6
5-Me-2,4-C$_2$B$_5$H$_6$	C$_{cage}$(4)	82.7	74.3	71.2
5-Me-2,4-C$_2$B$_5$H$_6$	Me	−3.5	−2.8	−3.1
1-Cl-2,4-C$_2$B$_5$H$_6$	C(2,4)	82.7	74.2	70.7
3-Cl-2,4-C$_2$B$_5$H$_6$	C(2,4)	81.2	73.3	69.1
5-Cl-2,4-C$_2$B$_5$H$_6$	C(2)	78.3	71.0	66.9
5-Cl-2,4-C$_2$B$_5$H$_6$	C(4)	83.7	78.1	73.5
5,6-Me$_2$-2,4-C$_2$B$_5$H$_5$	C$_{cage}$(2,4)	82.1	72.4	69.3
5,6-Me$_2$-2,4-C$_2$B$_5$H$_5$	Me	−3.7	−3.4	−4.0
1,5-Cl$_2$-2,4-C$_2$B$_5$H$_5$	C(2)	80.6	71.6	67.4
1,5-Cl$_2$-2,4-C$_2$B$_5$H$_5$	C(4)	85.7	78.6	73.7
3,5-Cl$_2$-2,4-C$_2$B$_5$H$_5$	C(2)	76.1	69.5	64.5
3,5-Cl$_2$-2,4-C$_2$B$_5$H$_5$	C(4)	83.0	77.2	71.9
5,6-Cl$_2$-2,4-C$_2$B$_5$H$_5$	C(2,4)	82.1	74.9	69.6

(Contd.)

TABLE 10-1 *(Continued)*

Compound	Carbon Position	δ, expt	δ, IGLO DZ// 6-31G*[a]	δ, GIAO 6-31G* //6-31G*
$1,5,6$-Me$_3$-$2,4$-C$_2$B$_5$H$_6$	C$_{cage}$(2,4)	83.2	73.0	70.1
$1,5,6$-Me$_3$-$2,4$-C$_2$B$_5$H$_6$	Me(1)	−4.7	−3.7	−4.5
$1,5,6$-Me$_3$-$2,4$-C$_2$B$_5$H$_6$	Me(5,6)	−4.3	−3.3	−3.9
$1,3,5,6,7$-Me$_5$-$2,4$-C$_2$B$_5$H$_2$	C$_{cage}$(2,4)	83.9	71.2	69.8
$1,3,5,6,7$-Me$_5$-$2,4$-C$_2$B$_5$H$_2$	Me(1,7)	−4.1	−4.3	−5.2
$1,3,5,6,7$-Me$_5$-$2,4$-C$_2$B$_5$H$_2$	Me(5,6)	−2.4	−3.0	−3.9
$1,3,5,6,7$-Me$_5$-$2,4$-C$_2$B$_5$H$_2$	Me(3)	−0.9	−0.4	−1.9
$1,7$-C$_2$B$_6$H$_8$	C(1,7)	72.9	65.4	63.0
[$1,3$-C$_2$B$_6$H$_9$]$^-$	C(1)	34.5	29.3	25.0
[$1,3$-C$_2$B$_6$H$_9$]$^-$	C(3)	76.0	71.5	64.8
[$2,6$-C$_2$B$_6$H$_{11}$]$^-$	C(2)	82.9	79.7	70.0
[$2,6$-C$_2$B$_6$H$_{11}$]$^-$	C(6)H$_2$	−22.4	−21.6	−23.9
$1,6$-C$_2$B$_7$H$_9$	C(1,6)	70.3[c]	64.6	61.7
$1,6$-C$_2$B$_8$H$_{10}$	C(1)	57.7[c]	40.6	40.4
$1,6$-C$_2$B$_8$H$_{10}$	C(6)	35.4[c]	35.1	32.6
$1,10$-C$_2$B$_8$H$_{10}$	C(1,10)	102.9	91.0	89.0
$2,3$-C$_2$B$_9$H$_{11}$	C(2,3)	88.8[c]	83.3	81.0
[$7,8$-C$_2$B$_9$H$_{12}$]$^-$	C(7,8)	ca. 45[e]	35.3	31.1
[$7,9$-C$_2$B$_9$H$_{12}$]$^-$	C(7,9)	ca. 35[f]	24.7	19.3
$1,2$-C$_2$B$_{10}$H$_{12}$	C(1,2)	55.5[b,c]	43.2	41.9
$1,7$-C$_2$B$_{10}$H$_{12}$	C(1,7)	55.1[b,c]	43.8	44.5
$1,12$-C$_2$B$_{10}$H$_{12}$	C(1,12)	63.5[c]	60.6	58.1

[a] The designation before the // marks represents the level of calculation carried out with the IGLO/NMR or GIAO/NMR methods. The designation after the // marks represents the level of calculation used for the geometry optimization procedure using the Gaussian code.

[b] Our experimentally measured values agree reasonably well with those reported earlier by: Olah, G. A., Prakash, G. K. S., Liang, G., Henold, K. L., and Haigh, G. B., *Proc. Natl. Acad. Sci. U.S.A.*, 1977, **74**, 5217: $1,5$-C$_2$B$_3$H$_5$, 103.3; $1,6$-C$_2$B$_4$H$_6$, 78.5; $2,3$-C$_2$B$_4$H$_8$, 122; $2,4$-C$_2$B$_5$H$_7$, 80; $1,2$-C$_2$B$_{10}$H$_{12}$, 56.4; $1,7$-C$_2$B$_{10}$H$_{12}$, 56.3.

[c] Our experimentally measured values for certain carboranes agree quite well with those reported earlier by: Todd, L. J., *Pure Appl. Chem.*, 1972, **30**, 587: $1,5$-C$_2$B$_3$H$_5$, 102.4; $1,6$-C$_2$B$_4$H$_6$, 77.2; $2,4$-C$_2$B$_5$H$_7$, 80; $1,6$-C$_2$B$_7$H$_9$, 70.3; $1,6$-C$_2$B$_8$H$_{10}$, C(1) 57.1, C(6) 33.4; $2,3$-C$_2$B$_9$H$_{11}$, 86.5; $1,2$-C$_2$B$_{10}$H$_{12}$, 55.5; $1,7$-C$_2$B$_{10}$H$_{12}$, 55.4. The value reported for $1,12$-C$_2$B$_{10}$H$_{12}$, $\delta = 63.5$, was not rechecked by us.

[d] Our experimentally measured value for 1-Me-B$_5$H$_8$ agrees well with that, $\delta = -12.2$, reported by Hall, L. W., Lowman, P. D., Ellis, J. D., Odom, J. D., *Inorg. Chem.* 1975, **14**, and 580.

[e] Our experimentally measured value for [$7,8$-C$_2$B$_9$H$_{12}$]$^-$ agrees reasonably well with that, $\delta = 47.8$, reported by Howe, O. V., Jones, C. J., Wiesema, R. J., and Hawthorne, M. F., *Inorg. Chem.*, 1971, **10**, 2516, but is somewhat different than that, $\delta = 38.0$, reported by Clouse, A. O., Doddrell, D., Kahl, S. B., and Todd, L. J., *Chem. Commun.*, 1969, 729.

[f] Our experimentally measured value for [$7,9$-C$_2$B$_9$H$_{12}$]$^-$ agrees reasonably well with that, $\delta = 38.0$, reported by Clouse, A. O., Doddrell, D., Kahl, S. B., and Todd, L. J., *Chem. Commun.*, 1969, 729, but is somewhat different than that, $\delta = 28.8$, reported by Howe, O. V., Jones, C. J., Wiesema, R. J., and Hawthorne, M. F., *Inorg. Chem.*, 1971, **10**, 2516.

[g] This value agrees with that reported earlier by Bühl, M. and Schleyer, P. v. R., *J. Am. Chem. Soc.*, 1992, **114**, 477–491.

sponding experimentally observed chemical shifts in Figure 10-6. The linear correlation between the IGLO/NMR//ab initio calculated ^{13}C NMR chemical shifts and the corresponding experimentally obtained chemical shifts is very good. From the findings in this study, ^{13}C IGLO/NMR calculations carried out at the DZ level on a 6-31G* optimized geometries (and even on geometries optimized at the 3-21G level of theory) of carboranes provide a reasonable estimate of where to expect the experimental ^{13}C chemical shift within the approximately 180 ppm range ($\delta = \approx + 150$ to ≈ -30) in which the reported $\delta(^{13}C)$ values are normally found. Considering that the calculations were

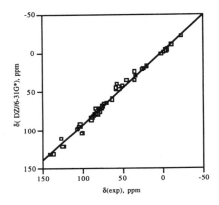

Figure 10-6   The ^{13}C NMR chemical shift comparisons between experimental and IGLO(DZ//6-31G*) calculated values for all the compounds in Table 10-1; $\delta(DZ//6\text{-}31G^*) = 0.941\delta(\text{expt}) - 1.897$ ($r^2 = 0.990$).

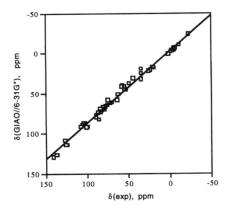

Figure 10-7   The ^{13}C NMR chemical shift comparisons between experimental and GIAO(6-31G*//6-31G*) calculated values for all the compounds in Table 10-1; $\delta(6\text{-}31G^*//6\text{-}31G^*) = 0.893\delta(\text{expt}) - 2.554$ ($r^2 = 0.991$).

carried out at these modest DZ/3-21G (IGLO/Gaussian geometry optimized) and DZ/6-31G* levels of theory, it is satisfying that the slope of each derived relationship $[\delta(DZ/3\text{-}21G) = 0.976\delta(\text{expt}) - 1.908$ $(r^2 = 0.986)$; $\delta(DZ/6\text{-}31G^*) = 0.941\delta(\text{expt}) - 1.897$ $(r^2 = 0.990)]$ is not far from unity, the intercept is not far from the ideal value of zero, and the r^2 value is close to unity. Similar results are found with the application of the GIAO/NMR method [20] to the same compounds (Figure 10-7).

When the splendid ^{13}C $\delta(\text{IGLO})/\delta(\text{expt})$ and ^{13}C $\delta(\text{GIAO})/\delta(\text{expt})$ correlations are considered alongside the previously very acceptable ^{11}B $\delta(\text{IGLO})/\delta(\text{expt})$ correlations among carborane compounds (Figures 10-6 and 10-7), it is clear that this calculational combination should significantly add to the confidence level in the assignment of specific polyhedral geometries for any newly discovered carborane compounds.

ACKNOWLEDGMENTS

I wish to thank the NSF and the MBRS-NIH program for partial support of this project.

REFERENCES

1. Williams, R. E., in *Electron Deficient Boron and Carbon Clusters* (Olah, G. A., Wade, K., and Williams, R. E., eds), Wiley, New York, 1991, Chap. 4, p. 91; Williams, R. E., *Chem. Rev.*, 1992, **92**, 177–207, and references therein. For many of the primary sources, consult also the references in the pertinent volumes of the Boron Compound series of the *Gmelin Handbook of Inorganic and Organometallic Chemistry*, Springer-Verlag, Berlin, in particular: pp. 249–288 of the Gmelin *Handbook of Inorganic Chemistry*, Springer-Verlag, Berlin, 1974, Borverbindungen 2, pp. 1–288; also see *Gmelin Handbook of Inorganic Chemistry*, Springer-Verlag, Berlin, 1975, Borverbindungen 6, pp. 1–150; ibid., 1977, Borverbindungen *11*, pp. 1–207; ibid., 1977, Borverbindungen 12, pp. 1–306; ibid., 1977, Borverbindungen 14, pp. 149–233; ibid., 1977, Borverbindungen 15; ibid., 1977, Borverbindungen 18; *ibid.*, 1978, Borverbindungen 20, pp. 1–239; ibid., 1980, Boron Compounds, 1st Suppl. Vol. 1, pp. 84–108; ibid., 1980, Boron Compounds, 1st Suppl. Vol. 3, pp. 105–256; ibid., 1983, Boron Compounds, 2nd Suppl. Vol. 1, pp. 84–204; ibid., 1982, Boron Compounds, 2nd Suppl. Vol. 2, pp. 223–335; ibid., 1987, Boron Compounds, 3rd Suppl. Vol. 1, pp. 90–240; ibid., 1988, Boron Compounds, 3rd Suppl. Vol. 4, pp. 153–254; T. Onak, ibid., 1993, Boron Compounds, 4th Suppl. Vol. 4, pp. 178–321.
2. Onak, T., Tran, D., Tseng, J., Diaz, M., Arias, J., and Herrera, S., *J. Am. Chem. Soc.*, 1993, **115**, 9210–9215, and references therein.
3. Onak, T., Tran, D., Tseng, J., Diaz, M., Herrera, S. and Arias, J., in *Current Topics in the Chemistry of Boron* (ed. G. W. Kabalka), Royal Society of Chemistry, Cambridge, U.K., 1994, 236–241.
4. McKee, M. L., *Inorg. Chem.*, 1988, **27**, 4241.

5. Lockman, B. and Onak, T., *J. Am. Chem. Soc.*, 1972, **94**, 7923–7924; Onak, T., Lockman, B., and Haran, G., *J. Chem. Soc., Dalton Trans.*, 1973, 2115–2118; Lew, L., Haran, G., Dobbie, R., Black, M. and Onak, T., *J. Organomet. Chem.*, 1976, **111**, 123–130.
6. Burg, A. B. and Reilly, T. J., *Inorg. Chem.*, 1972, **11**, 1962.
7. Onak, T., 1998, in preparation.
8. Onak, T., Tseng, J., Tran, D., Herrera, S., Chan, B., Arias, J., and Diaz, M., *Inorg. Chem.*, 1992, **31**, 3910–3913.
9. Siwapinyoyos, G. and Onak, T., *J. Am. Chem. Soc.*, 1980, **102**, 420.
10. Siwapinyoyos, G. and Onak, T., *Inorg. Chem.*, 1982, **21**, 156–163.
11. Onak, T. and Jaballas J., 1998, in preparation.
12. Onak, T. et al., unpublished studies.
13. Chowdhry, V., Pretzer, W. R., Rai, D. N., and Rudolph, R. W., *J. Am. Chem. Soc.*, 1973, **95**, 4560–4565.
14. Diaz, M., Jaballas, J., Tran, D., Lee, H., Arias, J., and Onak, T., *Inorg. Chem.*, 1996, 35, 4536–4540.
15. Olah, G. A., Prakash, G. K. S., Liang, G., Henold, K. L., and Haigh, G. B., *Proc. Natl. Acad. Sci. U.S.A.*, 1977, **74**, 5217.
16. Todd, L. J., *Pure Appl. Chem.*, 1972, **30**, 587.
17. Wrackmeyer, B., *Carbon-13 NMR Spectroscopy of Boron Compounds*, Progress in Nuclear Magnetic Resonance Spectroscopy, Vol. 12, 1978, pp. 227–259, Springer Verlag, Berlin.
18. Bühl, M. and Schleyer, P. v. R., *J. Am. Chem. Soc.*, 1992, **114**, 477–491.
19. See the many references in footnote 9 of Diaz, M., Jaballas, J., Arias, J., Lee, H., and Onak, T., *J. Am. Chem. Soc.*, 1996, 118, 4405–4410.
20. Wolinski, K., Hinton, J.F., and Pulay, P., *J. Am. Chem. Soc.*, 1990, **112**, 8251; Ditchfield, R., *Mol. Phys.*, 1974, **27**, 789.s

11

MECHANISTIC PATTERNS IN CARBORANE REACTIONS AS REVEALED BY AB INITIO CALCULATIONS

MICHAEL L. MCKEE

Department of Chemistry, Auburn University, 179 Chemistry, Alabama 36849

11.1 INTRODUCTION

Structural rationalization of electron-deficient compounds gave birth to the three-center two-electron bonding concept.[1] Rules for predicting and understanding structure and bonding in boron hydrides and carboranes were invented by Lipscomb (*styx*) [2], Williams [3] and Wade (Wade's Rules).[4] As opposed to electron-rich or electron-precise systems, these electron-poor systems have not received the same degree of scrutiny from the scientific community. This is particularly true for mapping out the conformational or reaction surfaces.

Steps have been taken to develop systematic reactions for cage expansion, cage coupling, and cage contraction. Although the field of synthesis for boron has not developed as far as for its neighbor in the Periodic Table, carbon, there are signs for optimism. In the future, it may be possible to use standard reactions to make cages of arbitrary size with a variety of cage vertices as well as substituents.

Soon after the discovery of carboranes and the nature of their cage structure, it was realized that the carbon vertices, initially adjacent, migrated to non-adjacent positions when the temperature was raised sufficiently.[5] For example, in 1,2-$C_2B_{10}H_{12}$ (or simply *ortho*-carborane), the carbons migrated to the 1,7-positions (*meta*-carborane) at 470°C, and the 1,12-positions (*para*-carborane) at 700°C (Figure 11-1).[6]

$1,2\text{-}C_2B_{10}H_{12}$ $1,7\text{-}C_2B_{10}H_{12}$ $1,12\text{-}C_2B_{10}H_{12}$

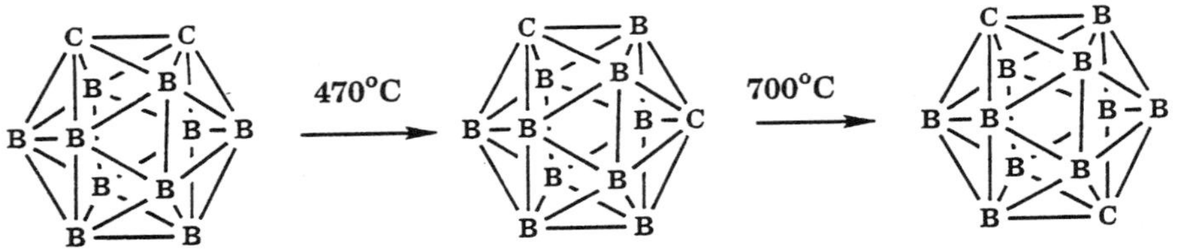

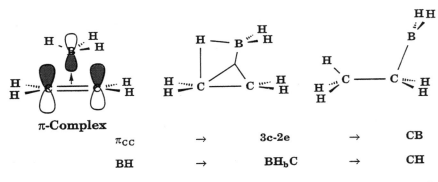

Figure 11-1 The $1,2\text{-}C_2B_{10}H_{12}$ isomer rearranges to more thermodynamically stable products: the 1,7-isomer at 470°C and the 1,12-isomer at 700°C.

These rearrangements and reactions take place by one or more mechanistic steps. The motivation behind this chapter is to discuss the mechanisms of the rearrangement and reaction of several carboranes to see if the reaction steps might fall into identifiable categories.

11.2 MECHANISTIC STEPS

Reactions of boranes and carboranes may involve the following steps.

11.2.1 Hydroboration

The interaction of the empty orbital on boron and the π bond of an alkene or alkyne is stabilizing. A stable adduct may not exist, however, due to the ease of hydrogen migration from boron to carbon (Figure 11-2). When the borane unit is complexed with a Lewis base, such as tetrahydrofuran (THF), the reaction is

Figure 11-2 The empty *p* orbital on a borane interacts with the filled π orbital of an alkene. A C=C π orbital is transformed into a three-electron two-center (3c2e) CBC orbital than into a C—B bond. At the same time, a BH bond is transformed into a BH_bC bridge and then into a C—H bond.

DSD

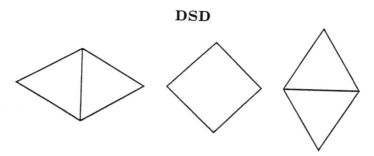

Figure 11-3 The diamond-square-diamond (DSD) rearrangement is a common step in inorganic cluster chemistry.

well understood and leads to anti-Markovnikov addition.[7] As a mechanistic step, hydroboration will be considered to be the addition of boron to one end of a multiple bond and hydrogen to the other.

11.2.2 Diamond-Square-Diamond

The diamond-square-diamond (DSD) step is one of the best-recognized steps in inorganic chemistry.[8,9] Two atoms, originally not connected, are transformed through a square transition state into a bonded pair, while the bond between the other two vertices is broken (Figure 11-3). It has been shown that the DSD step may be accompanied by a HOMO/LUMO crossing. Generally, if a mirror plane is preserved bisecting the DSD face during the single DSD rearrangement, then a HOMO/LUMO crossing will occur.[9]

11.2.3 Local Bond Rotation

A local bond rotation is a multiple DSD process that can be viewed as the local rotation of a bonded pair of atoms around an axis passing through the midpoint of the bonded pair and the center of the complex (Figure 11-4).[10] It has been shown that when a HOMO/LUMO crossing occurs for the single DSD path, and

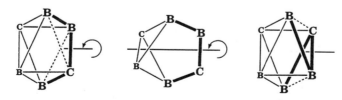

Figure 11-4 The local bond rotation step is a multiple concerted DSD step that can be viewed as a rotation of a local bonded pair of atoms.

if a C_2 axis is retained in the rearrangement process, then there is an avoided crossing with the double DSD rearrangement and the process becomes allowed.[9]

11.2.4 Avoid Crossing

The HOMO/LUMO crossings occur with some regularity in the rearrangements of boron hydride and carborane cages.[11] In organic reactions, "forbidden" reactions usually have high activation barriers due to the presence of large HOMO/LUMO gaps. Reactions tend to find alternative, lower energy pathways. In carborane and boron hydride rearrangements, "forbidden" reactions are sometimes still the lowest energy pathway.[12] In the rearrangement of $1,2\text{-}C_2B_4H_6$ to $1,6\text{-}C_2B_4H_6$, a HOMO/LUMO crossing is encountered between the reactant and a benzvalene-like intermediate (Figure 11-5).[10] However, rather than maintain a plane of symmetry throughout the reaction, the transition state is reached with a slight twist, reducing the C_{2v} symmetry to C_2, mixing the b_1 and b_2 orbitals, and avoiding the orbital crossing.

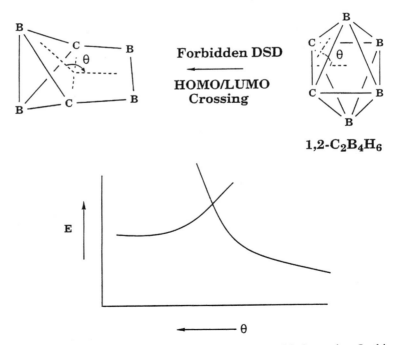

Figure 11-5 A least-motion step may be blocked by an orbital crossing. In this example, a crossing is encountered as θ increases from the $1,2\text{-}C_2B_4H_6$ reactant.

11.2.5 Nonclassic–Classic–Nonclassic

Lipscomb and coworkers [13] originally suggested that classic structures might be transition states. While boron atoms are involved in multicenter bonding in the most stable structure, an orbital vacancy may develop on one or more boron centers in the transition state. A trivial example would be the cleavage of diborane into two borane units. In B_2H_6, both borons have nearly sp^3 hybridization. As the cleavage occurs, an empty orbital develops on each center. In this example, there is not a transition state on the potential energy surface since dimerization occurs without barrier.

Classic structures can be stabilized by coordinating the empty orbital on boron with a Lewis base.[14] Numerous examples of Lewis base adducts exist. For example, Lewis bases preferentially stabilize the less stable (but more classic) structure of the boron hydrides B_3H_7 and B_4H_8. In addition, the substitution of an amino group for hydrogen on a boron center can stabilize the empty orbital by π-donation from nitrogen to boron.

11.2.6 *Closo–Nido–Closo*

If a rearrangement step opens a closed cage into a fragment that resembles a *nido*-structure, the step may be considered a *closo–nido–closo* mechanism.[15] There is no clear distinction between the classic–nonclassic–classic step and the *closo–nido–closo* step. In either case, the transition state is characterized by a structure having fewer multicenter bonds and more orbital vacancies on boron.

Edvenson and Gaines[16] considered an exhaustive list of rearrangements in $C_2B_{10}H_{12}$. They concluded that the isomerization mechanism in closest agreement with experiment involved a 12-vertex *nido*-intermediate.

11.2.7 Double-Bridged to Terminal or Triple-Bridged

The hydrogens of some boron hydrides become fluxional before vertex rearrangement occurs. The bridging hydrogen may rearrange to a new position by either decreasing its coordination number (terminal) or increasing its coordination number (triple-bridging) in the transition state (Figure 11-6). For example, consider the bridging and basal terminal hydrogens in B_6H_{10}, as well as the bridging and equatorial hydrogens in CB_5H_7.[17] In the former case (B_6H_{10}), the pathway is double-bridging to terminal to double-bridging, while in the latter case (CB_5H_7), the pathway is face-bridging to edge-bridging to face-bridging. The double to triple to double mechanism is encountered in the reaction of B_2H_6 and B_3H_7 with BH_3. In both reactions, the transition state has a triple-bridging hydrogen.

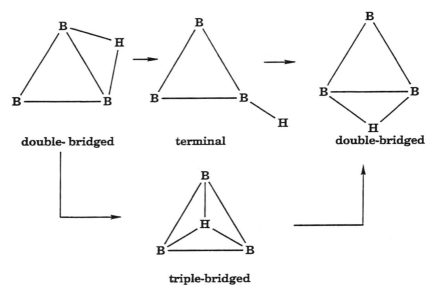

Figure 11-6 Hydrogen migrations may occur by either decreasing the coordination of hydrogen (bridging to terminal) or by increasing the coordination (bridging to triple-bridging).

11.3 COMPUTATIONAL DETAILS

The present generation of computers allows the routine use of reliable levels of theory for systems of six heavy atoms and smaller.[18] Relevant stationary points can be determined and characterized as minima (zero imaginary frequencies) or transition states (one imaginary frequency) and intrinsic reaction paths (IRP) connecting them can be constructed.[19]

To determine the level of theory sufficient for a survey of carborane reactions, various levels were used to calculate the relative energies of the early stages of the diborane pyrolysis (Figure 11-7).

$$B_2H_6 \rightleftharpoons 2BH_3 \tag{1}$$

$$BH_3 + B_2H_6 \rightarrow B_3H_9 \tag{2}$$

$$B_3H_9 \rightarrow B_3H_7 + H_2 \tag{3}$$

The minima and transitions states for Eqs (1)–(3) have recently been calculated at the MP2/6-31G(d,p) level by Stanton et al.[20] The authors commented that polarization functions on hydrogen had little effect on geometries. In this work, the geometries have been recomputed at the MP2/6-31G(d) level. Using these fixed geometries, relative energies were computed with a number of

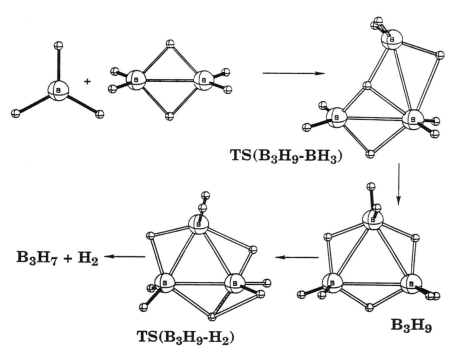

Figure 11-7 The initial step in the pyrolysis of diborane is the addition of BH_3 to B_2H_6, followed by the elimination of H_2 from B_3H_9. The relative energies of these species can be used to determine the reliability of computational methods.

different basis sets at the HF and MP2 levels. If the MP2/6-311 + G(3df,2p) level is assumed to be close to reality, the reliability of the various methods can be assessed (Table 11-1). The MNDO level severely overestimates the stability of three classic BH_3 units by 81.9 kcal/mol relative to nonclassic B_3H_9 (Table 11-1a). Clearly, the potential energy surface will be grossly distorted at the MNDO level, which would probably lead to unrealistic stationary points. Even at the HF/6-311 + G(3df,2p) level, the stability of three boranes is overestimated by 63.0 kcal/mol relative to B_3H_9. The MP2/3-21G level (Table 11-1b) is the first level to give qualitatively correct results. However, the MP2/6-31G(d) level is significantly better at reproducing classic/nonclassic energy differences. Since many mechanisms involve both classic and nonclassical intermediates, the additional reliability compensates for the additional computational expense. Therefore, the MP2/6-31G(d) method was chosen as the standard method for obtaining geometries.

Energies were further improved by making higher level calculations on fixed geometries. Electron correlation was improved by performing a single-point MP4/6-31G(d) calculation, and the basis set flexibility was improved with a

TABLE 11-1a Relative Energies (kcal/mol) at MP2/6-31G(d) Optimized Geometries

	MNDO	HF/ STO-3G	HF/ 3-21G	HF/ 6-31G(d)	HF/ 6-311 + G(3df,2p)
3BH$_3$	−26.0	−4.1	−3.4	3.9	−7.1
B$_2$H$_6$+ BH$_3$	−13.7	−19.2	−17.7	−16.3	−14.6
B$_3$H$_9$	0.0	0.0	0.0	0.0	0.0
TS(B$_3$H$_9$—H$_2$)	17.2	32.6	21.4	24.2	21.1
TS(B$_3$H$_9$—BH$_3$)	16.6	19.5	17.7	17.6	16.6

TABLE 11-1b Relative Energies Including Electron Correlation (kcal/mol) at MP2/6-31G(d) Optimized Geometries

	MP2/ STO-3G	MP2/ 3-21G	MP2/ 6-31G(d)	MP2/ 6-311 + G(3df,2p)
3BH$_3$	18.1	31.8	48.8	55.9
B$_2$H$_6$+ BH$_3$	−7.7	2.0	8.5	11.8
B$_3$H$_9$	0.0	0.0	0.0	0.0
TS(B$_3$H$_9$—H$_2$)	28.5	16.0	16.2	12.5
TS(B$_3$H$_9$—BH$_3$)	22.4	23.3	25.2	23.3

MP2/6-311 + G(d,p) calculation. An estimate of relative energies at the MP4/6-311 + G(d,p) level was made by using Eq. (4).[21]

$$\Delta E([\text{MP4/6-311} + \text{G(d,p)}]) = \\ \Delta E(\text{MP4/6-31G(d)}) + \Delta E(\text{MP2/6-311} + \text{G(d,p)}) - \Delta E(\text{MP2/6-31G(d)}) \tag{4}$$

Vibrational frequencies were determined at the MP2/6-31G(d) level to establish the nature of the potential energy surface and to make zero-point corrections (frequencies weighted by a 0.95 factor) and integrated heat capacity corrections to 298 K (unscaled frequencies). The "standard" level is [MP4/6-311 + G(d,p)] with zero-point and heat capacity corrections. All MP2 and MP4 calculations were made with the "frozen-core" approximation.

In cases where the reactant and product from a given transition structure were not clear, an IRC was constructed [22] at the MP2/6-31G(d) level. Molecular plots are made from calculated geometries. A boldface notation system is used for the species in the figures and text to aid in identification. For example, the bold notation **TS1/2** refers to the transition state between structures **1** and **2**, while the notation **TS6/7 + BH$_3$** refers to the transition state for loss of BH$_3$ from structure **6** to form structure **7**.

One complicating feature of the potential energy surface is that small activation barriers are calculated at the MP2/6-31G(d) level; these barriers disappear when corrections are included. Thus, when corrections are added, the energy of the transition state may be lower than the energy of the reactants or products, which implies either (1) that the location of the transition state is different at the higher level of theory or (2) that the reaction proceeds without barrier in the exothermic direction. In constructing the potential energy surface (PES), intermediates that have very small forward or reverse barriers may not be shown.

11.4   REACTIONS WITH ACETYLENE

11.4.1 $C_2H_2 + BH_3$ and $C_2H_2 + B_2H_6$

The reaction of borane with an unsaturated hydrocarbon, also known as hydroboration, is a reaction of tremendous synthetic utility. [7] The borane, R_2BH, adds hydrogen to the carbon with the least number of hydrogens (anti-Markovnikov addition) and the resulting organoborane can be easily hydrolyzed to an alcohol. There remains some question about the nature of the activated borane. If the borane is present as an ether–borane complex (e.g., $THF—BH_3$), to what extent is the ether interaction diminished in the transition state?

Attention is now turned to computational results at the "standard" level (see Section 11.3, Computational Details). The π-complex (structure **1**) is 3.3 kcal/mol more stable than BH_3 and C_2H_2, and is separated from $H_2B—CH=CH_2$ (structure **2**) by a 2.0 kcal/mol barrier (**TS1/2**).[23] If the BH_3 forms a dimer, B_2H_6, considerably more activation is required. At the MP2/6-31G(d) level, a transition state (**TS3/4**) was located for forming a π-complex intermediate (structure **4**), which was only 0.3 kcal/mol below the transition state. When zero-point corrections are added, the transition state is predicted to be more stable than the intermediate, which suggests that the reaction should proceed in the gas phase with no π-complex. The migration of the $B—H$ hydrogen to carbon (**TS4/5**) occurs with only 2.3 kcal/mol of additional activation.

11.4.2 $C_2H_2 + B_3H_7$

The reaction of B_4H_{10} with acetylene and substituted acetylenes has been studied by several groups.[24] The reaction produces different products depending on the reaction conditions. If moderate heat is used, a number of hydroboration products, such as alkylboranes, as well as *nido*-carboranes, such as CB_5H_9, $C_2B_4H_8$, and $C_3B_3H_7$, are formed. If an AC discharge is used, more compact species, such as *closo*-carboranes, are formed. The initial step in the reaction of B_4H_{10} with C_2H_2 is probably the decomposition of B_4H_{10} to form $B_3H_7 + BH_3$ or $B_4H_8 + H_2$. The next step in the reaction of C_2H_2 with B_3H_7 or B_4H_8 can occur via two main branches: addition and hydroboration.

In the initial decomposition of B_4H_{10} (structure **6**) to form $BH_3 + B_3H_7$ (structure **7**), a triple-bridging hydrogen is encountered in the transition state (**TS6/7+BH$_3$**).[23] The B_3H_7 product can exist in two low-energy minima, a species (structure **7**) with two hydrogen bridges (*styx*-2102) or a species (3.5 kcal/mol higher in energy; structure **7′**) with a single hydrogen bridge (*styx*-1103). The 1103 species (structure **7′**) forms a stronger π-complex with C_2H_2 than the 2102 species (structure **7**) because one boron has an orbital vacancy. A simple rotation of the C_2H_2 group (**TS adduct/8**) leads to a second minimum (structure **8**), essentially without barrier. From this second minimum (structure **8**), a hydroboration step (**TS 8/9**) leads to a number of vinyl-substituted B_3H_7 species (structure **9**, **9′**, and **9″**). The most stable form occurs when the terminal CH_2 group bends over and interacts with a carbon center (structure **9″**). A very

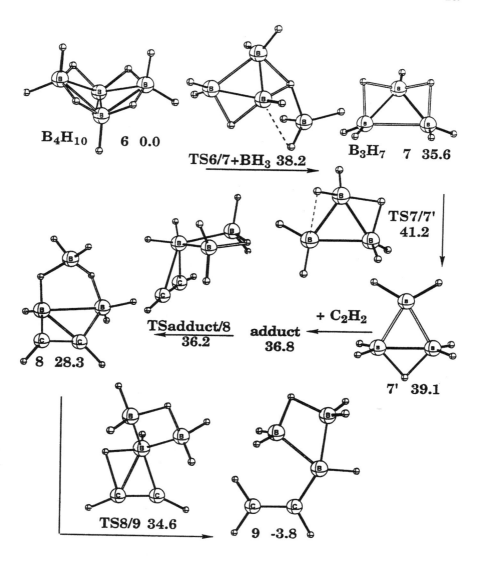

B_4H_{10} 6 0.0

TS6/7+BH$_3$ 38.2

B_3H_7 7 35.6

TS7/7'
41.2

TSadduct/8
36.2

adduct
36.8

+ C$_2$H$_2$

8 28.3

7' 39.1

TS8/9 34.6

9 -3.8

small further barrier (**TS9″/10**) exists for the transfer of a bridging hydrogen to the methylene center to form a terminal CH$_3$ group. A summary of the surface from structure **6** to structure **10** is given in Figure 11-8.

Methylated carboranes are known to be produced from the reaction of C$_2$H$_2$ with B$_4$H$_{10}$.[24n] The structure of the *nido*- compound CH$_3$—CB$_3$H$_6$ was correctly predicted by Williams using stability rules.[3e] It is also the most stable species produced in the hydroboration branch of the reaction of C$_2$H$_2$ with B$_3$H$_7$.[23] The reaction below [Eq. (5)] indicates how one of the observed

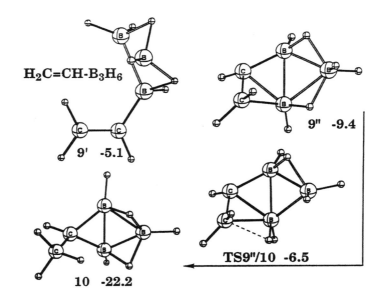

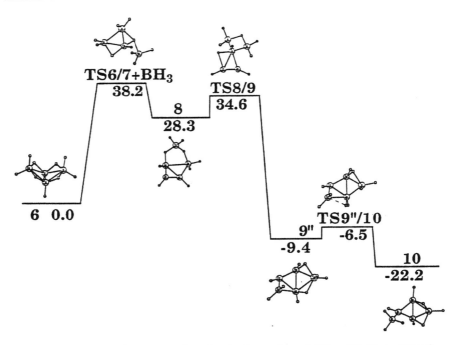

Figure 11-8 Potential energy surface for the formation of $CH_3-CB_3H_6$ (+ BH_3) from B_4H_{10} (+ C_2H_2). Energies are in kcal/mol at the MP4/6-311 + G(d,p) level with zero-point and heat capacity (298 K) corrections.

products, CH_3—$C_3B_3H_6$, [24n] could be produced by the addition of another unit of C_2H_2 and elimination of H_2.

$$CH_3-CB_3H_6 + C_2H_2 \rightarrow CH_3-C_3B_3H_8 \rightarrow CH_3-C_3B_3H_6 + H_2 \qquad (5)$$

When the first step of reaction (5) was modeled with CB_3H_7 rather than CH_3—CB_3H_6, the calculated activation energy was 17.3 kcal/mol and the reaction was exothermic by 21.5 kcal/mol,[25] indicating that the reaction is feasible.

11.4.3 $C_2H_2 + B_4H_8$

Since the elimination of H_2 from B_4H_{10} has a lower barrier than the elimination of BH_3 from B_4H_{10} (**TS6/11 + H_2** versus **TS6/7 + BH_3**),[26,27] the main branch of the C_2H_2 reaction is likely to be with B_4H_8 rather than B_3H_7. This appears to be true since almost all of the observed products (with the exception of CH_3—$C_3B_3H_7$) can be accounted for on this branch of the mechanism. The only minimum found for B_4H_8 at the MP2/6-31G(d) level was a C_1-symmetry structure (**11**) with three bridging hydrogens. However, the B_4H_8 moiety in the transition states for addition of C_2H_2 (**TS11/12** and **TS11/13**) resembles a higher energy stationary point (quadruple-bridged or double-bridged forms). The interaction of C_2H_2 with B_4H_8 stabilizes the higher-energy form through donation of electronic charge to the boron centers. In the absence of the stabilizing influence, the species with four bridging hydrogens or two bridging hydrogens spontaneously distorts to a structure with three bridging hydrogens. The symmetric addition of C_2H_2 to two opposite boron atoms (**TS11/12**) results in the formation of a "basket with a handle" (structure **12**). While this intermediate is unknown, a similar structure, where the handle is a CH_2—CH_2 group rather than CH=CH, is known.[28]

A second mode of C_2H_2 addition, where the C≡C π-bond interacts with one boron rather than two (**TS11/13**), was found to have a lower activation barrier than for addition to two different borons (**TS11/12**); 13.7 versus 21.6 kcal/mol.[27] The product, a *nido*-pentagonal pyramid with two bridging hydrogens around the perimeter (structure **13**), can rearrange to the "basket with a handle" (structure **12**) via a modest activation barrier (**TS13/12**, 22.4 kcal/mol).

The "basket" intermediate (structure **12**) can eliminate H_2 in a concerted step (**TS12/14 + H_2**) to yield the well-known *nido*-$C_2B_4H_8$ carborane (structure **14**). The reaction proceeds in two phases. In the first phase, two bridge hydrogens are converted into H_2 with little change in the remainder of the molecule. In the second phase, leading from the transition state (**TS12/14 + H_2**) to $C_2B_4H_8$ product (structure **14**), the C=C bond lengthens considerably as it interacts with the boron to which both of the departing hydrogens were bridged.[27]

Alternatively, the "basket" intermediate (structure **12**) can lose BH_3 to form a $C_2B_3H_7$ carborane with a five-membered open face and two bridging hydrogens

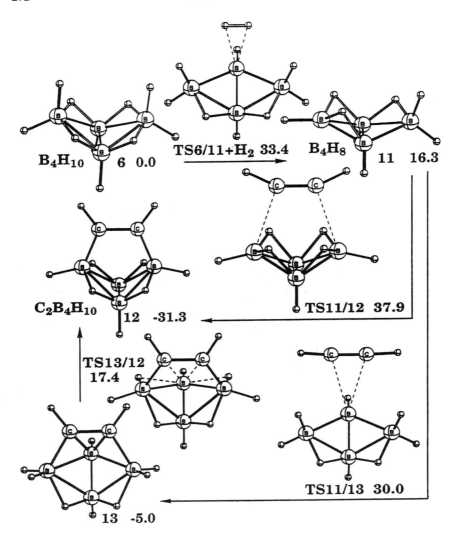

B_4H_{10} **6**   0.0

TS6/11+H_2 33.4

B_4H_8 **11**   16.3

TS11/12 37.9

$C_2B_4H_{10}$ **12**   -31.3

TS13/12 17.4

TS11/13 30.0

13   -5.0

(structure **15**). While many transition metal complexes of this ligand (and the related $C_2B_3H_5^{2-}$ ligand) are known,[29] the free carborane is not known. The rearrangement from the five-membered open ring to the more stable square-pyramidal form of $C_2B_3H_7$ will be presented in Section 11.5.2.

From $C_2B_4H_8$ (structure **14**), there are, again, two pathways: one for loss of H_2 to form *closo*-1,2-$C_2B_4H_6$ (structure **17**) and one for loss of BH_3 to form *closo*-1,2-$C_2B_3H_5$ (structure **18**). As the first step in both pathways, $C_2B_4H_8$

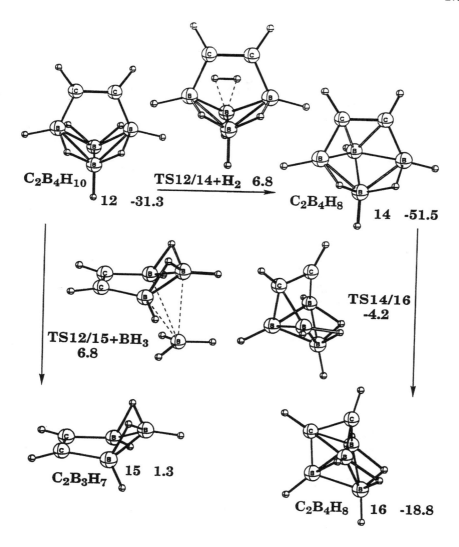

$C_2B_4H_{10}$ **TS12/14+H_2**   **6.8**

12   **-31.3**

$C_2B_4H_8$

14   **-51.5**

TS12/15+BH_3   **6.8**

TS14/16   **-4.2**

$C_2B_3H_7$ **15**   **1.3**

$C_2B_4H_8$ **16**   **-18.8**

rearranges via a 47.3 kcal/mol barrier (**TS14/16**) into another isomer (structure **16**), which is 32.7 kcal/mol higher in energy. The mechanistic step for rearrangement is a "local bond rotation" where the C—C bond pair rotates around an imaginary axis passing through the C—C bond and the remainder of the cage. From this intermediate (structure **16**), a second transition state (**TS 16/17 + H_2**) leads to *closo*-1,2-$C_2B_4H_6$. Elimination of H_2 from the intermediate (structure **16**) would appear straightforward. However, the least-motion path is blocked by a HOMO/LUMO crossing. An alternative transition state (**TS 16/17 + H_2**) was located where a bridging and terminal hydrogen are eliminated

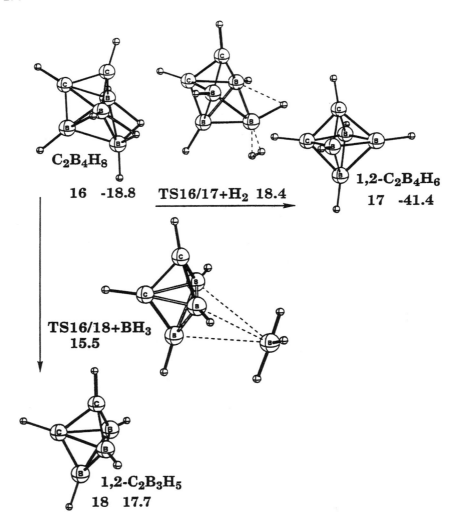

C₂B₄H₈

16   -18.8   **TS16/17+H₂ 18.4** $\longrightarrow$

1,2-C₂B₄H₆

17   -41.4

TS16/18+BH₃
15.5

1,2-C₂B₃H₅

18   17.7

to form molecular hydrogen. The barriers for the two pathways are very similar
(**16 → 17 + H₂** versus **16 → 18 + BH₃**). This section of the PES is summarized
in Figure 11-9.

The hydroboration branch of the reaction of C_2H_2 with B_4H_8 starts off with a
13.7 kcal/mol barrier (**TS 11/19**) to form a vinyl-substituted B_4H_8 species (structure **19**). Similar to the situation found in the $BH_3 + B_3H_7$ reaction,[27] the most
stable $H_2C{=}CH{-}B_4H_7$ species (structure **19′**) is formed when the methylene
group interacts with a boron center. However, unlike the methylene-bridged
$C_2B_3H_9$ case, the formation of a methyl group in $C_2B_4H_{10}$, by migration of a
bridging hydrogen to the methylene group, is not favorable (**19′ → 20**).

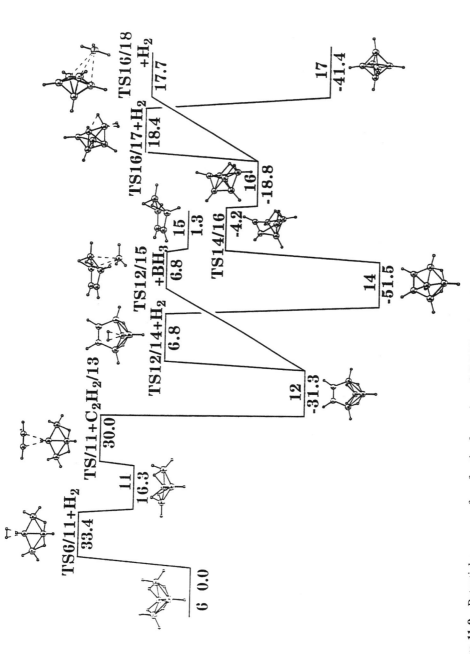

Figure 11-9 Potential energy surface for the formation of $C_2B_4H_6$ (+3H_2) from B_4H_{10} (+C_2H_2). Energies are in kcal/mol at the MP4/6-311+G(d,p) level with zero-point and heat capacity (298 K) corrections.

TS16/18 +H_2 17.7

TS16/17+H_2 18.4

16 -18.8

TS14/16 -4.2

17 -41.4

TS12/15 +BH_3 6.8

15 1.3

14 -51.5

TS12/14+H_2 6.8

TS/11+C_2H_2/13 30.0

TS6/11+H_2 33.4

11 16.3

12 -31.3

6 0.0

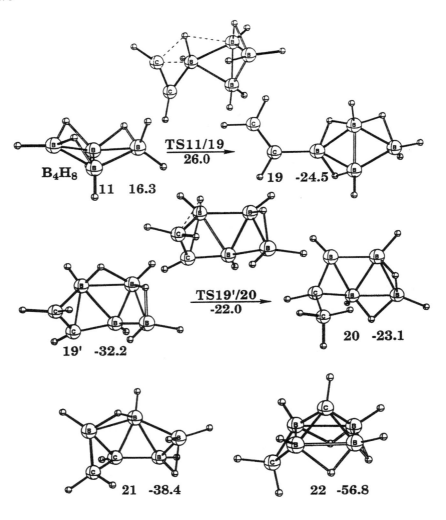

The most stable species on the $C_2B_4H_{10}$ surface is a square- pyramidal structure with a CH in the apical position and three bridging hydrogens and a bridging methylene group around the base (structure **22**). The C—C bond from acetylene has then finally broken. While a continuous reaction path to the square- pyramidal structure has not been found, a possible intermediate may be structure **21**, which is only 18.4 kcal/mol less stable than structure **22** and has a rather long C—C bond. In the study of $C_2H_2 + B_3H_7$, it was found that the C—C bond gradually increases in length through a series of intermediates.

11.4.4 $C_2H_2 + C_2B_3H_7$

Acetylenes are also known to react with stable carboranes such as $C_2B_3H_7$.[30] In section 11.4.2, CH_3—CB_3H_6, a *nido*-carborane with the same electron count as B_4H_8, was found to add C_2H_2 to a boron vertex with a modest activation barrier. In contrast to CB_3H_7, $C_2B_3H_7$ does not have a reactive boron center (i.e., a boron with an empty *p* orbital). The first step in the reaction is elimination of BH_3 from $C_2B_3H_7$ to form $C_2B_2H_4$ (**23** → **24**; no reverse barrier). The addition of C_2H_2 to $C_2B_2H_4$ takes place with little or no activation since (after corrections)

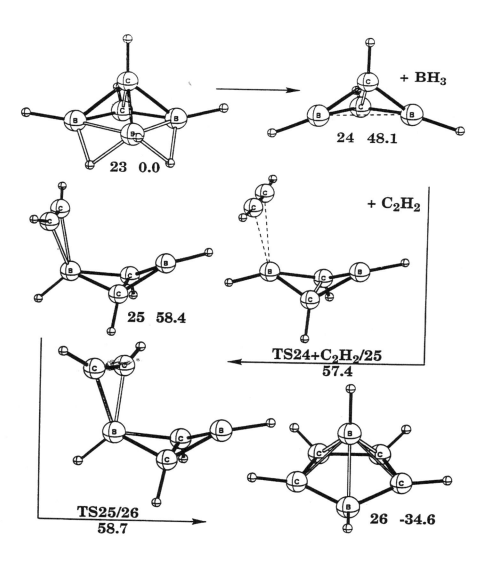

TABLE 11-2   Product Distribution Using Substituted Alkynes in the Reaction of C_2R_2 with $C_2B_3H_7{}^a$

Alkyne Reactant	Product	%
C_2H_2	$C_4B_2H_6$	100
C_2D_2	$2,3\text{-}D_2C_4B_2H_4$	93
	$3,4\text{-}D_2C_4B_2H_4$	7
$CH_3C{\equiv}CH$	$3\text{-}CH_3C_4B_2H_5$	88
	$2\text{-}CH_3C_4B_2H_5$	12
$CH_3C{\equiv}CCH_3$	$2,3\text{-}(CH_3)_2C_4B_2H_4$	87
	$3,4\text{-}(CH_3)_2C_4B_2H_4$	13

^{a} Ref. 30.

the transition state ($TS24 + C_2H_2/25$) is calculated to be lower in energy than the product (structure **25**). In turn, the π-complex rearranges with little or no barrier to form $C_4B_2H_6$ (**25** → **26**). Following the reaction path from the transition state (**TS25/26**) to the $C_4B_2H_6$ product (structure **26**) reveals that the boron to which the acetylene is complexed becomes an equatorial vertex, while the more distant boron swings up to assume the apical position. This rearrangement has consequences when labeled acetylenes are used, as shown in Table 11-2. With dideutero-acetylene, the major product is predicted to be $2,3\text{-}D_2\text{-}2,3,4,5\text{-}C_4B_2H_4$ in agreement with experiment. Similarly, dimethylacetylene results in 87% $2,3\text{-}(CH_3)_2\text{-}2,3,4,5\text{-}C_4B_2H_4$.

11.5   REARRANGEMENTS

The initially formed *closo*-carborane in the reaction of acetylene with a boron hydride often retains the C—C bond intact. For many carboranes (e.g., $1,2\text{-}C_2B_4H_6$, $1,2\text{-}C_2B_{10}H_{12}$), this 1,2-isomer is a kinetic product and rearranges, in a clean reaction, to a more stable isomer with the carbons in nonadjacent positions.[10] Considerable progress has been made by studying rearrangements in which the vertex positions are permutated. One particular approach is to construct a graph table that relates pairs of isomers which can be interconverted by simple DSD mechanisms.[9] By following the graph, one can figuratively follow a reactant through a series of steps to a final product. Orbital crossings may block some steps which may help to explain why some carboranes are fluxional and others are not.

Several carboranes, such as $1,2\text{-}C_2B_4H_6$[10] and $1,2\text{-}C_2B_{10}H_{12}$,[16,31] are known to undergo thermal rearrangement to more stable isomers. In contrast, the $1,2\text{-}C_3B_3H_5$ carborane does not undergo thermal rearrangement to the more stable $1,5\text{-}C_2B_3H_5$ because of the existence of an orbital crossing and high thermal barriers.[32,33]

11.5.1   1,2-C$_2$B$_4$H$_6$ → 1,6-C$_2$B$_4$H$_6$

The 1,2-isomer (structure **17**) is known to rearrange to the 1,6-isomer (structure **28**) with a barrier of about 42–45 kcal/mol.[34] In an earlier ab initio study at the MP4/6-31G(d)//6-31G(d) + ZPC/Cp(298 K) level, an activation barrier of 44.7 kcal/mol was calculated.[10b] For the purposes of this chapter, the rearrangement potential energy surface has been recalculated at the standard level of [MP4/6-311 + G(d,p)]//MP2/6-31G(d) + ZPC/MP2/6-31G(d). The rearrangement was found to occur in two steps with a benzvalene-like intermediate. In the

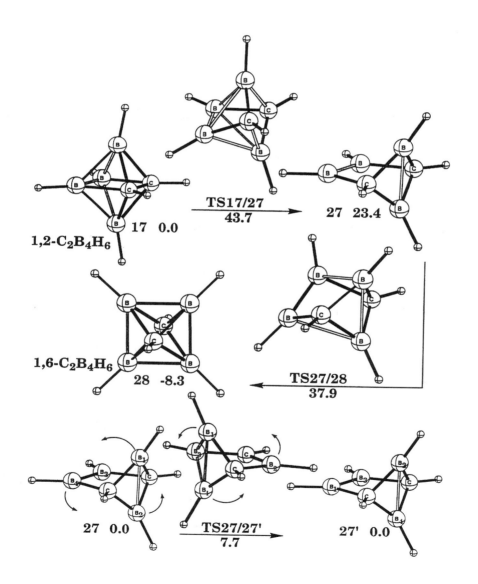

first step, 1,2-$C_2B_4H_6$ (structure **17**) undergoes a modified DSD rearrangement. A direct DSD step that preserves C_{2v} symmetry along the reaction path is blocked by a HOMO/LUMO crossing. The reactant has four b_1 and five b_2 orbitals occupied, while the product has three b_1 and five b_2 orbitals occupied. At the transition state (**TS17/27**), a slight twist lowers the symmetry from C_{2v} to C_2 and mixes the b_1 and b_2 orbitals. The intermediate (structure **27**), which is 23.4 kcal/mol less stable than 1,2-$C_2B_4H_6$, has two boron atoms with vacant p orbitals. An electron-donating group attached to these borons would greatly stabilize the intermediate. In fact, this intermediate has been isolated and characterized when N(i-Pr)$_2$ groups replace hydrogens on boron.[35] The reaction to form 1,6-$C_2B_4H_6$ (structure **28**) occurs by a local bond rotation (**TS27/28**). A second transition state (**TS27/27′**), 7.7 kcal/mol higher than the intermediate, exchanges boron positions.

11.5.2 $C_2B_3H_7$

The $C_2B_3H_7$ carborane (structure **15**) has a surprisingly complex potential energy surface. The 1,2-$C_2B_3H_7$ square pyramid (structure **23**) is the only isomer

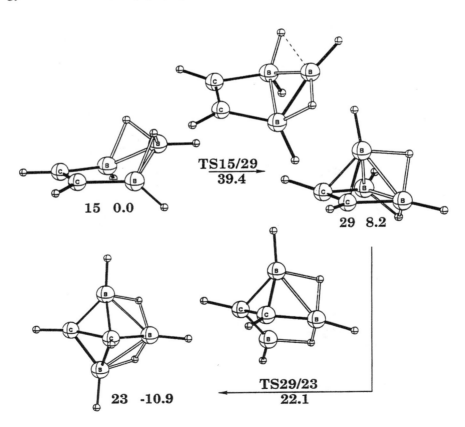

TS15/29
39.4

15 0.0

29 8.2

TS29/23
22.1

23 -10.9

characterized in the free form. It can be recovered in a 10% yield from the reaction of C_2H_2 with B_4H_{10} under mild conditions.[24f,24g] The five-membered open-face $C_2B_3H_7$ isomer (structure **15**) is known, when stabilized with a transition metal.[29]

A pathway for interconverting the two $C_2B_3H_7$ isomers was found earlier;[23] it involved the formation of two intervening bicyclic intermediates. A new pathway has been found that is quite different from the previous one. The highest energy barrier in the alternative pathway is 39.4 kcal/mol, 7.2 kcal/mol lower than the original pathway. An intermediate square pyramid (structure **29**) was found, with the two carbons in the base, which was 19.1 kcal/mol higher in energy than the $1,2$-$C_2B_3H_7$ isomer (structure **23**).

Perhaps most unusual is the pathway leading to a spiro system (structure **32**) in which the central planar boron has lost its hydrogen. A 20.3 kcal/mol activation barrier leads from structure **15** to an intermediate (structure **31**), where the boron at the formerly apical position has a bridging hydrogen but no terminal hydrogen. With further activation of 6.1 kcal/mol, a transition state (**TS31/32**) is reached that leads to the spiro product (structure **32**). It is interesting that the

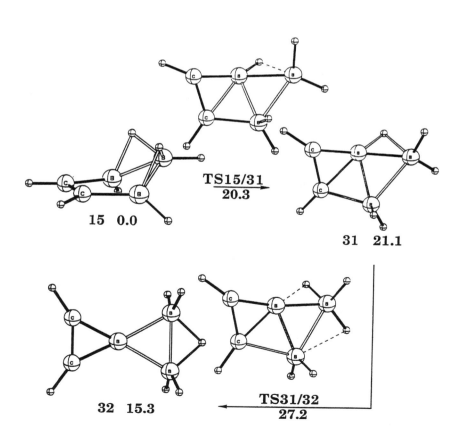

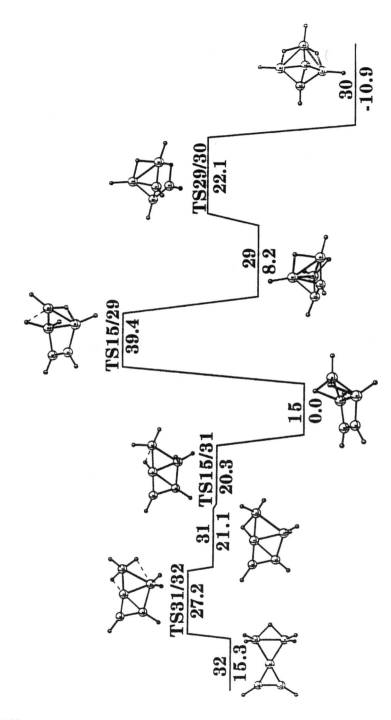

Figure 11-10 Potential energy surface for the rearrangements of $C_2B_3H_7$. Energies are in kcal/mol at the MP4/6-311+G(d,p) level with zero-point and heat capacity (298 K) corrections.

lowest energy pathway for unimolecular reaction of structure **15** forms the novel compound **32** rather than the *nido-* compound **23**. Experimentally, it may be possible to form compound **32** by photolytically cleaving a transition-metal complex of compound **15**, thus allowing free compound **15** to rearrange to compound **32**. A summary of the PES is given in Figure 11-10.

11.6   DECOMPOSITION

11.6.1 $C_2B_4H_{10} \rightarrow C_2B_3H_5 + BH_3 + H_2$

Recently, Köster et al.[24k] reported the formation a derivative of $C_2B_4H_{10}$ (structure **22**) from the condensation of two R_3C—Et units ($R = Et_2B$) that could be converted into a derivative of *closo*-1,5-$C_2B_3H_5$ (structure **35**), either directly or via a $C_2B_4H_8$-type intermediate. While the actual system studied had numerous alkyl substituents, the overall mechanism might be revealed in the parent system (all hydrogen substituents), where the mechanism below is proposed [Eqs (6) and (7)]:

$$C_2B_4H_{10} \rightarrow C_2B_4H_8 + H_2 \tag{6}$$

$$C_2B_4H_8 \rightarrow C_2B_3H_5 + BH_3 \tag{7}$$

In the experimental system, the first step in the indirect path to the derivative of $C_2B_4H_8$ takes place at 110°C (no solvent, low pressure) for a short time (1 h), while the second step to the $C_2B_3H_5$ derivative takes place with additional heating at 150°C. From $C_2B_4H_{10}$ (structure **22**), a transition state (**TS22/33 + H₂**) was located with an activation barrier of 51.8 kcal/mol to form a $C_2B_4H_8$ species (structure **33**). While the first intermediate formed (structure **33**) retains the methylene carbon, only a small activation barrier (**TS33/34**, 1.4 kcal/mol) separates it from a symmetric $C_2B_4H_8$ derivative (structure **34**) with two BHB bridges. From this intermediate (structure **34**), a least-motion symmetry-allowed path is available to form the 1,5-$C_2B_3H_5$ isomer (structure **35**) and BH_3.

Thus, in the study by Köster et al.[24k], a reasonable candidate for the $C_2B_4H_8$ intermediate would be structure **34**. Unfortunately, the identity of the $C_2B_4H_8$ intermediate was revealed, in a private communication from Wrackmeyer, to be not structure **34** but a completely different intermediate. In the parent system, the observed structure (**36**, Figure 11-11) was 4.3 kcal/mol higher than the first $C_2B_4H_8$ intermediate (structure **33**) and 24.1 kcal/mol higher than structure **34**. To determine whether the higher energy of the observed structure (**36**) relative to structure **33/34** might be due to a combination of electronic and steric effects of the ethyl substituents, single-point MP2/6-31G(d) calculations were made by freezing the carbon, boron, and bridging-hydrogen positions of structure **33** and the observed structure (**36**), while replacing the remaining seven hydrogens with methyl groups (optimized at the AM1 level). At this level, **Me₇-36** was 21.0 kcal/mol lower in energy than **Me₇-33**, a reversal of 25.3

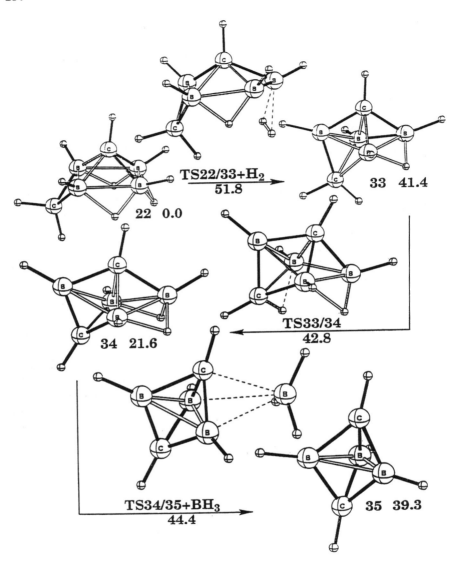

kcal/mol because of the methyl groups! The magnitude of the calculated energy difference also suggests that alkyl substitution would favor structure **36** over structure **34,** and thus explains the experimental observation of substituted structure **36** rather than structure **34.**

While the ethyl substituents may dictate the nature of the $C_2B_4H_8$ intermediate, the mechanism calculated for the parent system may still apply if the rearrangement barrier **36** → **33/34** for the system with ethyl substituents is small.

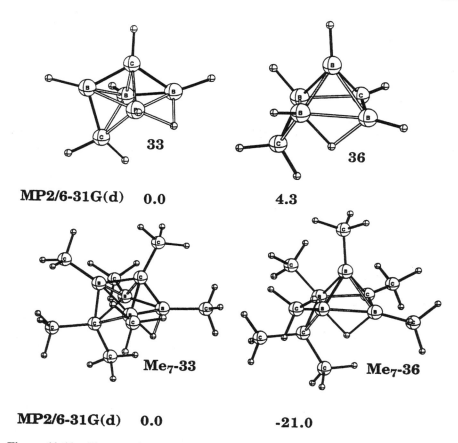

33

36

MP2/6-31G(d)   **0.0**   **4.3**

Me$_7$-33

Me$_7$-36

MP2/6-31G(d)   **0.0**   **-21.0**

Figure 11-11   The experimental intermediate (based on NMR data) in the decomposition of an alkyl-substituted derivative of $C_2B_4H_{10}$ is a derivative of structure **36**. Computational results suggest a different intermediate (structure **34**) for the parent system. While structure **36** is 4.3 kcal/mol less stable than structure **33** for the parent system, single-point calculations on a hepta-methyl-substituted model indicate substantial stabilization in favor of structure **36**.

11.7   CONCLUSIONS

Williams has developed methods for understanding, and predicting, the preferred structure in the boranes and carboranes. These "patterns" can be understood in terms of the vertex connectivity, charge smoothing, and related concepts. The next stage would be to predict mechanistic patterns — predicting reaction steps. At this stage, stabilities of transition states and intervening intermediates will also have to be predicted by heuristic rules. It is hoped that ab initio studies, such as those presented here, will constitute the raw material upon which to develop predictive models.

ACKNOWLEDGMENTS

Computer time for this study was made available by the Alabama Supercomputer Network and the NSF-supported Pittsburgh Supercomputer Center. I thank Dr. Nico J. R. van Eikema Hommes for making the Molecule program available, which was used for drawing the structures. A special acknowledgment is reserved for Paul v. R. Schleyer, who has inspired and encouraged my work over the years.

REFERENCES

1. Longuet-Higgens, H. C., *J. Chem. Phys.*, **46**, 275, 1949.
2. (a) Lipscomb, W. N., *Adv. Inorg. Chem. Radiochem.*, **1**, 117, 1950; (b) Lipscomb, W. N., *Boron Hydrides,* Benjamin, New York, 1963.
3. (a) Williams, R. E., *Adv. Inorg. Chem. Radiochem.*, **18**, 67, 1976; (b) Williams, R. E., *Inorg. Chem.*, **10**, 210, 1971; (c) Williams, R. E., in *Electron Deficient Boron and Carbon Clusters* (G. Olah, K. Wade, and R. E. Williams, eds), Wiley, New York, 1991, pp. 11–93; (d) Williams, R. E., *Chem. Rev.*, **92**, 177, 1992; (e) Williams, R. E., in *Advances in Organometallic Chemistry* (F. G. A. Stone and R. West, eds), Vol. 36, Academic Press, New York, 1994, pp. 1–55.
4. (a) Wade, K., *Chem. Commun.*, 791, 1971; (b) Wade, K., *Adv. Inorg. Chem. Radiochem.*, **18**, 1, 1976.
5. (a) Grimes, R. N., *Carboranes*, Academic Press, New York, 1970, pp. 23–31; (b) Stíbr, B., *Chem. Rev.*, **92**, 225, 1992.
6. King, R. B., *Inorganic Chemistry of Main Group Elements*, VCH Publishers, New York, 1995.
7. Brown, H. C., *Hydroboration*, Benjamin, New York, 1962.
8. Lipscomb, W. N., *Science* (Washington, DC), **153**, 373, 1966.
9. (a) Gimarc, B. M. and Ott, J. J., in *Graph Theory and Topology in Chemistry* (R. B. King and D. H. Rouvray, eds), Elsevier, Amsterdam, 1987, pp. 285–301; (b) Mingos, D. M. P. and Johnston, R. L., *Polyhedron*, **7**, 2437, 1988; (c) Gimarc, B. M. and Ott, J. J., *Main Group Met. Chem.*, **12**, 77, 1989; (d) Mingos, D. M. P. and Wales, D. J., *Introduction to Cluster Chemistry*, Prentice Hall, New Jersey, 1990, pp. 218–248; (e) Mingos, D. M. P. and Wales, D. J., in *Electron Deficient Boron and Carbon Clusters*, (G. A. Olah, K. Wade, and R. E. Williams, eds), Wiley, New York, 1991, pp. 143–163; (f) King, R. B., in *Graph Theoretical Approaches to Chemical Reactivity*, (D. Bonchev and O. Mekenyan, eds), Kluwer Academic Publishers, Dordrecht, 1994, pp. 109–135.
10. (a) McKee, M. L., *J. Am. Chem. Soc.*, **110**, 5317, 1988; (b) McKee, M. L., *J. Am. Chem. Soc.*, **114**, 879, 1992.

11. (a) Gimarc, B. M. and Ott, J. J., *Inorg. Chem.*, **25**, 83, 1986; (b) Gimarc, B. M. and Ott, J. J., *Inorg. Chem.*, **25**, 2708, 1986; (c) Gimarc, B. M. and Ott, J. J., *J. Comput. Chem.*, **7**, 673, 1986.

12. McKee, M. L. and Lipscomb, W. N., *Inorg. Chem.*, **20**, 4148, 1981.

13. Camp, R. N., Marynick, D. S., Graham, G. D., and Lipscomb, W. N., *J. Am. Chem. Soc.*, **100**, 6781, 1978.

14. McKee, M. L., *Inorg. Chem.*, **110**, 4241, 1988.

15. (a) Graham, G. D., Marynick, D. S., and Lipscomb, W. N., *J. Am. Chem. Soc.*, **102**, 2939, 1980; (b) Wong, H. S. and Lipscomb, W. N., *Inorg. Chem.*, **14**, 1350, 1975.

16. Edvenson, G. M. and Gaines, D. F., *Inorg. Chem.*, **29**, 1210, 1990.

17. McKee, M. L., *J. Phys. Chem.*, **93**, 3426, 1989.

18. Langhoff, S. R. (ed.), *Understanding Chemical Reactivity, Quantum Mechanical Structure Calculations with Chemical Accuracy*, Vol. 13, Kluwer Academic Publishers, Boston, 1995.

19. For a description of basis sets see: Hehre, W. J., Radom, L., Schleyer, P. v. R., and Pople, J. A., *Ab Initio Molecular Orbital Theory*, Wiley, New York, 1986.

20. (a) Stanton, J. F., Lipscomb, W. N., Bartlett, R. J., and McKee, M. L., *Inorg. Chem.*, **29**, 109, 1989; (b) Stanton, J. F., Lipscomb, W. N., and Bartlett, R. J., *J. Am. Chem. Soc.*, **111**, 5165, 1989.

21. (a) McKee, M. L. and Lipscomb, W. N., *J. Am. Chem. Soc.*, **103**, 4673, 1981; (b) Nobes, R. H., Bouma, W. J., and Radom, L., *Chem. Phys. Lett.*, **89**, 497, 1982; (c) McKee, M. L., and Lipscomb, W. N., *Inorg. Chem.*, **24**, 762, 1985.

22. McKee, M. L. and Page, M., in *Reviews in Computational Chemistry* (K. B. Lipkowitz and D. B. Boyd, eds), Vol. 4, VCH, New York, 1993, pp. 35–65.

23. McKee, M. L., *J. Am. Chem. Soc.*, **117**, 8001, 1995.

24. (a) Grimes, R. N., *J. Am. Chem. Soc.*, **88**, 1895, 1966; (b) Grimes, R. N. and Bramlett, C. L., *J. Am. Chem. Soc.*, **89**, 2557, 1967; (c) Grimes, R. N., Bramlett, C. L., and Vance, R. L., *Inorg. Chem.*, **7**, 1066, 1968; (d) Grimes, R. N., Bramlett, C. L., and Vance, R. L., *Inorg. Chem.*, **8**, 55, 1969; (e) Bramlett, C. L. and Grimes, R. N., *J. Am. Chem. Soc.*, **88**, 4269, 1966; (f) Franz, D. A. and Grimes, R. N., *J. Am. Chem. Soc.*, **92**, 1438, 1970; (g) Franz, D. A., Miller, V. R., and Grimes, R. N., *J. Am. Chem. Soc.*, **94**, 412, 1972; (h) Franz, D. A. and Grimes, R. N., *J. Am. Chem. Soc.*, **93**, 387, 1971; (i) Greatrex, R., Greenwood, N. N., and Kirk, M., *J. Chem. Soc., Chem. Commun.*, 1510, 1991; (j) Fox, M. A., Greatrex, R., Greenwood, N. N., and Kirk, M., *Polyhedron*, **12**, 1849, 1993; (k) Köster, R., Boese, R., Wrackmeyer, B., and Schanz, H.-J., *J. Chem. Soc., Chem. Commun.*, 1691, 1995; (l) Köster, R., Günter, S., Wrackmeyer, B., and Schanz, H.-J., *Angew. Chem. Int. Ed. Engl.*, **33**, 2294, 1994; (m) Fox, M. A., Greatrex, R., Hofmann, M., and Schleyer, P. v. R., *Angew. Chem. Int. Ed. Engl.*, **33**, 2298, 1994; (n) Fox, M. A. and Greatrex, R., *J. Chem. Soc., Chem. Commun.*, 667 (1995).

25. McKee, M. L., unpublished work.

26. (a) Greenwood, N. N. and Greatrex, R., *Pure Appl. Chem.*, **59**, 857, 1987; (b) Gibbs, T. C., Greenwood, N. N., Spalding, T. R., and Taylorson, D., *J. Chem. Soc., Dalton Trans.*, 1392, 1979; (c) Greenwood, N. N., in *Electron Deficient Boron and Carbon Clusters* (G. A. Olah, K. Wade, and R. E. Williams, eds), Wiley, New York, 1991, pp. 165–181; (d) Greatrex, R., Greenwood, N. N., and Potter, C. D., *J. Chem. Soc.*,

Dalton Trans., 2435, 1984; (e) Greatrex, R., Greenwood, N. N., and Potter, C. D., *J. Chem. Soc., Dalton Trans.*, **81**, 1986; (f) Greenwood, N. N., *Chem. Soc. Rev.*, **49**, 1992.

27. McKee, M. L., *J. Am. Chem. Soc.*, **118**, 421, 1996.

28. (a) Hnyk, D., Brain, P. T., Rankin, D. W. H., Robertson, H. E., Greatrex, R., Greenwood, N. N., Kirk, M., Bühl, M., and Schleyer, P. v. R., *Inorg. Chem.*, **33**, 2572, 1994; (b) Harrisom, B. C., Solomon, I. J., Hites, R. D., and Klein, M., *J. Inorg. Nucl. Chem.*, **14**, 195, 1960; (c) For a methyl-substituted derivative see: Brain, P. T., Bühl, M., Fox, M. A., Greatrex, R., Leuschner, E., Picton, M. J., Rankin, D. W. H., and Robertson, H. E., *Inorg. Chem.*, **34**, 2841, 1995.

29. Many complexes of $C_2B_3H_7$ or $C_2B_3H_5^{2-}$ with different transition metals are known. The following are two examples: (a) Sneddon, L. G., Beer, D. C., and Grimes, R. N., *J. Am. Chem. Soc.*, **95**, 6623, 1973; (b) Spencer, J. T. and Grimes, R. N., *Organometallics*, **6**, 323, 1987.

30. Miller, V. R. and Grimes, R. N., *Inorg. Chem.*, **11**, 862, 1972.

31. (a) Gimarc, B. M., Warren, D. S., Ott, J. J., and Brown, C., *Inorg. Chem.*, **30**, 1598, 1991; (b) Wu, S. and Jones, M., *J. Am. Chem. Soc.*, **111**, 5373, 1989; (c) Wales, D. J., *J. Am. Chem. Soc.*, **115**, 1557, 1993.

32. McKee, M. L., *THEOCHEM*, **168**, 191, 1988.

33. Hofmann, M., Fox, M. A., Greatrex, R., Schleyer, P. v. R., Bausch, J. W., and Williams, R. E., *Inorg. Chem.*, **35**, 6170, 1996.

34. (a) Halgren, T. A., Pepperberg, I. M., and Lipscomb, W. N., *J. Am. Chem. Soc.*, **97**, 1248, 1975; (b) Onak, T., Drake, R. P., and Dunks, G. B., *Inorg. Chem.*, **3**, 1686, 1964.

35. Krämer, A., Pritzkow, H., and Siebert, W., *Angew. Chem., Int. Ed. Engl.*, **29**, 292, 1990.

12

REACTIONS OF UNSATURATED HYDROCARBONS WITH SMALL BORANES: NEW INSIGHTS AND RECENT ADVANCES

Robert Greatrex and Mark A. Fox

School of Chemistry, University of Leeds, Leeds LS2 9JT, U.K.

12.1 INTRODUCTION

This chapter reviews some aspects of recent work in Leeds on gas-phase reactions of binary boranes with unsaturated hydrocarbons and other small molecules. It concentrates on structural rather than mechanistic aspects, and deals, in particular, with a number of cases where, with the benefit of modern instrumentation and techniques, we have been able to resolve inaccuracies that existed in the earlier literature. Several of these were cases for which the reported results were at odds with Williams' structural rules, as discussed in Chapter 1 and elsewhere.[1,2]

The volatile products of the gas-phase reactions were separated by cold-column methods, and the fractions were studied by ^{11}B and ^{1}H NMR spectroscopy. Unstable intermediates were trapped by means of hot–cold reaction techniques or, in selected cases, by the quenching of mass-spectrometrically (MS) monitored gas-phase reactions. The apparatus and the MS techniques were similar to those developed in our earlier kinetic studies of borane decompositions and thermal interconversions,[3,4] but no attempt was made to develop a rigorous quantitative method of analyzing these extremely complex gaseous mixtures, and no detailed kinetic measurements have been made. The value of the semiquantitative technique in detecting fugitive species is illustrated in Figure 12-1, which shows mass spectra of the gaseous mixture at various stages of the reaction of B_4H_{10} and HC≡CH at 70°C. The concentration of the

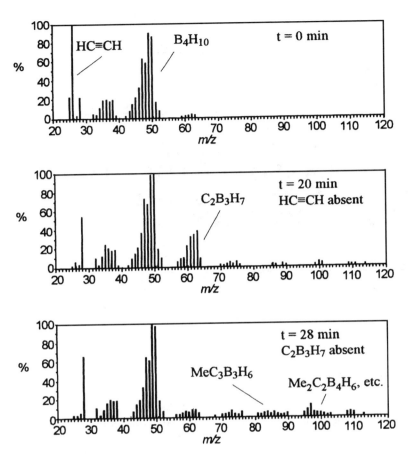

Figure 12-1   Mass spectra of the gaseous mixture of B_4H_{10} and HC≡CH at various stages of the reaction at 70°C. The concentration of *nido*-1,2-$C_2B_3H_7$ is a maximum after 20 min, at which stage all the ethyne has been consumed. The peak at *m/z* 28 is from background N_2. The other products from this reaction are discussed elsewhere.[8]

unstable intermediate *nido*-1,2-$C_2B_3H_7$, structure **1**,[5–7] is seen to reach a maximum when the alkyne (ethyne) concentration has just fallen to zero, and the yield was optimized by quenching the reaction at this stage. Other products from this reaction are discussed elsewhere [8]; note that the raft of peaks near *m/z* 100 arises from derivatives of 2,3-$C_2B_4H_8$, and not, as reported originally [9] and in subsequent papers and reviews,[7,10–12] from derivatives of 2,3,4-$C_3B_3H_7$.

This chapter begins with some comments on the ab initio/IGLO(or GIAO)/NMR technique, which has been an integral part of many aspects of the work. It then deals, in turn, with the following topics:

1. New *arachno*-carboranes from reactions of B_4H_{10} with alkynes.
2. Some controversial *closo*-carboranes.
3. "Basket-like" derivatives from reactions of B_4H_{10} with alkenes.

12.2 THE AB INITIO/IGLO/NMR METHOD

Much of our current work on small carboranes and related systems is benefiting considerably from the use of the combined ab initio/IGLO(or GIAO)/NMR approach. The method was initiated by Kutzelnigg in 1980,[13] and has subsequently been developed and used extensively by Schleyer [14] and others [15] as a powerful tool for the determination of molecular structure. The technique involves three stages. First, geometries of trial structures are optimized using ab initio computations; then, the resulting geometries are used as input for an IGLO(GIAO)/NMR chemical shift calculation; and, finally, the best fit with experimental NMR chemical shifts is deemed to be the best representation for the molecule in solution. The procedure is particularly useful in situations where, for whatever reason, the compound is unsuitable or unavailable for diffraction methods.

Bühl and Schleyer have shown that there is excellent agreement between experimental and calculated IGLO ^{11}B chemical shift values for a wide range of boranes and carboranes.[14] This is indicated in Table 12-1, in which experimentally observed values determined recently in Leeds by the present authors are compared with the calculated ^{11}B NMR chemical shifts. In many cases, the newly determined experimental data give even better agreement than those reported in the original comparison. Two compounds, *nido*-1,2-$C_2B_3H_7$ (struc-

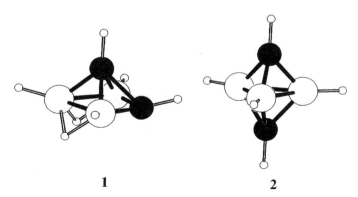

1   **2**

Figure 12-2 *Nido*-1,2-$C_2B_3H_7$ (structure **1**) and *closo*-1,5-$C_2B_3H_5$ (structure **2**), two compounds for which the reported experimental shifts were initially difficult to reconcile with the computed values.[16]

TABLE 12.1   Boron NMR Chemical Shifts of Some Boranes and Carboranes

Compound	ab initio / IGLO[a]	Experimental[b]	Difference
B_2H_6	15.4	18.2	−2.8
B_4H_{10}	−5.3	−6.6	1.3
	−40.0	−41.2	1.2
B_5H_9	−12.8	−13.0	−0.2
	−55.1	−52.6	2.5
B_5H_{11}	av. 8.0	7.3	0.7
	av. −1.6	0.4	−2.0
	−53.9	−54.8	0.9
B_6H_{10}	av. 14.0	av. 14.3	−0.3
	−51.5	−51.4	0.1
B_6H_{12}	23.5	24.0	−0.5
	10.1	10.3	−0.2
	−22.8	−21.1	−1.7
B_8H_{12}	av. 9.4	7.8	1.6
	av. −16.8	−19.3	2.5
	av. −22.3	−22.2	0.1
$1,5\text{-}C_2B_3H_5$	11.4	3.5	7.9
	$(1.9)^c$		$(-1.6)^c$
$1,2\text{-}C_2B_3H_7$	−13.6	−13.4	−0.2
	−15.7	−15.1	−0.6
$1,6\text{-}C_2B_4H_6$	−18.6	−17.5	−0.9
$2,3\text{-}C_2B_4H_8$	0.4	−0.2	0.6
	−1.6	−1.5	0.1
	−54.8	−53.1	−1.7
$2,4\text{-}C_2B_5H_7$	8.1	7.2	0.9
	3.3	3.9	−0.6
	−21.7	−21.5	−0.2
$2,3,4\text{-}C_3B_3H_7^d$	$(-0.3)^e$	0.0	$(-0.3)^e$
	$(-57.9)^e$	−55.2	$(-2.5)^e$
$2,3,4,5\text{-}C_4B_2H_6$	10.9	10.8	0.1
	−62.1	−60.6	−1.5

[a] MP2/6-31G*/II′; [b] In $CDCl_3$, 233 K, 128 MHz ^{11}B NMR; [c] GIAO-MP2/6-31G*/TZP′; [d] Ref. 7; [e] MP2/6-31G*/DZ.

ture **1**) and *closo*-1,5-$C_2B_3H_5$ (structure **2**) are of particular interest (Figure 12-2). Both gave IGLO results that did not match the experimental shifts, until it was shown [16] that the chemical shifts of −21.7 and −23.7 ppm reported [3] for structure **1** were in error by several ppm, and that higher level (GIAO-MP2) calculations were required for structure **2**. Data for the parent tricarbahexaborane *nido*-2,3,4-$C_3B_3H_7$, recently synthesized and characterized unequivocally for the first time in Leeds [17] are also included in Table 12-1.

12.3 NEW *ARACHNO*-CARBORANES FROM REACTIONS OF TETRABORANE(10) WITH ALKYNES

Reactions of binary boranes with alkynes were first studied in the early 1960s, and were found to produce a wide range of products, many in very small yields. The precise nature of the products was shown to depend crucially on the conditions of the reaction.[10] High-energy thermal, electric discharge, and flash reactions were found to generate *closo*-carboranes,[18–22] whereas typical products from reactions under mild conditions included derivatives of the *nido*-carboranes, such as CB_5H_9, $C_2B_4H_8$, and $C_3B_3H_7$ [5–7,10,11,23,24], No *arachno*- or *hypho*-species were observed in this early work, but it seemed likely that such species might play an important role in the mechanisms of these complex reactions as unstable intermediates. Accordingly, the hot–cold reaction between B_4H_{10} and propyne was found to give an unstable compound of composition $C_3B_4H_{12}$ in low yield.[25] Initially, the spectroscopic data were interpreted in terms of a *hypho*-structure, but this proposal was ruled out by subsequent ab initio calculations by Schleyer and his group at Erlangen. Later, it was found that analogous compounds were formed at 70°C in monitored reactions of B_4H_{10} with the alkynes $EtC{\equiv}CH$ and $MeC{\equiv}CMe$, but not with $HC{\equiv}CH$.[26]

The compound of molecular formula $C_4B_4H_{14}$ from the $B_4H_{10}/EtC{\equiv}CH$ reaction gave a ^1H NMR spectrum similar to that of $C_3B_4H_{12}$, except that the presence of an ethyl group in the compound was clearly indicated by the observation of a quartet and a triplet in the spectrum. This suggested that the compound was *arachno*-$EtC_2B_4H_9$, and that $C_3B_4H_{12}$ was probably the methyl analog, *arachno*-$MeC_2B_4H_9$. Subsequent ab initio/IGLO/NMR calculations confirmed that the latter compound was the methylene-bridged species *arachno*-1-Me-2,5-μ-CH_2-1-CB_4H_7 (structure **3**) (Figure 12-3), the smallest known *arachno*-carborane, and the first fully characterized carborane related to *arachno*-B_5H_{11}.

$$B_4H_{10} + MeC{\equiv}CH \xrightarrow{70°C} \textbf{3} + \text{other carboranes}$$

The structure is interesting from a mechanistic point of view in that it represents a snapshot of the cleavage of a $C{\equiv}C$ triple bond by the reactive borane intermediate $\{B_4H_8\}$, itself produced by the initial decomposition of B_4H_{10}.[3] It

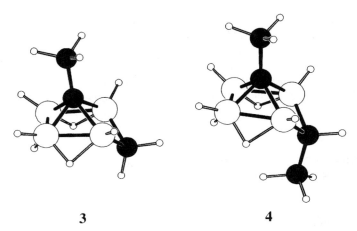

3   **4**

Figure 12-3 Optimized structures of the most stable methyl and dimethyl derivatives, $2,5\text{-}\mu\text{-}CH_2\text{-}1\text{-}Me\text{-}1\text{-}CB_4H_7$ (structure **3**) and $2,5\text{-}\mu\text{-}Me_{endo}CH\text{-}1\text{-}Me\text{-}1\text{-}CB_4H_7$ (structure **4**).[26]

appears that the alkyne undergoes hydroboration, and that, in the process, the borane intermediate transfers a hydrogen atom to the carbon atom that ultimately becomes part of the bridging methylene group. The other carbon atom is incorporated into the apical position of the pyramidal *arachno*-cluster.

A minor isomer, in which the alkyl substituent is located in the methylene bridge rather than at the apical carbon site, is also observed. The ratio of the two isomers was found to be ca. 6:1 for R = Me, and 31:1 for R = Et, and the computed relative energies of the various isomers were found to be consistent with these observations. Analogous $2,5\text{-}\mu\text{-}RCH\text{-}1\text{-}R'\text{-}1\text{-}CB_4H_8$ derivatives were obtained from quenched reactions of B_4H_{10} with the substituted alkynes $RC{\equiv}CR'$, and these are included in Figure 12-4, which summarizes all the known compounds. The optimized structure of the most stable dimethyl derivative, $2,5\text{-}\mu\text{-}Me_{endo}CH\text{-}1\text{-}Me\text{-}1\text{-}CB_4H_7$ (structure **4**) is included in Figure 12-3.

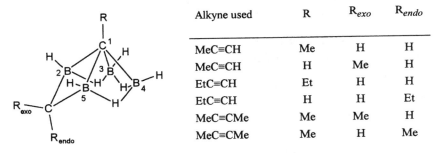

Alkyne used	R	R_{exo}	R_{endo}
MeC≡CH	Me	H	H
MeC≡CH	H	Me	H
EtC≡CH	Et	H	H
EtC≡CH	H	H	Et
MeC≡CMe	Me	Me	H
MeC≡CMe	Me	H	Me

Figure 12-4 Derivatives of *arachno*-$2,5\text{-}\mu\text{-}CH_2\text{-}1\text{-}CB_4H_8$ obtained from quenched reactions of B_4H_{10} with substituted alkynes.[26]

Interestingly, the parent compound, $2,5-\mu-CH_2-1-CB_4H_8$, was not observed among the products of the reaction between B_4H_{10} and $HC{\equiv}CH$. Neither was there any evidence of a "basket-like" derivative analogous to that obtained in the reaction of B_4H_{10} and ethene (see later).

Structures related to compound 3 were first proposed by Köster et al. in 1968 for peralkyl products derived from dehalogenation reactions,[27] and, shortly after that, Matteson and Mattschei reported a compound believed to be CB_4H_{10} (m/z 65, M-1),[28] for which Williams predicted an *arachno*-B_5H_{11}-like structure with carbon at the apex.[1] More recently, Wrackmeyer and colleagues have synthesized polyalkylated derivatives analogous to structure 3,[29] and Siebert and coworkers have reported a system with an *ortho*-phenylene bridge instead of the $CHR_{exo/endo}$ moieties.[30]

12.4 SOME CONTROVERSIAL *CLOSO*-CARBORANES

The first *closo*-carboranes, $1,5-C_2B_3H_5$ (structure 2) (see Figure 12-2), 1,2- and $1,6-C_2B_4H_6$ (structure 5), and $2,4-C_2B_5H_7$ (structure 6) (Figure 12-5), were discovered over 35 years ago [19] and this led Williams to note the following structural patterns [1,2]: (1) skeletal carbon atoms tend to occupy the lowest connected vertices available, and (2) the carbon atoms tend to be separated from each other in the thermodynamically most stable configurations.

However, in 1966 Grimes and coworkers reported a product from high-energy electric discharge and flash reactions of B_2H_6 and B_4H_{10} with alkynes [10,20–22], for which the proposed structures were inconsistent with Williams' empirical rules.[1,2] The compound had a cut-off in the mass spectrum at m/z

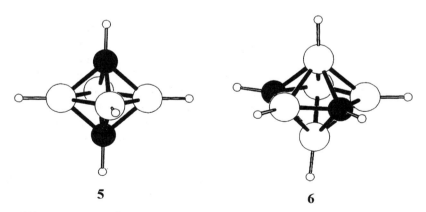

5 **6**

Figure 12-5 $1,6-C_2B_4H_6$ (structure 5) and $2,4-C_2B_5H_7$ (structure 6), two of the first *closo*-carboranes synthesized by Williams et al. (1961–1963).[19] The 1,2-isomer of structure 5 and $1,5-C_2B_3H_5$ (structure 2) (see Figure 12-2) were also synthesized at the time.

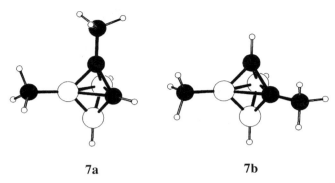

7a   7b

Figure 12-6 Possible structures for the compound formulated by Grimes et al. as *closo*-C,3-Me$_2$-1,2-C$_2$B$_3$H$_3$.[10,20–22] This compound has not been detected in any attempts to repeat the work, and its existence is refuted by ab initio/IGLO computations.[40]

90, and was formulated on the basis of NMR and IR spectra as C,3-Me$_2$-1, 2-C$_2$B$_3$H$_3$ (structure **7a, b**); i.e., one methyl group was believed to be attached to B(3), and the other to one of the carbon atoms (Figure 12-6). Surprise was expressed about the stability of the compound, which, upon heating, did not rearrange into a derivative of the 1,5-carborane (i.e., structure **2**). The ^{11}B NMR data for the compound were also problematic. First, the resonance assigned to the B(3) atom, which was claimed to carry a methyl group, was at higher field than the resonance of B(4) (attached to hydrogen). This was contrary to the relationship that had been found in all other similar pairs of boron atoms.[32,33] Second, the ^{1}H shifts did not parallel the ^{11}B shifts. This differs from the situation that is now recognized as normal, not only for other *closo*-carboranes,[34] but also for other boranes.[35] Third, the ^{1}H shifts of the methyl groups attached to boron and carbon were reported to be the same — a most unlikely occurrence.

In 1969, the same compound was identified by Grimes as being a significant component of a mixture of carboranes that had been generated from the thermal reaction of B$_5$H$_9$ with HC≡CH, and supplied by Williams as part of a scientific exchange.[36] However, Williams' analysis of a similar mixture, albeit by a different GLC apparatus, revealed only known methyl carborane derivatives,[37] among which was 2,3-Me$_2$-1,5-C$_2$B$_3$H$_3$ (structure **8**) [32] (see later). Unfortunately, this particular controversy was not resolved at the time, and the problem was not reinvestigated. In the ensuing years, "*closo*-C,3-Me$_2$-1,2-C$_2$B$_3$H$_3$" became accepted in the literature [38] as a genuine species, and therefore represented an important exception to the generality embodied in the rules referred to earlier.

In 1988, McKee's computational studies on the *closo*-C$_2$B$_3$H$_5$ system showed that the 1,5-isomer is thermodynamically more stable than the 1,2-isomer (by 35.1 kcal/mol),[39] and this was later confirmed by the Erlangen group, not only for the parent carboranes, but also for the dimethyl derivatives.[40]

Furthermore, it was shown that the barrier for the isomerization of the 1,2- to the 1,5-isomer was low (ca. 20 kcal/mol), and that the 1,2-isomer, even if formed, would be most unlikely to survive the high-energy conditions of the flash reaction. In addition, ab initio / IGLO / NMR studies of Bausch, Prakash and Williams,[41] and of the Erlangen group, yielded no evidence to support either the $C,3$-Me_2-$1,2$-$C_2B_3H_3$ structure, or any alternative structure that might have chemical shift values similar to those reported.[20,21] This led to the repetition of some of the original experiments in Leeds in an attempt to clarify the issue.

We attempted to repeat the synthesis of "$C,3$-Me_2-$1,2$-$C_2B_3H_3$" (structure **7a, b**) by a variety of "flash reactions" of B_2H_6 or B_4H_{10} with $HC\equiv CH$, $MeC\equiv CH$, or $MeC\equiv CMe$, but the electric discharge reactions of B_2H_6 and ethyne have not been repeated. Volatile fractions collected from our low-temperature column were obtained, with cut-offs at m/z 90, in comparable yields and giving comparable ^1H NMR spectra to those reported, but no peaks were observed in the ^{11}B NMR spectra with chemical shifts corresponding to the reported values (53.1, 26.0, and 24.4 ppm relative to $BF_3 \cdot Et_2O$). Indeed, the major component of the m/z 90 fraction in each case was identified as $2,3$-Me_2-$1,5$-$C_2B_3H_3$ (structure **8a**) [32] (see Figure 12-7), the same compound that had been implicated in the controversy mentioned earlier.[37] Therefore, it seems likely to us that the compound reported by Grimes and coworkers as $C,3$-Me_2-$1,2$-$C_2B_3H_3$ (structures **7a, b**) is actually structure **8a**. This would explain the unexpectedly high stability of the compound. The $1,2$-Me_2- and $1,5$-Me_2-isomers (structures **8b, c**) were also observed in our experiments, but in much smaller yields than structure **8a**. The latter was identified by Grimes in the reaction mixture, but was indicated to be present in trace amounts only.[21,31] The ^{11}B NMR spectrum reported by Grimes [20,21] remains unexplained.

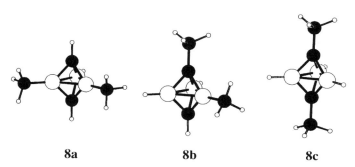

8a   **8b**   **8c**

Figure 12-7   Dimethyl carboranes from the B_4H_{10} / $HC\equiv CH$ flash reaction.[40] The isomer $2,3$-Me_2-$1,5$-$C_2B_3H_3$ (structure **8a**) was identified as the major component of the m/z 90 volatile fraction. Earlier reports suggested that the major component of this fraction was $C,3$-Me_2-$1,2$-$C_2B_3H_3$ (structures **7a, b**) (see Figure 12-6).[10,22]

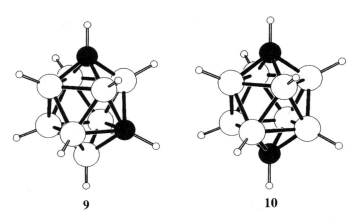

9   **10**

Figure 12-8   The dicarbadecaborane from the B_4H_{10}/HC≡CH flash reaction, originally reported as $1,6\text{-}C_2B_8H_{10}$ (structure **9**),[10,22] is now identified as $1,10\text{-}C_2B_8H_{10}$ (structure **10**).[42]

$$B_2H_6 + RC≡CR' \xrightarrow{\text{flash } 100°C} \textbf{8a} + \text{other carboranes}$$

$$B_4H_{10} + RC≡CR' \longrightarrow \textbf{8a} + \text{other carboranes}$$

$$R, R' = H, Me; R = Me, R' = H$$

The flash reaction of B_4H_{10} with HC≡CH was also reported to give 1, 6-$C_2B_8H_{10}$ (structure **9**) (Figure 12-8) and its derivative 2-Me-1,6-$C_2B_8H_9$.[10,22] These formulations do not fit Williams' "rule" that the most stable *closo*-carboranes contain carbon atoms at the lowest connected vertices. In our own investigations,[42] we have found no evidence for the presence of these two compounds, but have identified, instead, the closely related compound $1,10\text{-}C_2B_8H_{10}$ (structure **10**) [43] and its derivative 2-Me-1,10-$C_2B_8H_9$, which do fit the rule. It would appear that these two compounds were incorrectly identified in the earlier work.

The only other polyhedral carborane that disobeyed Williams' structural patterns was one that had been reported as the first *closo*-tricarbaoctaborane (7), $C_3B_5H_7$ (structure **11**), a most unusual structural proposal in which a "bare" carbon atom was thought to be present.[44] To explain the ^{11}B NMR spectrum of structure **11**, which consisted of three overlapping doublets in an approximate area ratio of 2:2:1, a fluxional system involving dodecahedral equilibrium structures was proposed (Figure 12-9).

In common with C,3-Me_2-1,2-$C_2B_3H_3$ (structures **7a, b**), this unusual species also gained acceptance in the borane literature.[45] However, it was recently subjected to ab initio/IGLO/NMR studies independently by two groups, but no structure could be found that fitted the experimental data.[46,47] We then noted [48] that the NMR data reported for the compound were essentially the same as

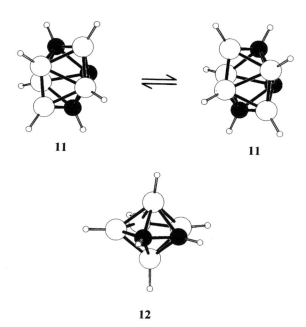

Figure 12-9 Upper: proposed dodecahedral equilibrium structures for the "bare"-carbon carborane "$C_3B_5H_7$" (structure **11**).[44] Lower: the true identity of this species, *closo*-2,3-$C_2B_5H_7$ (structure **12**).[48]

those of 2,3-Me$_2$-2,3-C$_2$B$_5$H$_7$ [49] and 2,3-Et$_2$-2,3-C$_2$B$_5$H$_7$,[50] and when the parent compound, *closo*-2,3-$C_2B_5H_7$ (structure **12**) (see Figure 12-9), was eventually reported,[51] its NMR data were found to give an even closer match. This observation, coupled with the fact that the cut-off in the electron ionization mass spectrum pointed to a compound with this molecular composition, left little doubt that the compound reported as $C_3B_5H_7$ was actually *closo*-2,3-$C_2B_5H_7$ (structure **12**).[48]

It is curious that the vapor density measurements pointed to a molecular weight of ca. 98, and that the cut-off in the chemical ionization mass spectrum, in which methane was used as the reagent gas, occurred at m/z 99.1089. This corresponds very closely to the ion $[C_3B_5H_8]^+$ (calculated m/z 99.1092), which was thought to have arisen from protonation of the parent carborane "$C_3B_5H_7$" by $[CH_5]^+$ produced from CH_4 in the ion source. The excellent agreement between the experimentally determined m/z value and the calculated relative molecular mass is difficult to explain in terms of the presence, in the mass spectrometer, of any species other than $[C_3B_5H_8]^+$, but there is, at present, no convincing explanation for the origin of this ion. The vapor density data also remain a mystery.

This compound, which we now know to be *closo*-2,3-$C_2B_5H_7$ (structure **12**), was produced from the pyrolysis of either *nido*-4- or μ-H$_3$Si-2,3-C$_2$B$_4$H$_7$ at

$$nido\text{-}H_3SiC_2B_4H_7 \xrightarrow{250°C} closo\text{-}2,3\text{-}C_2B_5H_7 \text{ (12)} \xrightarrow{100\%} closo\text{-}2,4\text{-}C_2B_5H_7$$
$$[\text{not } closo\text{-}C_3B_5H_7 \text{ (11)}]$$

$$nido\text{-}2,3\text{-}C_2B_4H_8 \xrightarrow{290–300°C} \{closo\text{-}2,3\text{-}C_2B_5H_7\} \longrightarrow closo\text{-}2,4\text{-}C_2B_5H_7$$
$$\mathbf{12}$$

Scheme 1 A reappraisal of the role of the "bare"-carbon carborane "$C_3B_5H_7$" (structure **11**) [now known to be $closo\text{-}2,3\text{-}C_2B_5H_7$ (structure **12**)] as an intermediate in the conversion of $nido\text{-}2,3\text{-}C_2B_4H_8$ to $closo\text{-}2,4\text{-}C_2B_5H_7$, first reported by Onak, Gerhart, and Williams in 1963.[23]

250°C.[44,52,53] When structure **12** was heated, it was observed to give the isomer $closo\text{-}2,4\text{-}C_2B_5H_7$ in 100% yield. This conversion is now seen to parallel the transformation of $nido\text{-}2,3\text{-}C_2B_4H_8$ to $closo\text{-}2,4\text{-}C_2B_5H_7$, which is initiated at the somewhat higher temperature of 290–300°C.[54]. Apparently, silylation facilitates the transformation to the $closo$-system, and $closo\text{-}2,3\text{-}C_2B_5H_7$ (structure **12**) is probably an intermediate in the transformation of $nido\text{-}2,3\text{-}C_2B_4H_8$ to $closo\text{-}2,4\text{-}C_2B_5H_7$, as indicated in Scheme 1.

In summary, the elimination of these incorrect structural assignments removes all known exceptions to the patterns noted by Williams in the very early days of carborane chemistry, and, similarly, all exceptions to two important trends in the NMR data of $closo$-carboranes. These generalizations, which now resume their rightful undisputed status, are as follows [40]: (1) carbon atoms tend to occupy sites with the lowest connectivity, (2) carbon atoms tend to be nonadjacent in the thermodynamically most stable $closo$-carboranes, (3) the ^{11}B NMR shifts of boron atoms with exo-terminal alkyl groups are always found at lower field than those of otherwise identical boron atoms with exo-terminal hydrogens, and (4) the ^{1}H NMR shifts of exo-terminal hydrogens normally parallel the ^{11}B NMR shifts of the boron nuclei to which they are bound.

12.5 "BASKET-LIKE" COMPOUNDS FROM REACTIONS OF TETRABORANE(10) WITH ALKENES

In contrast to the reactions of B_4H_{10} with alkynes, which generate a wide variety of products in which carbon is incorporated into the cluster (carboranes), the reactions of alkenes are known to give products of a quite different type. Thus, ethene affords the classic "basket-like" compound, 2,4-ethanotetraborane(8), 2,4-$(CH_2CH_2)B_4H_8$ (structure **13**). This was synthesized originally by the hot–cold reaction of B_4H_{10} with ethene,[55] but we now find that it can be isolated in high yield simply by quenching the "normal" gas-phase reaction.[42]

$$B_4H_{10} + CH_2{=}CH_2 \xrightarrow{100°C/0°C} \mathbf{13}$$

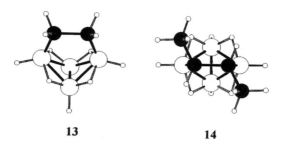

13   **14**

Figure 12-10   The molecular structures of 2,4-$(CH_2CH_2)B_4H_8$ (structure **13**)[56] and 2,4-(*trans*-MeCHCHMe)B_4H_8 (structure **14**)[59] as determined by gas-phase electron diffraction.

The geometry of structure **13** was proposed on spectroscopic evidence,[55] and this has subsequently been endorsed by a gas-phase electron diffraction (GED) study.[56] The molecule consists of a tetraborane(10) cluster, substituted at the "wing-tip" boron atoms by a bridging dimethylene, C_2H_4, moiety (Figure 12-10). All structural parameters and experimental ^1H, ^{11}B, and ^{13}C chemical shifts agreed well with the ab initio (MP2/6-31G*) optimized molecular geometry and the computed IGLO/NMR values. A suggestion [57] that structure **13** might be an intermediate in carborane formation was addressed by a careful study of the gas-phase thermolysis of this compound, but no evidence was found for any interesting carborane species among the volatile products.[42]

Methyl derivatives of structure **13** have been synthesized in the past from reactions of B_4H_{10} with propene and with both *cis*- and *trans*-but-2-ene, but the question of *cis*/*trans* isomerism was not addressed for 2,4-(MeCHCHMe)B_4H_8.[58] We have found that the quenched reaction of B_4H_{10} with *trans*-MeCH=CHMe takes place with retention of conformation to generate exclusively the basket compound 2,4-(*trans*-MeCHCHMe)B_4H_8 (structure **14**) (Figure 12-10).[59]

$$B_4H_{10} + trans\text{-MeCH=CHMe} \xrightarrow{70°C} \mathbf{14}$$

The geometry of structure **14** was confirmed in detail by a GED study. The complete absence of the *cis*-isomer indicates that the incoming ethene moiety interacts simultaneously at B(2) and B(4) of the tetraborane cluster. Likewise, the reaction of B_4H_{10} with *cis*-MeCH=CHMe gives only the *cis*-isomer 2,4-(*cis*-MeCHCHMe)B_4H_8.[60] These observations are consistent with the concerted mechanism proposed originally by Williams and Gerhart for the formation of structure **13**.[61]

The reaction of B_4H_{10} with propene was found to give three "basket-like" derivatives: 2,4-(MeCHCH$_2$)B_4H_8 (structure **15**), 2-Pr-2,4-(MeCHCH$_2$)B_4H_7 (structure **16**), and 4-Pr-2,4-(MeCHCH$_2$)B_4H_7 (structure **17**) (see Figure

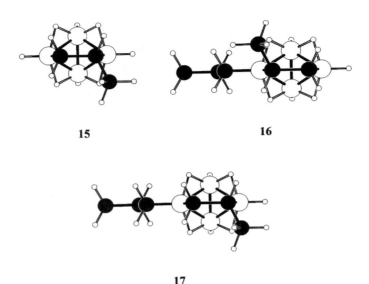

15   16

17

Figure 12-11   Basket-like derivatives from the reaction between B_4H_{10} and propene
[59]: 2,4-$(MeCHCH_2)B_4H_8$ (structure **15**), 2-Pr-2,4-$(MeCHCH_2)B_4H_7$ (structure **16**), and
4-Pr-2,4-$(MeCHCH_2)B_4H_7$ (structure **17**).

12-11).[59] The structure of the major product (compound **15**) was determined
by GED, and all three compounds (**15–17**) have been examined by ab
initio/IGLO/NMR calculations. Interestingly, compound **15** was found not to
react with propene to give compounds **16** or **17**, which suggests that the hydroboration process that introduces the substituents at the wing-tip boron atom must
occur before or during the formation of the dimethylene bridge.

$$B_4H_{10} + MeCH{=}CH_2 \xrightarrow{\text{70°C}} 15 + 16 + 17$$

ACKNOWLEDGMENTS

We thank Dr. Robert E. Williams for his continuing encouragement and interest
in this work, and Professor G. A. Olah for inviting us to participate in the mem-
orable symposium, at the Loker Hydrocarbon Research Institute, on which this
book is based. We also thank Professor Paul Schleyer and Dr Matthias Hofmann
at Erlangen and Dr Michael Bühl at Zurich for ab initio/IGLO/NMR computations, Professor David Rankin and Dr. Paul Brain at Edinburgh for gas-phase
electron diffraction studies, and Professor Russell Grimes for extensive discussions and exchanges of information. We acknowledge the contributions made by
Mr. Darshan Singh, Dr. Martin Kirk, Dr. Alireza Nikrahi, and Ellen Leuschner at

Leeds, and are grateful for financial support from the EPSRC and the Royal Society.

REFERENCES

1. Williams, R. E., *Adv. Inorg. Chem. Radiochem.*, 1976, **18**, 67.
2. Williams, R. E., *Progress in Boron Chemistry*, Vol. 2, Pergamon Press, Oxford, 1970; Williams, R. E., *Chem. Rev.*, 1992, **92**, 177; Williams, R. E., *Adv. Organometal. Chem.*, 1994, **36**, 1.
3. Greenwood, N. N. and Greatrex, R., *Pure Appl. Chem.*, 1987, **59**, 857, and references therein.
4. Greatrex, R., Greenwood, N. N., and Waterworth, S. D., *J. Chem. Soc., Dalton Trans.*, 1991, 643; Cranson, S. J., Greatrex, R., Greenwood, N. N., and Whitehouse, M., *Current Topics in the Chemistry of Boron* (ed. G. W. Kabalka), Royal Society of Chemistry, Cambridge, 1994, p. 392, and references therein.
5. Franz, D. A. and Grimes, R. N., *J. Am. Chem. Soc.*, 1970, **92**, 1438.
6. Franz, D. A., Miller, V. R., and Grimes, R. N., *J. Am. Chem. Soc.*, 1972, **94**, 412.
7. Franz, D. A. and Grimes, R. N., *J. Am. Chem. Soc.*, 1971, **93**, 387.
8. Fox, M. A. and Greatrex, R., *J. Chem. Soc., Chem. Commun.*, 1995, 667.
9. Bramlett, C. L. and Grimes, R. N., *J. Am. Chem. Soc.*, 1966, **88**, 4269.
10. Grimes, R. N. and Bramlett, C. L., *J. Am. Chem. Soc.*, 1967, **89**, 2557.
11. Grimes, R. N., Bramlett, C. L., and Vance, R. L., *Inorg. Chem.*, 1968, **7**, 1066.
12. Franz, D. A., Howard, J. W., and Grimes, R. N., *J. Am. Chem. Soc.*, 1969, **91**, 4010; Howard, J. W. and Grimes, R. N., *Inorg. Chem.*, 1972, **11**, 263; Grimes, R. N., *Carboranes*, Academic Press, London, 1970, pp. 24–31; *Gmelin Handbuch der Anorganische Chemie*, Vol. 15, Borverbindungen 2, Springer-Verlag, New York, 1974, pp. 153–155.
13. Kutzelnigg, W., *Isr. J. Chem.*, 1980, **19**, 193; Kutzelnigg, W., *J. Chem. Phys.*, 1982, **76**, 1919; Kutzelnigg, W., Schlindler, M., and Fleischer, U., *NMR Basic Princ. Prog.*, 1990, **23**, 165.
14. Bühl, M. and Schleyer, P. v. R., *J. Am. Chem. Soc.*, 1992, **114**, 477.
15. For the many examples of this method on boron compounds see Ref. 7 in: Diaz, M., Jaballas, J., Arias, J., Lee, H., and Onak, T., *J. Am. Chem. Soc.*, 1996, **118**, 4405.
16. Schleyer, P. v. R., Gauss, J., Bühl, M., Greatrex, R., and Fox, M. A., *J. Chem. Soc., Chem. Commun.*, 1993, 1766.
17. Fox, M. A., Greatrex, R., and Nikrahi, A., *Chem. Commun.*, 1996, 175.
18. Ditter, J. F., Klusmann, E. B., Oakes, J. D., and Williams, R. E., *Inorg. Chem.*, 1970, **9**, 889.
19. Shapiro, I., Good, C. D., and Williams, R. E., *J. Am. Chem. Soc.*, 1962, **84**, 3837; Shapiro, I., Keilin, B., Williams, R. E., and Good, C. D., *J. Am. Chem. Soc.*, 1963, **85**, 3167; Williams, R. E., Good C. D., and Shapiro, I., 140th Meeting of the American Chemical Society, Chicago, Sept. 1961, 14N, p. 36.
20. Grimes, R. N., *J. Am. Chem. Soc.*, 1966, **88**, 1070.
21. Grimes, R. N., *J. Am. Chem. Soc.*, 1966, **88**, 1895.

22. Grimes, R. N., Bramlett, C. L., and Vance, R. L., *Inorg. Chem.*, 1969, **8**, 55.

23. Onak, T., Williams, R. E., and Weiss, H. G., *J. Am. Chem. Soc.*, 1962, **84**, 2830; Onak, T., Gerhart, F. J., and Williams, R. E., *J. Am. Chem. Soc.*, 1963, **85**, 3378.

24. Onak, T., Dunks, G. B., Spielman, J. R., Gerhart, F. J., and Williams R. E., *J. Am. Chem. Soc.*, 1966, **88**, 2061.

25. Greatrex, R., Greenwood, N. N., and Kirk, M., *J. Chem. Soc., Chem. Commun.*, 1991, 1510; Fox, M. A., Greatrex, R., Greenwood, N. N., and Kirk, M., *Polyhedron*, 1993, **12**, 1849.

26. Fox, M. A., Greatrex, R., Hofmann, M., and Schleyer, P. v. R., *Angew. Chem.*, 1994, **106**, 2384; *Angew. Chem. Int. Ed. Engl.*, 1994, **33**, 2298.

27. Köster, R., Benedikt, G., and Grassberger, M. A., *Justus Liebigs Ann. Chem.*, 1968, **719**, 187.

28. Matteson, D. S. and Mattschei, P. K., *Inorg. Chem.*, 1973, **12**, 2472.

29. Köster, R., Seidel, G., and Wrackmeyer, B., *Angew. Chem.*, 1994, **106**, 2380; *Angew. Chem. Int. Ed. Engl.*, 1994, **33**, 2294; Köster, R., Boese, R., Wrackmeyer, B., and Schanz, H.-J., *J. Chem. Soc., Chem. Commun.*, 1995, 1961.

30. Gangnus, B., Stock, H., Siebert, W., Hofmann, M., and Schleyer, P. v. R., *Angew. Chem.*, 1994, **106**, 2383; *Angew. Chem. Int. Ed. Engl.*, 1994, **33**, 2296.

31. Grimes, R. N., *J. Organometal. Chem.*, 1967, **8**, 45.

32. Dobbie, R. C., DiStefano, E. W., Black, M., Leach, J. B., and Onak T., *J. Organometal. Chem.*, 1976, **114**, 233.

33. See, for example: Ditter, J. F., Klusmann, E. B., Williams, R. E., and Onak, T., *Inorg. Chem.*, 1976, **15**, 1063; Onak, T., Fung, A. P., Siwapinyoyos, G., and Leach J. B., *Inorg. Chem.*, 1979, **18**, 2878; Wilczynski, R. and Sneddon, L. G., *Inorg. Chem.*, 1982, **21**, 506.

34. See, for example: Gotcher, A. J., Ditter, J. F., and Williams, R. E., *J. Am. Chem. Soc.*, 1973, **95**, 7514; Onak, T. and Wan, E., *J. Chem. Soc., Dalton Trans.*, 1974, 665.

35. Kennedy, J. D., in *Multinuclear NMR* (Mason, J., ed.), Plenum Press, New York, 1987, p. 250.

36. Williams, R. E., personal communication.

37. Williams, R. E., *Pure Appl. Chem.*, 1972, **29**, 569.

38. Burg, A. B. and Reilly T. J., *Inorg. Chem.*, 1972, **11**, 1962; Grimes, R. N., *Carboranes*, Academic Press, London, 1970, pp. 34–36; Köster, R. and Grassberger, M. A., *Angew. Chem. Int. Ed. Engl.*, 1967, **6**, 224; *Gmelin Handbuch der Anorganische Chemie*, Vol. 15, Borverbindungen 2, Springer-Verlag, New York, 1974, pp. 165–166; Eaton, G. R. and Lipscomb, W. N., *NMR Studies of Boron Hydrides and Related Compounds*, W. A. Benjamin, New York, 1969, pp. 307–310; Onak, T., in *Comprehensive Organometallic Chemistry* (Wilkinson, G., Stone, F. G. A., and Abel, E. W., eds.) Vol. 1, Pergamon, Oxford, 1982, p. 413; Grassberger, M. and Köster, R., *Methoden der Organischen Chemie* (Köster, R., ed.) Vol. 2, Organoborverbindungen III, Georg Thieme Verlag, Stuttgart, 1984, p. 164; Štíbr, B., *Chem. Rev.*, 1992, **92**, 234.

39. McKee, M. L., *J. Mol. Struct. (THEOCHEM)*, 1988, **168**, 191.

40. Hofmann, M., Fox, M. A., Greatrex, R., Schleyer, P. v. R., Bausch, J. W., and Williams, R. E., *Inorg. Chem.*, 1996, **35**, 6170.

41. Bausch, J. W., Prakash, G. K. S., and Williams, R. E., presented at the BUSA-II Meeting, Research Triangle, NC, June 1990.

42. Fox, M. A. and Greatrex, R., unpublished results.

43. Tebbe, F. N., Garrett, P. M., Young, D. C., and Hawthorne, M. F., *J. Am. Chem. Soc.*, 1966, **88**, 609; Tebbe, F. N., Garrett, P. M., and Hawthorne, M. F., *J. Am. Chem. Soc.*, 1968, **90**, 869; Garrett, P. M., Smart, J. C., Ditta, G. S., and Hawthorne, M. F., *Inorg. Chem.*, 1969, **8**, 1907.

44. Thompson, M. L. and Grimes, R. N., *J. Am. Chem. Soc.*, 1971, **93**, 6677.

45. Onak, T., *Organometal. Chem.*, 1972, **1**, 106; Onak, T., in *Comprehensive Organometallic Chemistry* (Wilkinson, G., Stone, F. G. A., and Abel, E. W. eds), Vol. 1, Pergamon, Oxford, 1982, pp. 415–417; *Gmelin Handbuch der Anorganische Chemie*, Boron Compounds 1st Suppl., Vol. 3, Springer-Verlag, New York, 1981, p. 154.

46. Bausch, J. W., Prakash, G. K. S., and Williams, R. E., *Inorg. Chem.*, 1992, **31**, 3763.

47. Schleyer, P. v. R., personal communication.

48. Fox, M. A. and Greatrex, R., *J. Chem. Soc., Dalton Trans.*, 1994, 3197.

49. Rietz, R. R. and Schaeffer, R., *J. Am. Chem. Soc.*, 1971, **93**, 1263; ibid., 1973, **95**, 6254.

50. Beck, J. S., Kahn, A. P., and Sneddon, L. G., *Organometallics*, 1986, **5**, 2552; Beck, J. S. and Sneddon, L. G., *Inorg. Chem.*, 1990, **29**, 295.

51. Bausch, J. W., Matoka, D. J., Carroll, P. J., and Sneddon, L. G., *J. Am. Chem. Soc.*, 1996, **118**, 11423.

52. Thompson, M. L. and Grimes, R. N., *Inorg. Chem.*, 1972, **11**, 1925.

53. Compound **11** was also reported from thermolysis at 150°C of μ,μ'-$SiH_2(C_2B_4H_7)_2$: Tabereaux, A. and Grimes, R. N., *Inorg. Chem.*, 1973, **12**, 792.

54. Onak, T., Drake, R. P., and Dunks G. B., *Inorg. Chem.* 1964, **3**, 1686.

55. Harrison, B. C., Solomon, I. J., Hites, R. D., and Klein M. J., *J. Inorg. Nucl. Chem.*, 1960, **14**, 195.

56. Hynk, D., Brain, P. T., Rankin, D. W. H., Robertson, H. E., Greatrex, R., Greenwood, N. N., Kirk, M., Bühl, M., and Schleyer, P. v. R., *Inorg. Chem.*, 1994, **33**, 2572.

57. DeKock, R. L., Fehlner, T. P., Housecroft, C. E., Lubben, T. V., and Wade, K., *Inorg. Chem.*, 1982, **21**, 25.

58. Onak, T., Gross, K., Tse, J., and Howard, J., *J. Chem. Soc., Dalton Trans.*, 1973, 2633.

59. Brain, P. T., Bühl, M., Fox, M. A., Greatrex, R., Leuschner, E., Picton, M. J., Rankin, D. W. H., and Robertson, H. E., *Inorg. Chem.*, 1995, **34**, 2841.

60. Brain, P. T., Bühl, M., Fox, M. A., Greatrex, R., Hnyk, D., Nikrahi, A., Rankin, D. W. H., and Robertson, H. E., *J. Mol. Struct.*, accepted for publication.

61. Williams, R. E. and Gerhart, F. J., *J. Organometal. Chem.*, 1967, **10**, 168.

13

RECENT ADVANCES IN METALLACARBORANE SANDWICH CHEMISTRY: CONTROLLED LINKING, STACKING, AND COOL FUSION

RUSSELL N. GRIMES

Department of Chemistry, University of Virginia, Charlottesville, Virginia 22901

13.1 BACKGROUND

Boron has played a central role in multidecker sandwich chemistry, almost from its inception. Although this area was launched by the synthesis [1] of an NMR- and IR-characterized $Cp_3Ni_2^+$ cation whose proposed triple-decker structure was later established via X-ray crystallography,[2] the first stable (and the first structurally characterized) triple-decker complexes were $CpCo(C_2B_3)CoCp$ metallacarboranes.[3] These compounds were prototypes of a broad boron-based multidecker chemistry that extensively overlaps the carborane and organoborane fields and involves most of the transition elements [4]. Beyond the triple-decker level, the presence of boron as a stabilizing element has thus far been indispensable: at present, all known "linear" molecular sandwiches (those in which the metal centers are aligned) of four or more decks feature boron-containing heterocycles, such as C_2B_3, C_3B_2, or C_2B_2S rings.[4a,b,d] These can be combined in the same molecule, as in the "hybrid" sandwiches that incorporate both diborolyl (C_3B_2) and carborane (C_2B_3) ring ligands.[5]

Figure 13-1 outlines the general routes that we have employed to generate a wide range of carborane-bridged multideckers, the details of which have been outlined in several recent reviews.[4d,6] In this chapter, I summarize some recent findings in our laboratory and show how they shed new light on earlier work and suggest new directions in synthesis.

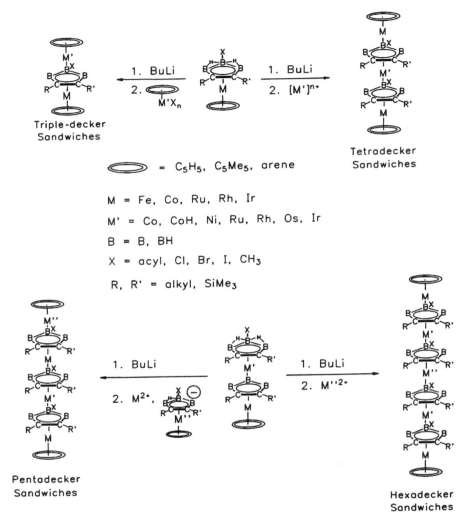

Figure 13-1   Synthetic approaches to multidecker sandwiches.

13.2   IRON-CENTERED TETRADECKERS AND OXIDATIVE FUSION

The approach that we have employed to generate tetradecker sandwich complexes (Figure 13-1, top) utilizes double-decker synthons of the type $LM(RR'C_2B_3H_4-X)$ (where L is a hydrocarbon ring, M is a transition metal, X is an electron-withdrawing substituent, and R, R' are alkyl or $SiMe_3$). These complexes are readily bridge-deprotonated to generate mono- or dianions that are

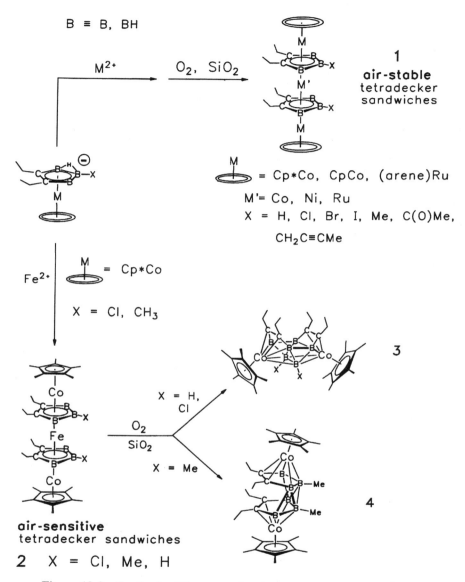

Figure 13-2 Synthesis of Fe-centered and other tetradecker sandwiches.

reacted with transition metal cations and the products separated on silica in air, as shown in Figure 13-2, top.[4d,6] The X group can be, for example, Cl, Br, acetyl, or methyl (which converts to CH_2^- via tautomerism and thus becomes electron-withdrawing [7]); its role is to activate the open C_2B_3 ring toward coordination to metal ions, and to stabilize the multidecker complexes once they have been formed. The effect of the X substituent has been observed quantitatively via the ^1H NMR shift of the BHB bridging proton in neutral $LM(R_2C_2B_3H_4-X)$ complexes: the more electron-withdrawing X is, the further downfield the NMR signal and the greater the reactivity of the complex toward metal ions.[8]

The general utility of this route has been demonstrated by the synthesis of a variety of tetradecker complexes in which M, M′, L, and X can be systematically varied, as shown.[7–9] However, we noted a curious limitation in this approach: attempts to prepare tetradeckers in which the central metal M′ is Fe invariably failed, giving, instead, fusion products of the type $L_2M_2(R_4C_4B_6H_4X_2)$, which contain open 12-vertex $M_2C_4B_6$ clusters (structures **3** and **4**, Figure 13-2). We suspected that the desired Fe-centered tetradecker complexes were, in fact, forming in solution, but were undergoing oxidative fusion during chromatographic separation on silica columns in air. This type of reaction, first observed in our laboratory in 1974 [10] and subsequently encountered by several groups,[6a,11] involves metal-promoted conjoining of two borane, carborane, metallaborane, or metallacarborane cage units under low-energy conditions (room temperature or below). In the present case, recent work has demonstrated that the Fe-centered tetradeckers (compounds **2**) do indeed form as expected (see Figure 13-2), and can be isolated from the reaction medium provided that exposure to silica in air is avoided.[12]

The fusion products, compounds **3** and **4**, are of interest in their own right. Compound **4** is structurally identical to a $Co_2C_4B_6$ cluster isolated some years ago from a different reaction system [13] (Figure 13-3, Type I), but the cage geometry of compound **3** (Figure 13-3, Type II) is novel, consisting of a 12-vertex fragment of an ideal 16-vertex *closo*-polyhedron of T_d symmetry.[14] Since this geometry is formally derived via removal of four vertices from a parent *closo*-framework, it is in the *klado*-class. However, according to Wade's Rules,[15] such a structure requires $2n + 10$ skeletal electrons, and thus seems grossly incompatible with the $2n + 4$ electrons actually present in compound **3** (X = H or Cl). This electron count is consistent with neither Type I nor II, but rather with a *nido*-geometry based on extraction of one vertex from a 13-vertex *closo*-polyhedron (Figure 13-3, Type III); indeed, an isomer having this cage structure has been prepared.[13,16] Figure 13-3 shows schematically how the three isomeric $Co_2C_4B_6$ cages may be intercoverted via minimal bond breaking and bond formation, and may help to account for the observed formation of Types I and II in the same reaction system, with Type II proposed as an intermediate species [13] 14 years prior to its actual discovery [14]).

The fact that the products (compounds **3**) adopt Type II (klado) structures, despite their $2n + 4$ electron counts, underlines the importance of *kinetic* stabilization in boron cluster chemistry and demonstrates, again, that so-called "violations" of Wade's Rules are often of kinetic rather than thermodynamic origin, i.e., a low-energy pathway to the thermodynamically favored structure is not available under reaction conditions. In such cases, esoteric electron/orbital-based rationales intended to explain unexpected cage geometries may be irrelevant.

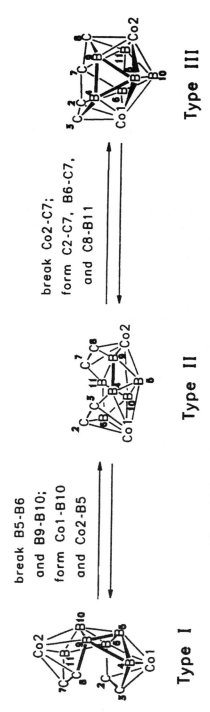

Figure 13-3 Possible cage rearrangement pathways for $Co_2C_4B_6$ isomers.

13.3   COMPLEXES OF SECOND AND THIRD TRANSITION SERIES METALS AS SYNTHONS FOR MULTIDECKER SANDWICHES

The chemistry of carborane-bridged multidecker sandwiches developed thus far [4d,6b,6c] has been largely based on $nido$-Cp′Co(RR′C$_2$B$_4$H$_6$) double-decker complexes (Cp′ = Cp or Cp* [η^5-C$_5$Me$_5$]) and their B-substituted derivatives, which afford ready access to triple- and higher-decker complexes having Cp′Co end units (Figure 13-1). It is desirable, however, to develop routes to a broader range of multidecker sandwich families, including complexes of heavy transition metals. Not only would one anticipate somewhat different electronic properties for these compounds in comparison with their first-row transition metal counterparts, but, in addition, synthetic advantages could be envisioned in having hydrocarbon end ligands, such as arenes, that are more reactive than the relatively inert Cp and Cp* units. Accordingly, we undertook to prepare and study several new classes of tetradeckers, including (arene)Ru- and Cp*Ir-capped species, some of which also feature second- or third-row metals in the central position.

Figure 13-4 shows the use of $nido$-(cymene)Ru(Et$_2$C$_2$B$_3$H$_4$-5-Y) complexes (compound **5**), originally prepared several years ago,[17] as precursors to (arene)Ru-capped tetradeckers (compound **6**) containing Co or Ni metal centers.[7] These syntheses parallel those employing $nido$-Cp′Co(carborane) complexes [4d,6b,8] and afford reasonable yields of isolated products, which, for M = Co^{IV} or Co^{III}H, are air-stable, colored crystals. Although the Co^{IV}- and Ni^{IV}-centered complexes are paramagnetic, the Co^{IV} compound was characterized via correlated NMR spectroscopy. In this technique, a paramagnetic

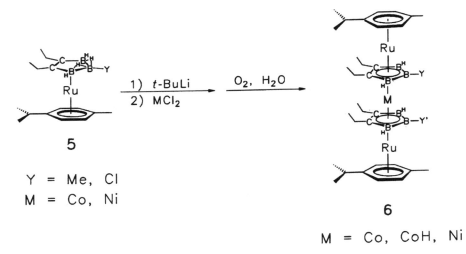

Figure 13-4   Synthesis of [(cymene)Ru(C$_2$B$_3$)]$_2$M tetradecker sandwiches.

species is reduced or oxidized, in stepwise fashion, to a diamagnetic product with the NMR data recorded at each stage, thereby allowing interpretation of the paramagnetic spectrum.[18] Crystal structure determinations on several type **6** tetradeckers [7] show molecular geometries very similar to those of their Cp*Co-capped analogs, i.e., the stacks are slightly bent and the metals are nearly centered over their carborane and hydrocarbon ring ligands (Figure 13-5).

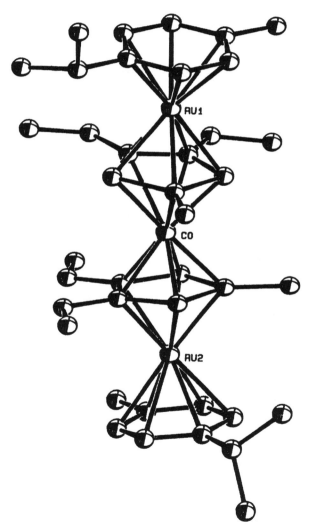

Figure 13-5   Structure of [(MeC$_6$H$_4$CHMe$_2$)Ru(2,3-Et$_2$C$_2$B$_3$H$_2$-5-Me)]$_2$Co.[7a]

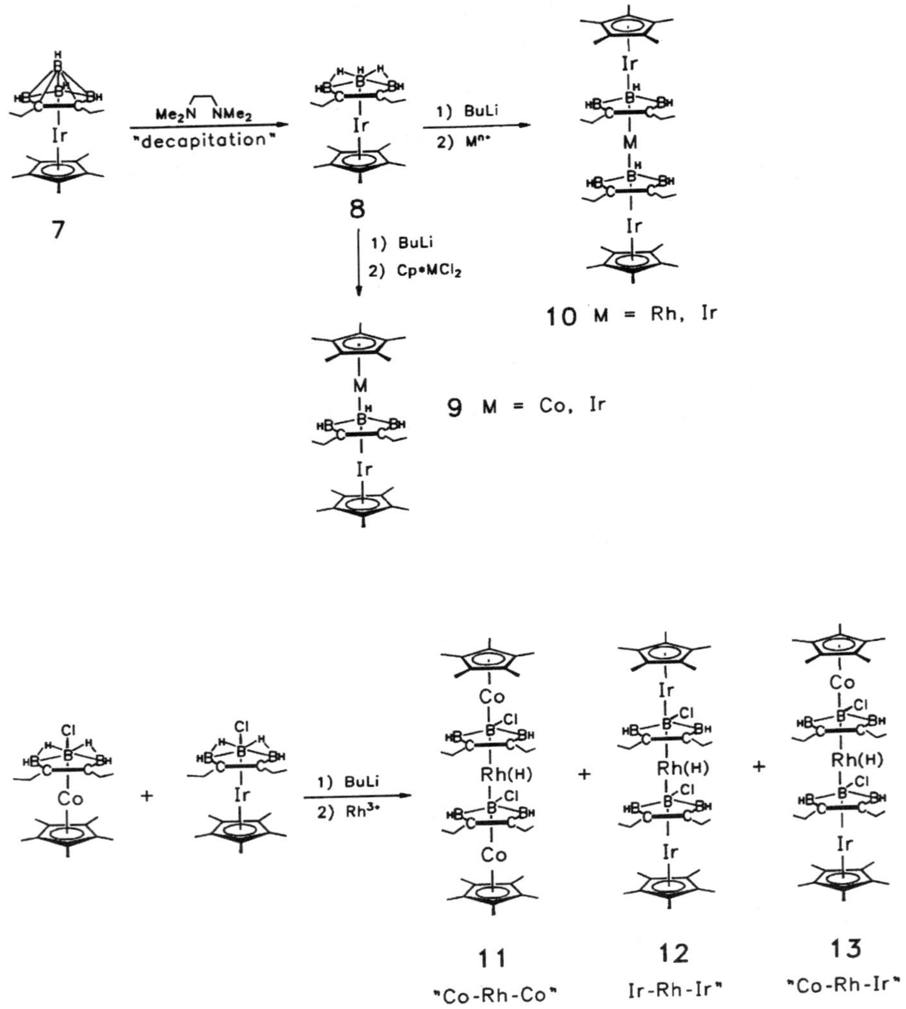

Figure 13-6 Preparation of tetradecker sandwiches incorporating Group 9 metals.

In very recent work,[19] the *nido*-iridacarborane Cp*Ir(Et$_2$C$_2$B$_3$H$_5$), compound **8**, has been employed as a building-block unit to prepare a series of triple- and tetradecker sandwiches featuring the Group 9 transition metals (Figure 13-6, complexes **10–13**).[19] Compound **13** ("Group 9 ocene") is unusual in that it incorporates the three elements, Co, Rh, and Ir, in sequence.

13.4   CLUSTER LINKING VIA B–B, Cp*–Cp*, AND METAL–METAL BONDS

In addition to metal–ligand stacking to create multidecker sandwich structures, other modes of connecting metallacarborane units have been investigated. One such approach involves linking the individual clusters via covalent bonds, which may involve carbon, boron, or metal atoms, and systematic methods for achieving each of these types of linkage have been developed. Figure 13-7

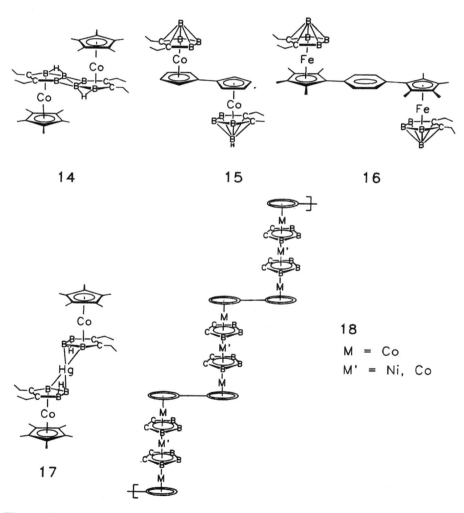

Figure 13-7   Examples of linked-cage and linked-sandwich poly(metallacarborane) complexes. The connecting groups in **18** contain C_5 or C_6 rings.

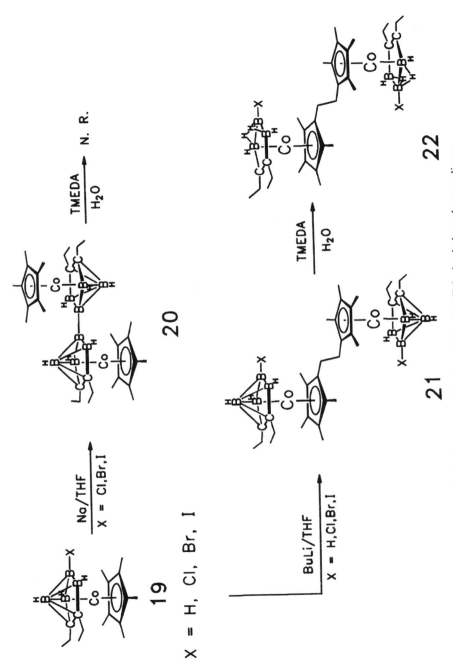

Figure 13-8 Synthesis of B–B and Cp*–Cp* linked cobaltacarborane dimers.

depicts a few examples of dimetallic (structures **14–16**), trimetallic (structure **17**), and polymetallic (structure **18**) complexes, that we have prepared in earlier work.[20,21] Recently, we found that B–B linked bis(cobaltacarboranes) (structure **20**) are readily obtained via Wurtz-type reactions of Cp*Co(Et₂C₂B₄H₃-5-X) halogenated complexes (structure **19**), as shown in Figure 13-8.[20] Treatment of the same monomeric clusters with butyllithium gave a surprising result: formation of Cp*-Cp* linked dimers (structure **21**), which can be decapped to give structure **22**. Activation of Cp* ligands is rare, having precedent only in a few reported reactions of Cp*IrL$_n$ and Cp*CoCp⁺.[22]

The molecular structures of compounds **20** and **21** [20] are depicted in Figure 13-9. Dimers of type **22**, having open carborane rings that can be coordinated to metal ions, are potential building blocks for assembling poly(cobaltacarborane) chains.

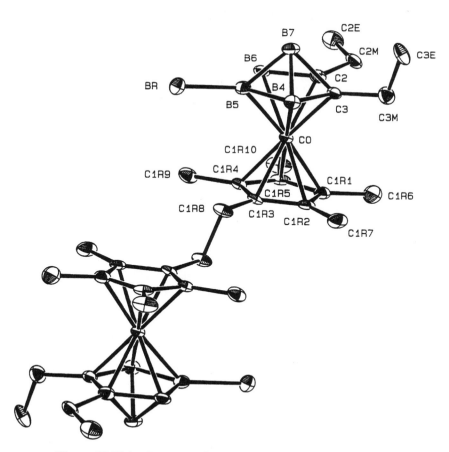

Figure 13-9(a)   Structure of [(–CH₂C₅Me₄)Co(Et₂C₂B₄H₃Br)]₂.

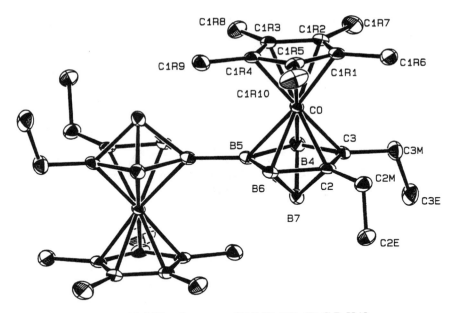

Figure 13-9(b)   Structure of $5,5\text{-}[Cp*Co(Et_2C_2B_4H_3)]_2$

Still another approach that affords linked dimers employs "recapitation" reactions as illustrated in Figure 13-10.[23] Treatment of a *nido*-cobaltacarborane dianion with a BX_3 or BX_2R reagent yields apically [B(7)] substituted *closo*-complexes (structures **23** or **24**); the bifunctional $Cl_2B\text{-}C_6H_4\text{-}BCl_2$ reagent gave the phenylene-linked bis(cobaltacarboranyl) compound **25,** whose structure has been established by X-ray crystallography.[23] This chemistry provides, for the first time, a systematic route to previously unavailable B(7)-derivatized MC_2B_4 clusters.

Small metallacarborane clusters that are connected by direct metal–metal bonds are unusual,[24] and $(B_4C_2)M\text{-}M(C_2B_4)$ dimers have not been reported. In a serendipitous discovery,[25] we found that halogen- and carbonyl-bridged dimolybdenum and ditungsten species (structure **27**) can be prepared from $[(CO)_3M(Et_2C_2B_4H_4)]^{2-}$ dianions (structure **26**) as shown in Figure 13-11. These dimers have been structurally characterized via spectroscopy and an X-ray crystallographic analysis of the bromine-bridged dimolybdenum complex (Figure 13-12). Double-decapitation of compound **27** affords open-ended species that are amenable to linkage via metal complexation to give linear polymers of type **28**, a prospect that is under investigation at the time of writing.

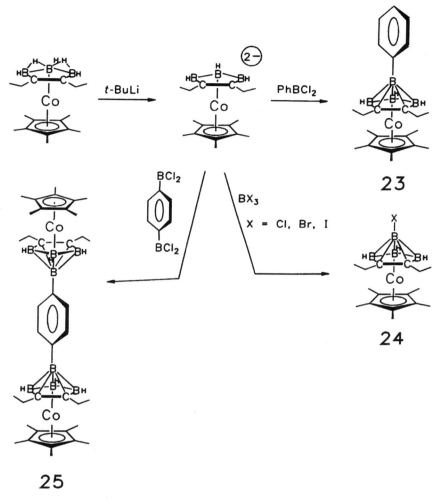

Figure 13-10   Synthesis of apically substituted cobaltacarboranes.

13.5   CARBORANE-ENDCAPPED SANDWICHES: IMPLICATIONS FOR POLYDECKER SYNTHESIS

Stacking of double-decker metallacarborane building-block complexes, such as $Cp*Co(R_2C_2B_3H_4)]^-$, with metal ions is a useful strategy for constructing triple- and tetradecker sandwich complexes (Figures 13-1 and 13-2). However, the hydrocarbon end ligands in these products are essentially inert to further metal stacking. Consequently, the synthesis of penta- and hexadecker complexes requires triple-decker sandwich precursors that have an open C_2B_3 end ring

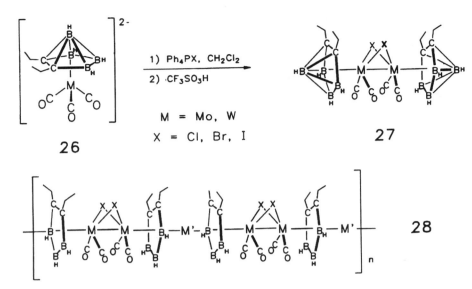

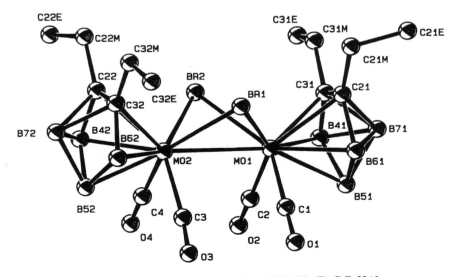

Figure 13-11 Synthesis of metal–metal bonded molybda- and tungstacarborane dimers (structure **27**) and proposed polymer (structure **28**).

Figure 13-12   Structure of $(\mu\text{-Br})_2[(CO)_2Mo(Et_2C_2B_4H_4)]_2$.

amenable to deprotonation and metal complexation, as shown at the bottom of Figure 13-1. The preparation of the carborane-capped triple-deckers, shown in Figure 13-13, proceeds in stages: first, C_2B_4-endcapped complexes (structure **29**) are obtained and these are decapped to generate structure **30**, which in turn, is deprotonated and air-oxidized to give the neutral paramagnetic complex **31**.[9]

Tetradecker sandwiches having carborane ligands at one or both ends have been prepared in an analogous manner, starting with the *nido,closo*-cobaltacarborane (structure **32**) as outlined in Figure 13-14.[9] Carborane-double-capped species such as structure **33** are potentially valuable synthetic agents, as decapitation and metal complexation at both ends could generate polydecker sandwiches. At this writing, double-decapitation of structure **33** has not yet been demonstrated, but the carborane-monocapped tetradecker structure **34** has been decapped to afford the open-ended tetradecker structure **35**, which is envisioned as a synthon for octadecker sandwiches (an as-yet unknown genre).[9]

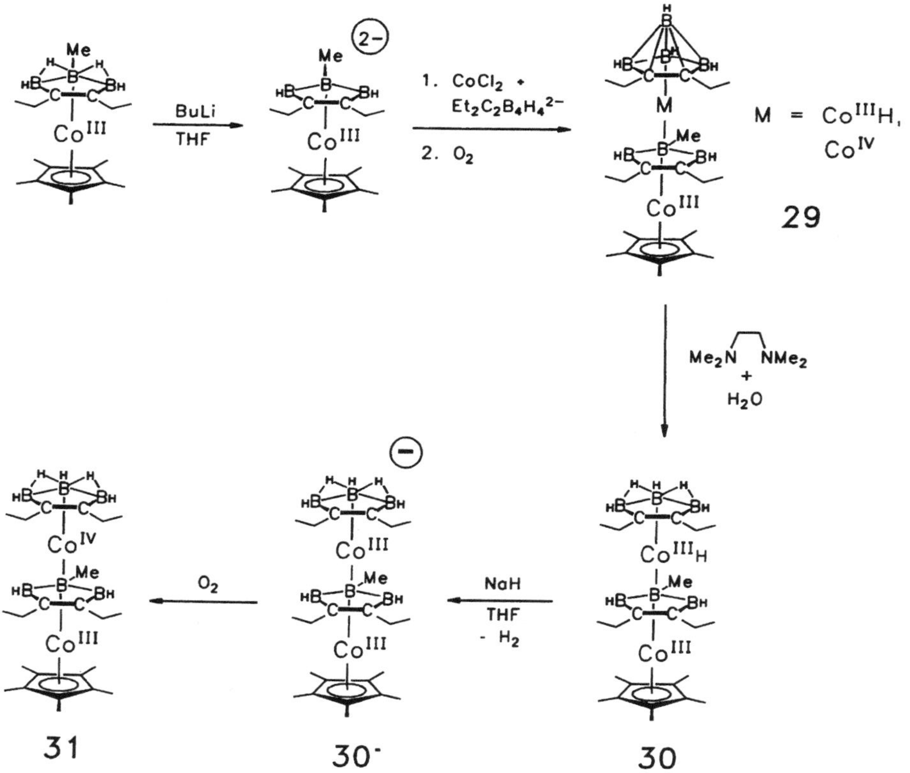

Figure 13-13 Synthesis of carborane-endcapped triple-decker complexes.

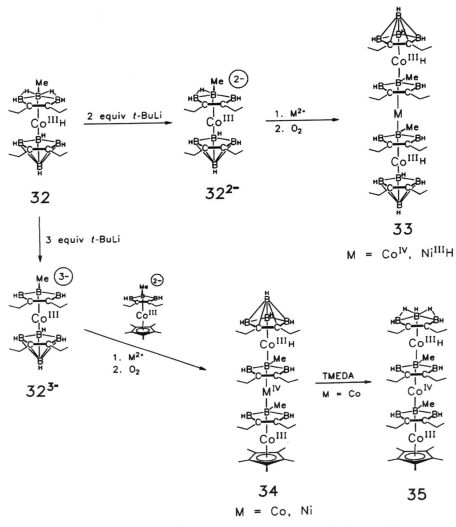

Figure 13-14 Synthesis of carborane-endcapped tetradecker complexes.

Open-ended triple-deckers of type **31** have been put to use in preparing the first C_2B_3-bridged pentadecker and hexadecker sandwiches (Figure 13-15),[7b] most of which are paramagnetic and exhibit limited electron-delocalization (see Section 13.6). A pentacobalt complex of type **37**, containing two $Co^{III}H$, two $Co^{III}Cp$, and one Co^{IV} center, has been characterized by X-ray crystallography [7b] and is the first hexadecker molecular sandwich to be structurally established, although, some time ago, Siebert's group reported an NMR-characterized C_3B_2-bridged hexadecker.[26]

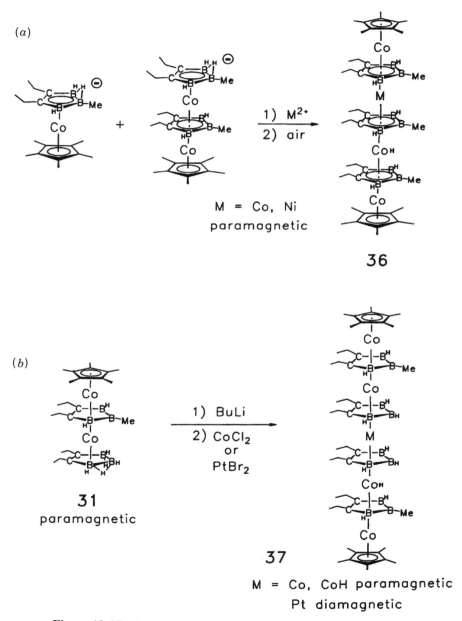

Figure 13-15 Synthesis of pentadecker and hexadecker complexes.

A point to be noted about the preparation of all these multidecker sandwich complexes is that they represent the application of directed synthesis — a time-honored concept in organic chemistry — to the assembly of a series of inorganic target complexes from small, specifically designed building-block units. This has been possible primarily because of the robustness, high solubility, ease of functionalization, and especially the versatility, of the metallacarborane reagents. These general properties afford significant advantages to small metallacarborane units in comparison with many types of organometallics (e.g., nonferrous metallocenes) as a basis for systematic construction of large, stable systems having specified composition and structure.

13.6 METAL–METAL COMMUNICATION IN BIMETALLIC AND POLYMETALLIC SYSTEMS

A focus of interest in many of the species discussed in this review is the degree and kind of electron delocalization between the transition metal centers. Our interest in this question is twofold. First, in terms of basic science, the stability and versatility of many of these systems allows detailed, systematic studies, not only of individual complexes, but also of entire families in which the electronic consequences of changing the metals, oxidation states, substituents, and other features can be examined. This contrasts with the situation presented by many interesting organometallic species, studies of which are precluded by chemical or electrochemical instability, or by an inability to extend a particular structural motif beyond one or two isolable examples. Second, the fact that most of the polymetallic metallacarboranes are either paramagnetic, or can be readily reduced or oxidized to a paramagnetic state, opens up myriad possibilities for practical application in technologically "hot" areas, such as molecular wires, molecular magnets, and molecular switches.[4d,21] Accordingly, we and our collaborators have investigated metal–metal electron delocalization in several types of bimetallic and polymetallic complexes, employing a range of physical techniques including ESR, NMR, electrochemistry, magnetic susceptibility, Mössbauer, UV-visible, and IR spectroscopy. Much of this work has been published in detail in recent years; here, I present just a few illustrative examples.

In one study,[27] the bis(ferracarboranyl)phenylene complex (structure **16**) mentioned earlier (see Figure 13-7) was examined in detail (Figure 13-16). The compound itself is diamagnetic with two $Fe^{II}H^+$ centers; monodeprotonation and one-electron oxidation yields a formal $Fe^{II}H^+$—Fe^{III} species, and repetition of this process affords an Fe^{III}—Fe^{III} system. As indicated in the figure, the $Fe^{II}H^+$—Fe^{III} complex was determined to be a trapped-valence (Class II) molecule, in which the unpaired electron is strictly localized on one metal. This finding is consistent with electrochemical data on phenylene-linked tetradecker complexes [21] that incorporate the same Me_4C_5—C_6H_4—C_5Me_4 connecting ligand as in structure **16**, and, similarly, exhibit no evidence of delocalization between stacks (i.e., across the phenylene rings), although there *is* metal–metal communication within individual stacks.

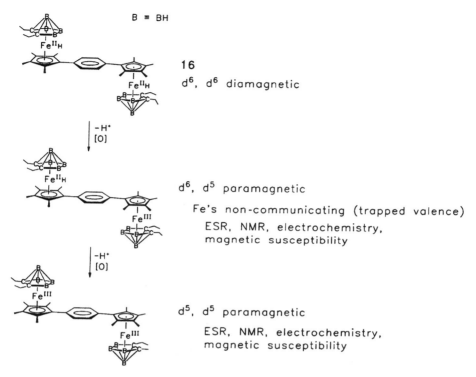

B ≡ BH

16
d⁶, d⁶ diamagnetic

d⁶, d⁵ paramagnetic

Fe's non-communicating (trapped valence)
ESR, NMR, electrochemistry,
magnetic susceptibility

d⁵, d⁵ paramagnetic

ESR, NMR, electrochemistry,
magnetic susceptibility

Figure 13-16 Stepwise oxidation of a phenylene-linked bis(ferracarboranyl) sandwich compound.

In contrast, the monoanion obtained by one-electron reduction of the fulvalene-bridged dicobalt complex **15** was shown via infrared spectroelectrochemistry to be a fully delocalized (Class III) mixed-valence system.[28] As shown in Figure 13-17 (bottom), the B–H stretching band in the neutral compound exhibits a drastic shift to lower frequency as it undergoes successive electrochemical reduction to the monoanion and the dianion. The values of ν_{BH} for the Co^{III}–Co^{III} neutral complex and the Co^{II}–Co^{II} dianion are almost identical to those of the corresponding Co^{III} and Co^{II} monocobalt complexes (Figure 13-17, top), while ν_{BH} for the Co^{III}–Co^{II} species, 2515 cm^{-1}, is midway between these, as expected for a delocalized Co$^{2.5}$–Co$^{2.5}$ system. Since the time scale for infrared spectroscopy is very short (10^{-13}–10^{-14} s), this method furnishes a very sensitive probe of electron delocalization. Electrochemical evidence also indicates that closely related fulvalene-linked oligomers (structure **18**, Figure 13-7) are similarly electron-delocalized.[29]

Communication between metal centers in multidecker sandwiches has also been explored, and the degree of intrastack delocalization varies widely in different systems. Thus, from ESR, electrochemical, and other evidence, it was

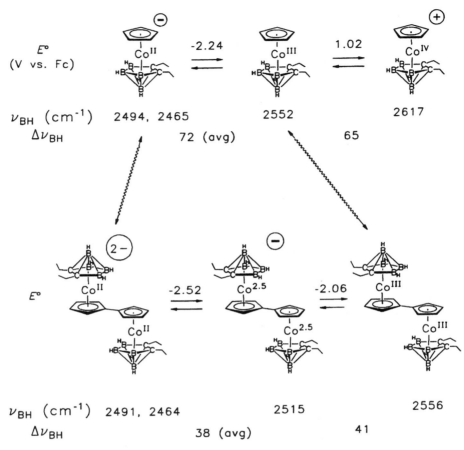

$$E°$$
(V vs. Fc)

ν_{BH} (cm^{-1})   2494, 2465   2552   2617

$\Delta\nu_{BH}$ 72 (avg)   65

$E°$

ν_{BH} (cm^{-1}) 2491, 2464   2515   2556

$\Delta\nu_{BH}$ 38 (avg)   41

Conclusion: The unpaired electron in the Co-Co monoanion is *fully delocalized* between metal centers

Figure 13-17   Stepwise oxidation/reduction of a fulvalene-bridged bis(cobaltacarboranyl) complex (bottom) and the corresponding monomer (top).

established that the 29-valence electron Co^{III}–Ru^{III} and Ru^{III}–Ru^{II} cationic triple-deckers (Figure 13-18a) are completely delocalized, with the unpaired electron in each case occupying a molecular orbital extending over both metals.[30] On the other hand, a family of Fe^{III}–Co^{III} 29-electron triple-decker sandwiches (Figure 13-18b) proved to be trapped-valence systems, with the unpaired electron localized on the iron atom in the neutral compounds [see Ref. 18a]. In this case, correlated NMR spectroscopy (see above) together with cyclic voltammetry, Mössbauer, and X-band ESR data were employed to probe the electronic structures.

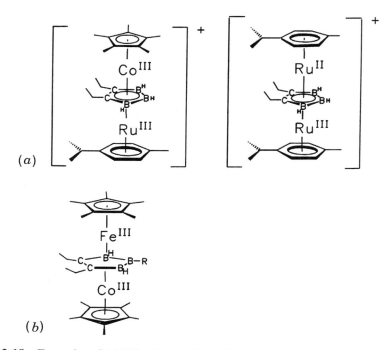

(a)

(b)

Figure 13-18 Examples of (a) fully electron-delocalized and (b) localized 29-electron triple-decker sandwiches.

Similar investigations of carborane-bridged paramagnetic $[Cp^*Co(C_2B_3)]_2M$ (M = Co, Ru, Ni) tetradecker [31] and $[Cp^*Co(C_2B_3)Co(C_2B_3)]_2M$ (M = Co, Pt) hexadecker sandwiches [7b,29] have led to the conclusion that the former compounds are essentially fully delocalized, while the hexadeckers exhibit partial delocalization that involves the inner metal centers but not those coordinated to the Cp* end ligands. In the latter systems (e.g., structure **37**, Figure 13-15), the end cobalts appear to have mainly diamagnetic character, corresponding to Co(III) d^6 centers, as evidenced by their normal Cp* proton NMR signals, which contrast with the paramagnetic nature of the remainder of the spectra; electrochemical and ESR data support this interpretation.[7b,29] These observations on the hexadeckers are tentative, pending further studies, which are continuing.

The fact that different C_2B_3-bridged multidecker sandwich assemblies exhibit widely different degrees of electron-delocalization attests to the versatility that is inherent in this area, and underlines the potential for developing materials that can be electronically tuned to have specified combinations of properties.

ACKNOWLEDGMENTS

I am pleased to acknowledge the contributions of numerous present and former graduate students and postdoctoral associates, named in the references, and of Dr. Michal Sabat in our department for crystal structure determinations. I am most grateful to Professors Walter Siebert (University of Heidelberg) and Bill Geiger (University of Vermont) and their students for continuing collaborations with their laboratories. Much of the work described was generously supported by the National Science Foundation and the U.S. Army Research Office.

REFERENCES

1. Werner, H. and Salzer, A., *Angew. Chem. Int. Ed. Engl.*, **11**, 930, 1972; *Synth. React. Inorg. Met.-Org. Chem.*, **2**, 239, 1972.

2. Dubler, E., Textor, M., Oswald, H.-R., and Salzer, A., *Angew. Chem. Int. Ed. Engl.*, **13**, 135, 1974.

3. Beer, D. C., Miller, V. R., Sneddon, L. G., Grimes, R. N., Mathew, M., and Palenik, G. J., *J. Am. Chem. Soc.*, **95**, 3046, 1973.

4. (a) Grimes, R. N., in *Comprehensive Organometallic Chemistry I* (Abel, E., Stone, F., and Wilkinson, G. A., eds), Pergamon Press, Oxford, 1982, Chap. 9, pp. 459–542; and *Comprehensive Organometallic Chemistry II* (Abel, E., Stone, F., and Wilkinson, G. A., eds), Pergamon Press, Oxford, 1995, Chap. 9, pp. 373–430; (b) Siebert, W., *Adv. Organometal. Chem.*, **35**, 187, 1993; (c) Hosmane, N. S. and Maguire, J. A., *J. Cluster Science*, **4**, 297, 1993; (d) Grimes, R. N., *Chem. Rev.*, **92**, 251, 1992; (e) Herberich, G. E., in *Comprehensive Organometallic Chemistry II* (Abel, E., Stone, F., and Wilkinson, G. A., eds), Pergamon Press, Oxford, 1995, Chap. 5, pp. 197–216.

5. (a) Fessenbecker, A., Attwood, M. D., Grimes, R. N., Stephan, M., Pritzkow, H., Zenneck, U., and Siebert, W., *Inorg. Chem.*, **29**, 5164, 1990; (b) Fessenbecker, A., Attwood, M. D., Bryan, R. F., Grimes, R. N., Woode, M. K., Stephan, M., Zenneck, U., and Siebert, W., *Inorg. Chem.*, **29**, 5157, 1990; (c) Attwood, M. A., Fonda, K. K., Grimes, R. N., Brodt, G., Hu, D., Zenneck, U., and Siebert, W., *Organometallics*, **8**, 1300, 1989.

6. (a) Grimes, R. N., *Coord. Chem. Rev.*, **143**, 71, 1995; (b) Grimes, R. N., in *Current Topics in the Chemistry of Boron* (Kabalka, G. W., ed.), Royal Society of Chemistry, 1994, p. 269. (c) Grimes, R. N., in *Advances in Boron Chemistry* (Siebert, W., ed.), Royal Society of Chemistry, 1997, p, 321–332.

7. (a) Greiwe, P., Sabat, M., and Grimes, R. N., *Organometallics*, **14**, 3683, 1995; (b) Wang, X., Sabat, M., and Grimes, R. N., *J. Am. Chem. Soc.*, **117**, 12227, 1995.

8. Piepgrass, K. W., Meng, X., Hoelscher, M., Sabat, M., and Grimes, R. N., *Inorg. Chem.*, **31**, 5202, 1992.

9. Wang, X., Sabat, M., and Grimes, R. N., *J. Am. Chem. Soc.*, **117**, 12218, 1995.

10. Maxwell, W. M., Miller, V. R., and Grimes, R. N., *J. Am. Chem. Soc.*, **96**, 7116, 1974.

11. (a) Grimes, R. N., *Adv. Inorg. Chem. Radiochem.*, **26**, 55, 1983; (b) Bould, J., Greenwood, N. N. and Kennedy, J. D., *Polyhedron.*, **2**, 1401, 1983; (c) Gaines,

D. F., Nelson, C. K., and Steehler, G. A., *J. Am. Chem. Soc.*, **106**, 7266, 1984; (d) Jun, C.-S., Powell, D. R., Haller, K. J., and Fehlner, T. P., *Inorg. Chem.*, **32**, 5071, 1993.

12. Wang, X., Sabat, M., and Grimes, R. N., *Inorg. Chem.*, **34**, 6509, 1994.

13. Pipal, J. R. and Grimes, R. N., *Inorg. Chem.*, **18**, 1936, 1979.

14. Piepgrass, K. W., Curtis, M. A., Wang, X., Meng, X., Sabat, M., and Grimes, R. N., *Inorg. Chem.*, **32**, 2156, 1993.

15. (a) Wade, K., *Adv. Inorg. Chem. Radiochem.*, **18**, 1, 1976; (b) Mingos, D. M. P., *Accounts Chem. Res.*, **17**, 311, 1984; (c) Mingos, D. M. P. and Wales, D. J., *Introduction to Cluster Chemistry*, Prentice Hall, Englewood Cliffs, NJ, 1990.

16. Wong, K.-S., Bowser, J. R., Pipal, J. R., and Grimes, R. N., *J. Am. Chem. Soc.*, **100**, 5045, 1978.

17. Davis, J. H. Jr, Sinn, E., and Grimes, R. N., *J. Am. Chem. Soc.*, **111**, 4776, 1989.

18. (a) Stephan, M., Mueller, P., Zenneck, U., Pritzkow, H., Siebert, W., and Grimes, R. N., *Inorg. Chem.*, **34**, 2058, 1995; (b) Stephan, M., Hauss, J., Zenneck, U., Siebert, W., and Grimes, R. N., *Inorg. Chem.*, **33**, 4211, 1994, and references therein.

19. Franz, D. A., Houser, E. J., Sabat, M., and Grimes, R. N., *Inorg. Chem.*, **35**, 7027, 1996.

20. Wang, X., Sabat, M., and Grimes, R. N., *Organometallics*, **14**, 4668, 1995, and references therein.

21. Meng, X., Sabat, M., and Grimes, R. N., *J. Am. Chem. Soc.*, **115**, 6143, 1993.

22. (a) Miguel-Garcia, J. A., Adams, H., Bailey, N. A., and Maitlis, P. M., *J. Organomet. Chem.*, **413**, 427, 1991; (b) Miguel-Garcia, J. A., Adams, H., Bailey, N. A., and Maitlis, P. M., *J. Chem. Soc., Dalton Trans.*, 131, 1992; (c) Gloaguen, B. and Astruc, D., *J. Am. Chem. Soc.*, **112**, 4607, 1990.

23. Curtis, M. A., Sabat, M., and Grimes, R. N., *Inorg. Chem.*, **35**, 6703, 1996.

24. Oki, A. R., Zhang, H., Hosmane, N. S., Ro, H., and Hatfield, W. E., *J. Am. Chem. Soc.*, **113**, 8531, 1991.

25. Curtis, M. A., Houser, E. J., Sabat, M., and Grimes, R. N., *Inorg. Chem.*, **37**, in press, 1998.

26. Kuhlmann, T. and Siebert, W., *Z. Naturforsch.*, **40**, 167, 1985.

27. Stephan, M., Davis, J. H. Jr, Meng, X., Chase, K. P., Hauss, J., Zenneck, U., Pritzkow, H., Siebert, W., and Grimes, R. N., *J. Am. Chem. Soc.*, **114**, 5214, 1992.

28. Chin, T. T., Lovelace, S. R., Geiger, W. E., Davis, C. M., and Grimes, R. N., *J. Am. Chem. Soc.*, **116**, 9359, 1994.

29. Geiger, W. E., Lovelace, S. R., and Grimes, R. N., unpublished results.

30. Merkert, J., Davis, J. H. Jr, Geiger, W., and Grimes, R. N., *J. Am. Chem. Soc.*, **114**, 9846, 1992.

31. Pipal, J. R. and Grimes, R. N., *Organometallics*, **12**, 4452 and 4459, 1993.

PART IV

NEW SPECIES OF BORANES AND CARBORANES

14

ISOELECTRONIC BORANE AND HYDROCARBON METAL COMPLEXES

THOMAS P. FEHLNER

Department of Chemistry and Biochemistry, University of Notre Dame, Notre Dame, Indiana 46556

14.1 INTRODUCTION

The typical approach to new chemistry proceeds from an empirical or molecular formula to a geometric structure to a chemical bonding model of electronic structure. Right from their first beginnings in 1953,[1] the species now known as carboranes were curiosities on all three counts. The elemental formulae led chemists astray as far as structure, and the geometric structures themselves, when finally defined, only seemed to make the necessary connection between chemical bonding, structure, and stoichiometry more obscure. It is a measure of Bob Williams' abilities that he was able to draw a beautifully simple relationship between borane and carborane stoichiometries and shapes in his important 1971 *Inorganic Chemistry* paper.[2] This was followed by a perceptive review of main-group cluster chemistry, which included the effects of heteroatoms and endo-hydrogens.[3] Although the connection between geometric structure and electronic structure was developed principally by Wade and Mingos,[4–9] the interrelationship between cluster geometries remains a fundamental expression of the electronic structure. Further, Williams not only pointed out one of the key relationships, but, over the years, has kept reminding us of the connections to spectroscopic and chemical properties. In the end, it is the properties of compounds that one wishes to manipulate.

A common thread that runs throughout the work of Williams is that of isoelectronic species. This is an old, but still very useful, method for the intercomparison of ostensibly very different compounds.[10] The isolobal concept, in which main-group element fragments and transition-metal fragments are said to

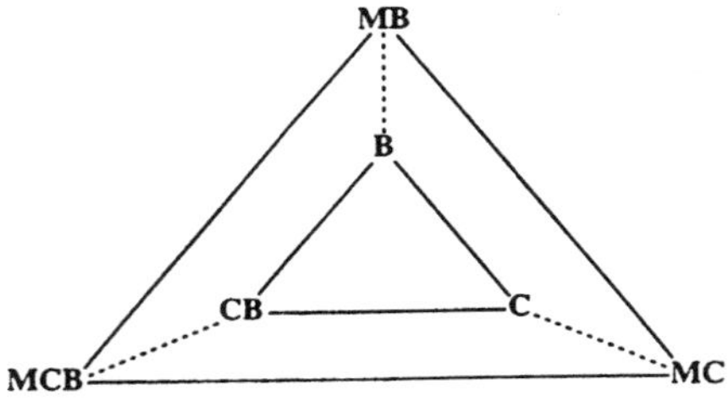

METALLABORANES

METALLACARBORANES   METALLACARBANES

Figure 14-1 Schematic relationship between boranes, carbaboranes, i.e., carboranes, and carbocations (inner triangle), and metallaboranes, metallacarboranes, and metallacarbanes, i.e., hydrocarbyl complexes (outer triangle).

be related in terms of bonding capabilities, is a powerful corollary.[11,12] It permits the interrelationship of organometallic compounds and main-group compounds, e.g., $(CO)_5Mn—Mn(CO)_5$ is said to be isolobal with $H_3C—CH_3$. It also gives us another dimension of comparison for cage compounds.

Thus, just as Williams compared series of boranes and carboranes leading up to carbocations by sequential replacement of $[BH]^-$ or BH_2 fragments with CH, so, too, one can compare series of boranes and carboranes in which the BH or CH fragments are replaced with isolobal metal fragments. This is illustrated schematically in Figure 14-1, in which the inner triangle represents the borane, carborane, carbocation comparison of Williams, and the outer triangle represents isolobal replacement by metal fragments leading, respectively, to metallaboranes, metallacarboranes, and organometallic complexes that may be charged or uncharged. The first compound type lies in the area of inorganometallic chemistry (compounds containing M—E bonds, E ≠ C),[13] the last in the area of organometallic chemistry (compounds containing M—C bonds), and the second can have either M—B, M—C, or both types of bonds, [14] Many of our interests lie in these areas and what follows is an account that interrelates structure and chemistry after the manner of Williams. Although the present discussion is focused on the work of myself and my colleagues, the reader should know that others have similar interests and, as shown in the referenced reviews, the scope of metallaborane chemistry has become significantly large.[15–19]

14.2   ELEMENTARY IDEAS

With our present knowledge and understanding of cluster systems, the pentagonal pyramidal series beginning with B_6H_{10} and ending with $C_4B_2H_6$, described by Williams in 1970,[1] appears simple and maybe even obvious. But, it is still

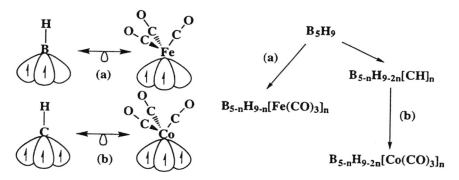

Figure 14-2 Main group fragment and metal fragment isolobal relationships and an example of the stoichiometric consequences.

a useful exercise of pedagogical value. With such an approach, a systematic enumeration of achievable compound stoichiometries and isomer counts for each composition becomes an easy exercise. In addition, the characteristics of a given cluster atom environment permits estimates of relative isomer stabilities. Further, correlations of NMR chemical shifts, as well as estimates of Brönsted acidities for the various types of hydrogens present, can be carried out.[3,20] Finally, when isolobal metal fragments are added to the problem, truly large numbers of compound possibilities are generated. The isolobal relationships pertinent to this account are shown in Figure 14-2.

However, the situation becomes somewhat more complex when metal fragments are introduced. First, the metal fragments can be used to replace either BH or CH, or, with two metal fragments, both BH and CH. Second, although there are clear connections between the cluster bonding of main-group element cages and that of transition-metal clusters, there are large differences as well,[3,8,9] Three differences of importance here are (1) the tendency of metal clusters to form more condensed, interconnected clusters, (2) the much larger steric demands of a metal fragment vs a BH or CH fragment, and (3) the much weaker M—M cluster bond vs B—B, B—C, C—C, M—B, or M—C bonds. Thus, although one can set up series analogous to those of Williams for main-group cages, one does not know to what extent the different tendencies of the main group and metal clusters will be competitive or complementary. Indeed, the most interesting situations will be those in which the number of main-group and metal fragments are similar, so that neither fragment dominates.

In the following, real examples are explored more fully in terms of (1) overall structure vs composition, (2) selected structural features vs composition, (3) heterometal effects, and (4) selected chemical properties. Of course, the compounds used as examples have to be made and characterized, and in any given series there will likely be missing compounds. It is tempting to interpret the absence of certain members of a chosen series as evidence of intrinsic instability, but, in the absence of additional information, no such meaning can be attached to the missing compounds. In my experience, all of the metallaboranes are thermodynamically unstable with respect to separate boron- and metal-containing fragments. Hence, the missing compounds can simply reflect a lack of a suitable synthetic pathway. On the other hand, partially characterized compounds, often intermediates in a reaction series, when compared with fully characterized relatives, do give a clear indication of the direction of greater kinetic or thermodynamic stability.

14.3 INTERCOMPARISON OF CLUSTER GEOMETRIC STRUCTURE

Three series of selected compounds will be considered. First, n atom clusters, $n = 4, 5$, as the number of boron atoms runs from n to the minimum value presently known, will be compared. Second, examples of monoboron compounds, BM_n, as n runs from 3 to 7 are given. Finally, monometal compounds, B_nM, as n runs from 2 to 5are discussed.

14.3.1 B_nM_{4-n}

A complete series of clusters is known for M = Fe and the structures are shown in Figure 14-3.[21–25] The compounds belong two structure types, i.e., an *arachno*-set, $B_nH_{n+6}[Fe(CO)_3]_{4-n}$ for $n = 3, 4$, and a *nido*- set, $B_nH_{n+4}[Fe(CO)_3]_{4-n}$ for $n = 0, 1, 2$. Evidence that the change in cluster type as the metal fraction increases truly reflects a change in stability is given by the fact that *nido*-B_4H_8

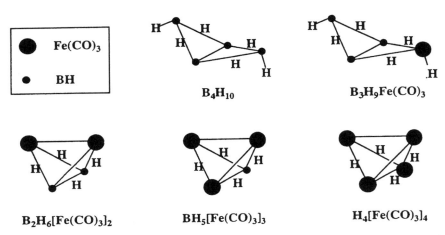

Figure 14-3 Structures observed for B_nM_{4-n} clusters showing the change from *arachno*- to *nido*-geometry as the metal content increases.

has been observed as an intermediate species with a short lifetime [26]; the fact that the Mn analog of *arachno*-$B_3H_9Fe(CO)_3$, $B_3H_8Mn(CO)_4$, can be converted reversibly into $B_3H_8Mn(CO)_3$ with a *nido*-geometry [22,27]; and the fact that reactions of $BH_5Fe_3(CO)_9$ [24] and a related Os metallaborane [28] with Lewis bases appear to proceed via addition to yield an intermediate with a presumed *arachno*-structure.

As the change in stoichiometry and associated structure does depend on *n*, the number of boron atoms, it must reflect differences in main-group and transition-metal fragment characteristics. Clearly, the switch in structure type is not steric in origin as the much larger $Fe(CO)_3$ fragment should be more easily accommodated in the open *arachno*-structure.[29] It is more likely associated with the tendency of a metal atom to prefer six neighbors in an octahedral geometry and the main-group atom to prefer four neighbors in a tetrahedral geometry. On the one hand, this limits the ability of the three coordinate $Fe(CO)_3$ fragments to accommodate hydrogen ligands and, on the other hand, limits the ability of a BH fragment to form BB bonds in a closed tetrahedron without the bonding flexibility provided by B—H—B bridges.

14.3.2 B_nM_{5-n}

No complete series is known for any one metal or even for different metals. The situation is summarized in Figure 14-4, where it is seen that, although there are

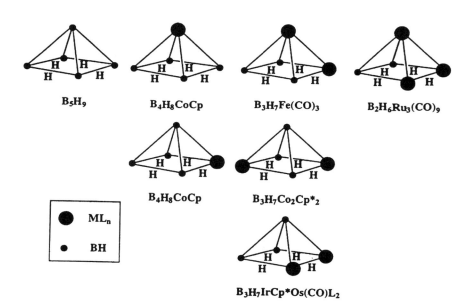

Figure 14-4 Structures observed for known B_nM_{5-n} clusters illustrating the geometric isomerism required by the cluster geometry.

many interesting examples for the boron-rich members, including all geometrical isomers for $n = 3, 4$, there is only one example for $n = 2$.[21,30–37] Two closely related compounds, $BH_3Fe_4(CO)_{12}$ [38] and $CFe_5(CO)_{15}$ [39], could be considered as examples for $n = 1, 0$, respectively; however, the former is viewed as an *arachno* four-atom cluster with an interstitial boron (see below) and the latter, although a five-atom *nido*-cluster, also possesses an interstitial carbon atom rather than four hydrogen atoms. In the context of the present discussion, the presence of interstitial atoms introduces additional parameters that will be dealt with below. In comparing the four- and five-membered cages, it is clear that the perturbation caused by replacing a BH fragment with an isolobal metal fragment also depends of the ability of the particular cluster geometry to accommodate the bonding and steric requirements of each fragment type. Up to and including three metals, the *nido* five-atom structure with four bridging hydrogens is a stable arrangement. Note that for the one-metal derivative, the *arachno*-geometry is known as well, e.g., $B_4H_{10}CoCp^*$, $Cp^* = \eta^5-C_5Me_5$.[34]

14.3.3 BM_n

This series is an interesting one. Pertinent examples are shown in Figure 14.5. In contrast to the preceding series, the examples are necessarily dominated by the metal-rich systems and, in fact, many more examples of the latter type might have been given.[17,24,38,40–46] An example of a monometal compound $(CO)_4CoBH_2\cdot THF$ [47] is known, but it is extremely labile. No example of a dimetal compound with a monoborane (BH_n) fragment has been reported. The

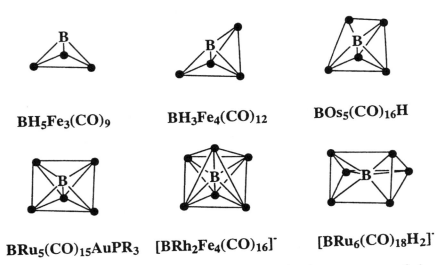

$BH_5Fe_3(CO)_9$ $BH_3Fe_4(CO)_{12}$ $BOs_5(CO)_{16}H$

$BRu_5(CO)_{15}AuPR_3$ $[BRh_2Fe_4(CO)_{16}]^-$ $[BRu_6(CO)_{18}H_2]^-$

Figure 14-5   Core structures observed for BM_n, illustrating the consequences of a large number of metal atoms. The filled circles represent the metal atom positions.

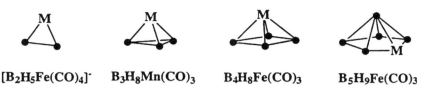

$[B_2H_5Fe(CO)_4]^-$ $B_3H_8Mn(CO)_3$ $B_4H_8Fe(CO)_3$ $B_5H_9Fe(CO)_3$

Figure 14-6 Examples of core structures observed for B_nM, $n = 2 - 5$, metallaboranes. The filled circles represent the boron atom positions.

three-metal compound can be considered a *nido*-structure, but all the compounds with four or more metal atoms contain boron as an interstitial atom, i.e., it contributes all three valence electrons to cluster bonding and the cluster itself no longer contains a BH fragment with a terminal hydrogen atom. Apparently, the metal-dominated cluster denudes the BH fragment of its hydrogen ligand and incorporates it totally into the bonding network. In a sense, we no longer have a mixed main-group/transition-metal atom cluster but rather a metal cluster alone. Hence, the cluster nuclearity is reasonably considered to be defined by the number of metal atoms alone.

14.3.4 B_nM

The other side of the coin is shown in Figure 14-6, where only a few of the many examples of monometal ferraboranes are illustrated.[22,30,48,49] This constitutes the largest set of compounds [18,19] and it is clear that a borane framework of practically any size can accommodate the requirements of a metal fragment isolobal with BH. Indeed, in these compounds, there are cogent reasons for considering the borane fragment as a neutral or anionic ligand. This constitutes a viable alternative to the view of these compounds as mixed main-group/transition-metal clusters.[50] It becomes a matter of convenience which model one uses in a given instance. Certainly, in the hands of Hawthorne, Grimes, Siebert, and others, the ligand model has been an extremely fruitful one.[51–53] For two or more metals, however, the ligand model places largely inappropriate restrictions on the structural interpretations.

14.4 ORGANOMETALLIC VERSUS INORGANOMETALLIC CLUSTERS: INFLUENCE OF THE MAIN-GROUP ATOM

Just as Williams compared carbocations with isoelectronic boranes, one can compare organometallic clusters (C_xM_y) with metallaboranes (B_xM_y). We have pointed out the differences between the closely related and isoelectronic clusters with the compositions $CH_4Fe_3(CO)_9$ and $BH_5Fe_3(CO)_9$.[54,55] Although the cluster frameworks possess the metal triangle capped by a main-group atom, the distribution of the nonterminal hydrogen atoms is very different. As the only difference between the clusters is C vs BH, the origin of the differences must lie in the properties of C vs B or in the number of hydrogen atoms.

It is instructive to review the situation. The compound $CH_4Fe_3(CO)_9$ exists in solution as an equilibrium mixture of three tautomers: $H_3Fe_3(CO)_9CH$, $H_2Fe_3(CO)_9(HCH)$, and $HFe_3(CO)_9(H_2CH)$ in the ratio of 86:13:1. The order of stability was unambiguously demonstrated by deprotonating the mixture to yield a single anion, $[HFe_3(CO)_9(HCH)]^-$. This anion, when protonated at low temperature, yields exclusively $HFe_3(CO)_9(H_2CH)$ as the sole kinetic product. Warming to room temperature resulted in conversion to the more abundant and more stable tautomers shown in Figure 14-7. These results imply that protonation takes place preferentially at an exposed C—Fe edge of the cluster, even though the Fe—Fe edge is clearly the more basic site. It is perhaps not surprising, then, that protonation of $H_3Ru_3(CO)_9CR$ to yield $[H_3Ru_3(CO)_9(HCR)]^+$ has been observed showing that, as expected, the difference in basicity between the two sites cannot be large.[56]

In the case of the isoelectronic ferraborane, $BH_5Fe_3(CO)_9$, only one, fluxional species is observed in the solid state and in solution.[24] This is $HFe_3(CO)_9(H_3BH)$. Clearly, this structural form is analogous to the least stable form of the carbon analog. Is this due to the difference between boron and carbon or simply due to the difference in the number of hydrogen atoms in the two species? It appears to be associated with the former in that deprotonation of the ferraborane leads to $[HFe_3(CO)_9(H_2BH)]^-$, which has the identical hydrogen atom distribution as the least stable tautomer of $CH_4Fe_3(CO)_9$ (Figure 14-7).

The difference between these isoelectronic clusters shows that, even though they are geometrically similar (the bridging hydrogens are often considered

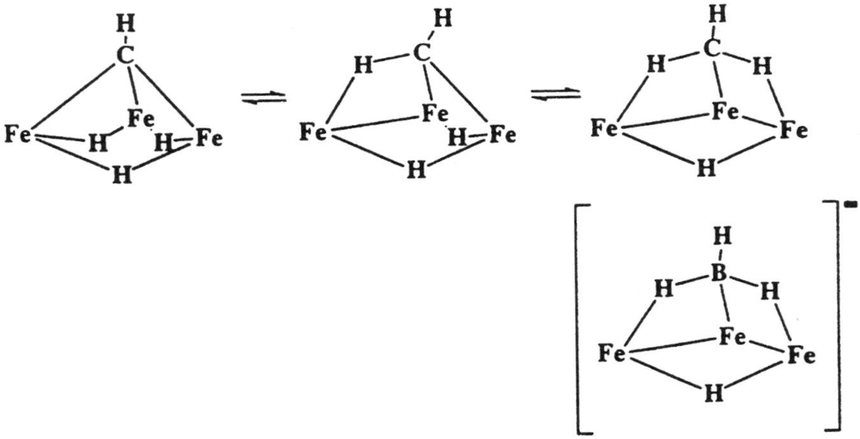

Figure 14-7 Proposed structures of the three tautomeric forms of $CH_4Fe_3(CO)_9$ and that of $[BH_4Fe_3(CO)_9]^-$ Three terminal CO ligands (not shown) are attached to each metal atom.

simply as contributing one electron to cluster bonding with little additional effect on the cluster core geometry), the cluster composition clearly affects cluster properties. Of course, this is expected but, even so, it is not a trivial point as it is in chemical properties that compounds find their unique value. I have described a simple explanation of the role of the main-group and transition-metal atoms in determining the most stable position of the bridging hydrogen atoms in the two compounds discussed, as well as in a considerable number of related compounds. The essence of the explanation lies in the difference in the effective electronegativities of the main-group atom and the metal atom. This model views the endohydrogen atom as simply a proton seeking maximum electronic charge, and the differing placements of the bridging atoms on the two tetrahedral frameworks as reflecting the differing skeletal charge distributions. As was pointed out in the work, this finding has relevance to processes in which H atoms bound to metal centers are added to or, removed from, main-group atoms.

One of the important findings in the above experimental work is that the differences between the energies of the various tautomeric forms for a given cluster are very small. Thus, it is not a surprise that when the terminal hydrogen on either the carbon or boron triiron cluster is replaced with a methyl group, the structural and reaction chemistry is considerably different. For example, the compound $(CH_3)CH_3Fe_3(CO)_9$ exists as a single form in solution and in the solid state, i.e., as the ethylidyne cluster, $H_3Fe_3(CO)_9CCH_3$.[57] On treatment with base, a proton is removed from the methyl group and the anion formed is very unstable with respect to H_2 elimination. The result is the formation of the vinylidene anion $[HFe_3(CO)_9CCH_2]^-$. The latter can be taken to a number of other products, as well as back to the initial ethylidyne cluster in the presence of protons and H_2. The dramatic effect of a simple methyl substituent in place of a hydrogen atom confirms the delicate balance both between different structural forms and also between the reaction pathways in these systems. Recent mechanistic studies show that the reaction pathways of trimetal akylidyne clusters are extremely sensitive to ostensibly small substituent changes.[58] In terms of designed chemistry, this is desirable as a small perturbation can be used to fine-tune chemistry. However, a large caveat still exists. One must understand the origin of the change and one must be able to modify the system chemically in an appropriate fashion.

14.5   ORGANOMETALLIC VERSUS INORGANOMETALLIC CLUSTERS: INFLUENCE OF THE METAL ATOM

The perturbation of cluster properties by the C vs BH interchange is clear. The same should be true of isolobal metal fragment interchange and, in fact, much has been written about heterometal effects in multinuclear organometallic compounds. Here, however, one must be careful to separate the differences in ML_x fragments due to nonidentical metals from differences caused by nonidentical

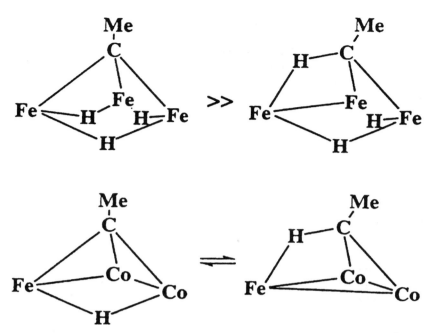

Figure 14-8 Comparison of the relative stabilities of the tautomeric forms of $H_3Fe_3(CO)_9CCH_3$ and $HFeCo_2(CO)_9CCH_3$. Three terminal CO ligands (not shown) are attached to each metal atom.

ligands. It is well known that the metal–ligand interactions are significantly larger energetically than metal–metal interactions within a cluster. The balance is more even for metal main-group atom interactions within the cluster but, as one is still looking at small effects, no conclusion is possible without dealing with compounds with identical ligands on the metal atoms. Thus, we have examined the interchange of $Co(CO)_3$ (isolobal with CH) for $HFe(CO)_3$ (isolobal with BH_2).

In the case of organometallic compounds, the primary comparison was between $H_3Fe_3(CO)_9CCH_3$ and $HFeCo_2(CO)_9CCH_3$.[59] The structures are shown in Figure 14-8 along with the key results. Substitution of the more electronegative Co for Fe (and H) increases the stability of the tautomer containing a M—H—C bridging hydrogen atom. By measuring the equilibrium constant as a function of temperature, we were able to show that the difference between the two tautomeric forms lies mainly in a small difference in enthalpies favoring the form with the M—H—C bridge. In this case, we were also able to define the effect of the substituent on the alkyne carbon atom. The equilibrium constant is only weakly affected by changes in the electronic effect of *para*-substituted phenyl groups, but changes markedly with the size of the substituent. There is a significantly larger preference for the tautomer with a M—H—C bridge as the

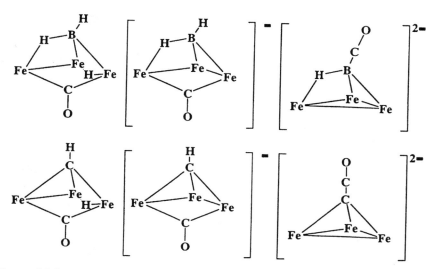

Figure 14-9 Comparison of the most stable tautomeric forms of $BH_3Fe_3(CO)_{10}$, $[BH_2Fe_3(CO)_{10}]^-$ (which also represents that of $BH_2Fe_2Co(CO)_{10}$), $[BHFe_3(CO)_{10}]^{2-}$, $CH_2Fe_3(CO)_{10}$, $[CHFe_3(CO)_{10}]^-$ and $[CFe_3(CO)_{10}]^{2-}$. Three terminal CO ligands (not shown) are attached to each metal atom.

substituent size is reduced. This is reasonable as introduction of a M—H—C interaction for a M—H—M interaction substantially increases the steric interaction between the equatorial CO ligands and the alkyne substituent. In this work, we were not able to separate the effects of H vs Fe in replacing Co with FeH (see also below).

The parallel comparison for boron is not as informative as no tautomeric equilibria have been observed. To be observed, the tautomers must have nearly equal energies as the NMR experiment has a dynamic range limited to ≈ 10^2. However, observations of metallaboranes and hydrocarbyl complexes in a closely related system are consistent with the above conclusions. This comparison is shown in Figure 14-9. It differs from the above trimetal cluster as there are now 10 CO ligands on the metal framework.[24,60–62] The high π-acceptor character of CO changes the overall effective electronegativity of the $Fe_3(CO)_{10}$ fragment relative to the $H_2Fe_3(CO)_9$ fragment and, in addition, the ligand requirements of CO vs 2 H are different. It is difficult to know precisely how these differences will affect the electronic situations and, consequently, frameworks that differ in numbers of CO ligands cannot be quantitatively compared.

The $BH_3Fe_3(CO)_{10}$ exhibits a single form having the same hydrogen atom distribution of $[HFe_3(CO)_9(HCH)]^-$. The hydrogen atom distribution of the ferraborane monoanion differs from that of the isoelectronic organometallic cluster in the manner that one would predict from the discussion above, i.e., the ferraborane has an Fe—H—B hydrogen whereas the hydrocarbyl cluster has an

Fe—H—Fe hydrogen. Replacement of one FeH of the former with Co in the neutral ferraborane leads to the same hydrogen atom distribution as observed for $[Fe_3(CO)_{10}(HBH)]^-$. However, the change in electronegativity is in the direction that favors Fe—H—B interactions, and so it drives the system even further from tautomeric equilibrium. Although the hydrogen atom distributions observed are consistent with C vs BH changes in Fe clusters and Co vs FeH changes in hydrocarbyl clusters, the result for Co vs FeH in ferraborane clusters is a null one. Note that, in the dianions, the unique bridging carbonyl ligand moves from the metals to the capping main-group atom.

The importance of the metal ancillary ligands in determining fragment properties becomes clear from a comparison of the isolobal, but not isoelectronic, $B_2H_6Fe_2(CO)_6$ and $B_2H_6Co_2Cp^*_2$ (where $Cp^* = \eta^5\text{-}C_5Me_5$) molecules. The former exhibits a single tautomeric form, albeit fluxional, with three Fe—H—B bridges and one B—H—B bridge (see Figure 14-10),[23] whereas the latter exhibits two tautomers. One is analogous to that of $B_2H_6Fe_2(CO)_6$, whereas the other has four Co—H—B bridges.[34] As the bridging-hydrogen location seems to be driven by available electronic charge,[63] this observation is consistent with a Cp*Co fragment being more electron rich than a $Fe(CO)_3$ fragment, i.e., the donor–acceptor properties of the ligands are more important than the intrinsic electronegativities of the metal atoms.

14.6   REACTION PROPERTIES: BRÖNSTED ACIDITY

Although hydrogen atom location proved to be a sensitive probe of the variation of certain cluster properties vs cluster composition, the approach only works if one can make the right compound pairs. Further, the least ambiguous results ensue from the observation of tautomeric equilibria, as one can then precisely specify the free energy, and sometimes enthalpy and entropy, differences between two or more defined structural forms.

Obviously, it would be more useful to examine a property containing similar information but one that can be measured quantitatively for all systems that one manages to make. The Brönsted acidity associated with B—H—B and M—H—M hydrogen atoms is well known and, in the case of the borane cages, a dependence on cluster size and shape has been established.[3,64] This property can be measured competitively in a relative sense and can, in principle, be measured quantitatively. Thus, it provides an alternative and more general approach to probe real cluster properties. In fact, Williams has connected trends in acidity with structural features in a semiquantitative fashion in order to demonstrate the existence of a structural pattern.[3] For the boranes, Shore has measured acidity orders as a function of cage size, including the role of substituents.[64] Thus, the increasing acidity of B—H—B hydrogens with cage size is now well established.

There has been considerable interest in the acidity of metal-bound hydrogen atoms over the same time period. For a given ligand set, it is clear that the acidity of the hydrogen atom in a mononuclear compound increases as one goes, e.g., from $HMn(CO)_5$ to $H_2Fe(CO)_4$ to $HCo(CO)_4$. In this case, pK_A values have been measured and the differences in acidity are sufficiently large that any difference between mono- and dihydrides is overwhelmed. For different ancillary ligands on the metal, e.g., phosphines, the pK_A values are substantially changed; however, the trend with metal identity remains the same [65–68]

Little quantitative data are available on metallaborane acidities vs boranes, on the one hand, and metal clusters, on the other hand. Here, the ligands on the metal fragment play a major role. For example, the $B_3H_7(ML_n)_2$ *nido-* cluster can be deprotonated with a weak base when ML_n is $Fe(CO)_3$, whereas B_5H_9 itself requires a very strong base. On the other hand, when ML_n is Cp*Co, we have been unable to prepare the anion. Weak bases are insufficient and strong bases result in degradation of the compound. In the case of this particular cluster system, the all-metal analog is unknown, i.e., the closest relative is $CFe_5(CO)_{15}$, which has no protons.

Despite not being the universal measurement it appears to be, when the proton transfer chemistry can be carried out, the approach gives substantial information of the type desired. In addition, the information is valuable as it also defines one important aspect of compound reactivity. Thus, we have examined some of the acid–base properties of metallaboranes as a function of metal identity.

The first set of compounds for which measurements have been completed are shown in Figure 14-10.[61,69,70] Again, each compound is strictly isoelectronic with the other two with identical ancillary ligands on the metal centers. As emphasized before, acidities for simply isolobal compounds would be difficult to interpret. Synthesis, then, could not be avoided, and it was necessary to make the cobalt analogs of $B_2H_6Fe_2(CO)_6$. The first, $B_2H_5FeCo(CO)_6$, results from metal fragment exchange with the diiron compound; however, the second, $B_2H_4CO_2(CO)_6$, required discovering the proper L in the reaction of $BH_3 \cdot L$ with $Co_2(CO)_8$.[61,70] A further complication arose when it was discovered that, although the anion $[B_2H_5Fe_2(CO)_6]^-$ is quite stable (more so, in fact, than its parent neutral molecule, $B_2H_6Fe_2(CO)_6$), the anions of the cobalt derivatives were highly unstable. Fortunately, $[B_2H_4FeCo(CO)_6]^-$ is stable in the solid state and can be quantitatively reprotonated to the neutral. However, in the presence of

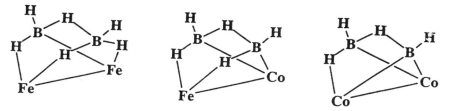

Figure 14-10 Structures of $B_2H_6Fe_2(CO)_6$, $B_2H_5FeCo(CO)_6$, and $B_2H_4Co(CO)_6$. Three terminal CO ligands (not shown) are attached to each metal atom.

polar solvents like THF, the anion rapidly decomposes. All attempts to deprotonate the dicobalt analog resulted in total decomposition. We attribute the latter to a very high reactivity of the anion via electron transfer processes that lead to rapid decomposition.

For this four-atom system, we were left with only two suitable compounds. However, the competition reactions were clean and definitive. Reaction (1) did not occur and reaction (2) was quantitative.

$$[B_2H_5Fe_2(CO)_6]^- + B_2H_5FeCo(CO)_6 \rightarrow B_2H_6Fe_2(CO)_6 + [B_2H_4FeCo(CO)_6]^- \quad (1)$$

$$B_2H_6Fe_2(CO)_6 + [B_2H_4FeCo(CO)_6]^- \rightarrow [B_2H_5Fe_2(CO)_6]^- + B_2H_5FeCo(CO)_6 \quad (2)$$

Therefore, contrary to our expectations, $B_2H_6Fe_2(CO)_6$ is a stronger Brönsted acid than $B_2H_5FeCo(CO)_6$.

Cluster size is not a variable and the metal ligands are identical. The only differences between the two compounds are the identities of the metal atoms and the number of M—H—B bridges. Based on the unambiguous results for mononuclear metal carbonyl hydrides, the change from Fe to Co should increase, rather than decrease, acidity. Thus, in contrast to the case for the mononuclear compounds where the number of hydrogen atoms had no qualitative effect on acidity, here, one is forced to associate a greater acidity with the larger number of bridging hydrogen atoms in $B_2H_6Fe_2(CO)_6$.

In replacing FeH with Co, one is effectively taking an unshielded proton and placing it in a metal nucleus where it is highly shielded by the metal core electrons. Its effect on the valence properties is clearly going to be larger in the first case. Thus, if the process by which a proton is lost is strongly affected by the other hydrogens present, then an effect on acidity might be expected.

Two factors make the cluster situation significantly different from the mononuclear compounds. First, in both $B_2H_6Fe_2(CO)_6$ and $B_2H_5FeCo(CO)_6$, it is a Fe—H—B bridging hydrogen that is removed as a proton, and the metal change must be an indirect one via cage bonding. Second, the extra hydrogen on $B_2H_6Fe_2(CO)_6$ is a bridging hydrogen, which must strongly perturb the cluster bonding network. It could very well play a role in the stabilization of the anion formed upon proton loss. Of course, we have not measured absolute acidities and the difference in acidity between the two compounds may well be small. However, the direction of reaction is established and, at least insofar as these two compounds are concerned, has consequences for their reactivity.

14.7   CONCLUSIONS

This brief account strongly emphasizes the fine balance between completing factors in determining structure and properties. Of course, this competition is present in all compounds but it appears to be a particularly delicate one in the case of cluster compounds containing metal atoms. The downside is that one must be extremely careful about drawing conclusions when comparing compounds. An isolobal comparison combined with the electron-counting rules is fine for interpreting basic structure. Even in this sense, there are a significant number of established exceptions, some of which can be explained on the basis of a closer examination of the details of the electronic structure. An isolobal comparison is useless, however, if one is interested in physical or reaction properties, as these may be more a function of the "ligands" present rather than of the cluster structure. The upside is that transition-metal systems permit fine control of properties. Thus, a detailed understanding of the origin of metallaborane cluster properties will, ultimately, permit the design of clusters with specified values of the desired properties. Just as Williams succeeded in revealing the unique aspects of the strange organoboron compounds that we know as carboranes, so, too, are we, and others, following in his footsteps in the exploration of strange metal–boron complexes known as metallaboranes. The fragments are different but the approach is the same.

ACKNOWLEDGMENTS

The results from Notre Dame constitute the work of the excellent set of coworkers, listed in the references, whom I would like to thank. The support of our research by the National Science Foundation is gratefully acknowledged.

REFERENCES

1. Williams, R. E., *Prog. Boron Chem.*, **2**, 37, 1970.
2. Williams, R. E., *Inorg. Chem.*, **10**, 210, 1971.
3. Williams, R. E., *Adv. Inorg. Chem. Radiochem.*, **18**, 67, 1976.
4. Wade, K., *Electron Deficient Compounds,* Nelson, London, 1971.
5. Wade, K., *New Scientist,* **62**, 615, 1974.
6. Wade, K., *Adv. Inorg. Chem. Radiochem.*, **18**, 1, 1976.
7. Mingos, D. M. P., *Acc. Chem. Res.*, **17**, 311–319, 1984.
8. Mingos, D. M. P. and May, A. S., in *The Chemistry of Metal Cluster Complexes* (D. F. Shriver, H. D. Kaesz, and R. D. Adams, eds) VCH, New York, 1990, p. 11.
9. Mingos, D. M. P. and Wales, D. J., *Introduction to Cluster Chemistry,* Prentice Hall, New York, 1990.
10. Aradi, A. A. and Fehlner, T. P., *Adv. Organomet. Chem.*, **30**, 189, 1990.
11. Hoffmann, R., *Science,* **211**, 995, 1981.
12. Mingos, D. M. P. and Johnston, R. L., *Structure and Bonding,* **68**, 29, 1987.
13. Fehlner, T. P., Ed., *Inorganometallic Chemistry,* Plenum, New York, 1992.
14. Elschenbroich, C. and Salzer, A., *Organometallics,* VCH, New York, 1989.
15. Grimes, R. N., in *Metal Interactions with Boron Clusters* (R. N. Grimes, ed.), Plenum, New York, 1982, p. 269.

16. Housecroft, C. E. and Fehlner, T. P., *Adv. Organomet. Chem.*, **21**, 57, 1982.

17. Housecroft, C. E., *Coordination Chem. Rev.*, **143**, 297, 1995.

18. Kennedy, J. D., *Prog. Inorg. Chem.*, **32**, 519, 1984.

19. Kennedy, J. D., *Prog. in Inorg. Chem.*, **34**, 211, 1986.

20. Williams, R. E. and Field, L. D., in *Boron Chemistry-4* (R. W. Parry and G. Kodama, eds) Pergamon, Oxford, 1980, p. 131.

21. Lipscomb, W. N., *Boron Hydrides,* Benjamin, New York, 1963.

22. Gaines, D. F. and Hildebrandt, S. J., *Inorg. Chem.*, **17**, 794, 1978.

23. Jacobsen, G. B., Andersen, E. L., Housecroft, C. E., Hong, F.-E., Buhl, M. L., Long, G. J., and Fehlner, T. P., *Inorg. Chem.*, **26**, 4040, 1987.

24. Vites, J. C., Housecroft, C. E., Eigenbrot, C., Buhl, M. L., Long, G. J., and Fehlner, T. P., *J. Am. Chem. Soc.*, **108**, 3304, 1986.

25. Chini, P., Longoni, G., Martinengo, S., and Ceriotti, A., *Adv. in Chem. Series* **167**, 1, 1978.

26. Hollins, R. E. and Stafford, F. E., *Inorg. Chem.*, **9**, 877, 1970.

27. Hildebrandt, S. J., Gaines, D. F., and Calabrese, J. C., *Inorg. Chem.*, **17**, 790, 1978.

28. Jan, D.-Y., Workman, D. P., Hsu, L.-Y., Krause, J. A., and Shore, S. G., *Inorg. Chem.*, **31**, 5123, 1992.

29. Mingos, D. M. P., *Inorg. Chem.*, **21**, 464, 1982.

30. Greenwood, N. N., Savory, C. G., Grimes, R. N., Sneddon, L. G., Davison, A., and Wreford, S. S., *J. Chem. Soc., Chem. Commun.*, 718, 1974.

31. Miller, V. R., Weiss, R., and Grimes, R. N., *J. Am. Chem. Soc.*, **99**, 5646, 1977.

32. Weiss, R., Bowser, J. R., and Grimes, R. N., *Inorg. Chem.*, **17**, 1522, 1978.

33. Andersen, E. L., Haller, K. J., and Fehlner, T. P., *J. Am. Chem. Soc.*, **101**, 4390, 1979.

34. Nishihara, Y., Deck, K. J., Shang, M., Fehlner, T. P., Haggerty, B. S., and Rheingold, A. L., *Organometallics,* **13**, 4510, 1994.

35. Bould, J., Pasieka, M., Braddock-Wilking, J., Rath, N. P., Barton, L., and Gloeckner, C., *Organometallics,* **14**, 5138, 1995.

36. Chipperfield, A. K., Housecroft, C. E., and Matthews, D. M., *J. Organomet. Chem.*, **384**, C38, 1990.

37. Housecroft, C. E., Matthews, D. M., and Rheingold, A. L., *J. Chem. Soc., Chem. Commun.*, 323, 1992.

38. Fehlner, T. P., Housecroft, C. E., Scheidt, W. R., and Wong, K. S., *Organometallics,* **2**, 825, 1983.

39. Braye, E. H., Dahl, L. F., Hubel, W., and Wampler, D. L., *J. Am. Chem. Soc.*, **84**, 4633, 1962.

40. Chung, J.-H., Knoeppel, D., McCarthy, D., Columbie, A., and Shore, S. G., *Inorg. Chem.*, **32**, 3391, 1993.

41. Housecroft, C. E., Matthews, D. M., and Rheingold, A. L., *Organometallics,* **11**, 2959, 1992.

42. Hong, F. E., Coffy, T. J., McCarthy, D. A., and Shore, S. G., *Inorg. Chem.*, **28**, 3284, 1989.

43. Draper, S. M., Housecroft, C. E., Keep, A. K., Matthews, D. M., Song, X., and Rheingold, A. L., *J. Organomet. Chem.*, **423**, 241, 1992.

44. Housecroft, C. E., Matthews, D. M., Rheingold, A. L., and Song, X., *J. Chem. Soc., Chem. Commun.*, 842, 1992.

45. Bandyopadhyay, A., Shang, M., Jun, C. -S., and Fehlner, T. P., *Inorg. Chem.*, **33**, 3677, 1994.

46. Housecroft, C. E., *Chem. Soc. Rev.*, 215, 1995.

47. Basil, J. D., Aradi, A. A., Bhattacharyya, N. K., Rath, N. P., Eigenbrot, C., and Fehlner, T. P., *Inorg. Chem.*, **29**, 1260, 1990.

48. Coffy, T. J., Medford, G., Plotkin, J., Long, G. J., Huffman, J. C., and Shore, S. G., *Organometallics*, **8**, 2404, 1989.

49. Shore, S. G., Raganini, D., Smith, R. L., Cottrell, C. E., and Fehlner, T. P., *Inorg. Chem.*, **18**, 670, 1979.

50. Housecroft, C. E. and Fehlner, T. P., *Inorg. Chem.*, **21**, 1739, 1982.

51. Hawthorne, M. F., *J. Organomet. Chem.*, **100**, 97, 1975.

52. Grimes, R. N., *Pure Appl. Chem.*, **54**, 43, 1982.

53. Siebert, W., *Pure Appl. Chem.*, **59**, 947, 1987.

54. Dutta, T. K., Vites, J. V., Jacobsen, G. B., and Fehlner, T. P., *Organometallics*, **6**, 842, 1987.

55. Lynam, M. M., Chipman, D. M., Barreto, R. D., and Fehlner, T. P., *Organometallics*, **6**, 2405, 1987.

56. Bower, D. K. and Keister, J. B., *Organometallics*, **9**, 1656, 1990.

57. Dutta, T. K., Meng, X., Vites, J. C., and Fehlner, T. P., *Organometallics*, **6**, 2191, 1987.

58. Duggan, T. P., Golden, M. J., and Keister, J. B., *Organometallics*, **9**, 1656, 1990.

59. Barreto, R. D., Puga, J., and Fehlner, T. P., *Organometallics*, **9**, 662, 1990.

60. Crascall, L. E., Thimmappa, B. H. S., Rheingold, A. L., Ostrander, R., and Fehlner, T. P., *Organometallics*, **13**, 2153, 1994.

61. Jun, C.-S., Bandyopadhyay, A. K., and Fehlner, T. P., *Inorg. Chem.*, **35**, 2189, 1996.

62. Kolis, J. W., Holt, E. M., and Shriver, D. F., *J. Am. Chem. Soc.*, **105**, 7307, 1983.

63. Fehlner, T. P., *Polyhedron*, **9**, 1955, 1990.

64. Shore, S. G., in *Boron Hydride Chemistry* (E. L. Muetterties, ed.), Academic Press, New York, 1975, p. 79.

65. Moore, E. J., Sullivan, J. M., and Norton, J. R., *J. Am. Chem. Soc.*, **108**, 2257, 1986.

66. Walker, H. W., Pearson, R. G., and Ford, P. C., *J. Am. Chem. Soc.*, **105**, 1179, 1983.

67. Kristjánsdóttir, S. S., Moody, A. E., Weberg, R. T., and Norton, J. R., *Organometallics*, **7**, 1983, 1988.

68. Weberg, R. T. and Norton, J. R., *J. Am. Chem. Soc.*, **112**, 1105, 1990.

69. Jun, C.-S. and Fehlner, T. P., *Organometallics*, **13**, 2145, 1994.

70. Jun, C.-S., Halet, J.-F., Rheingold, A. L., and Fehlner, T. P., *Inorg. Chem.*, **34**, 2101, 1995.

15

CYCLIC ORGANOHYDROBORATE ANIONS: HYDRIDE TRANSFER REACTIONS IN THE REDUCTION OF METAL CARBONYLS AND ORGANIC FUNCTIONAL GROUPS; REACTIONS WITH ZIRCONOCENE AND HAFNOCENE DICHLORIDES

SHELDON G. SHORE, GLENN T. JORDAN IV, JIANPING LIU, FU-CHEN LIU, EDWARD A. MEYERS, AND PAUL L. GAUS

Department of Chemistry, Ohio State University, 120 West 18th Avenue, Columbus, Ohio 43210

15.1 INTRODUCTION

Organoboranes are very useful chemical reagents in organic chemistry.[1] A product of the hydroboration of 1,3-butadiene is $(\mu\text{-H})_2B_2(\mu\text{-}C_4H_8)_2$, compound **1**, a transannular hydrogen-bridged ten-membered ring.[2] Our studies of this compound,[2] over 25 years ago, provided evidence for the structure proposed earlier by Zweifel and colleagues.[3] This compound is benign in character compared with other organodiboranes. It is stable in air and in the presence of water for extended periods of time. On the other hand the product of the hydro-

| 1 | 2 |

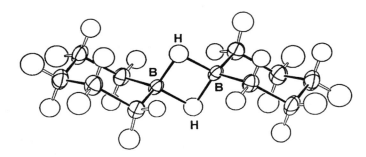

Figure 15-1   An ORTEP drawing of the molecular structure of $(\mu\text{-H})_2B_2(C_5H_{10})_2$ with 25% probability thermal ellipsoids.

boration of 1,4-pentadiene is $(\mu\text{-H})_2B_2(C_5H_{10})_2$, compound **2**.[4a,4b] It has a structure that consists of two boracyclopentane rings that are joined by a double hydrogen bridge. This structure was first proposed by Köster.[4a,4b] Recently, we confirmed this structure by means of a single crystal X-ray analysis (Figure 15-1).[4c] Unlike complex **1**, complex **2** is very sensitive to air and moisture.

Alkali metal tetrahydroborates have been studied extensively and have been found to be very useful reagents for organic and organometallic reactions,[5] and alkali metal organohydroborates are attracting increasing attention as potentially useful reagents.[5,6] Here, we summarize our recent research that is focused on syntheses and reactivities of alkali metal organohydroborates derived from complexes **1** and **2**. These represent the first detailed studies of cyclic organohydroborate anions. Also reported here are the subsequent reactions and structures of products derived from these reactions.

15.2   FORMATION OF ALKALI METAL CYCLIC ORGANOHYDROBORATE ANIONS M[(μ-H)B$_2$H$_2$(μ-C$_4$H$_8$)$_2$], COMPOUND 3, (M = Na, K) AND K[H$_2$BC$_5$H$_{10}$], COMPOUND 4

While structure **1** is relatively inert with respect to its sensitivity to air and moisture, it readily accepts hydride ion from alkali metal hydrides to form the transannular singly hydrogen bridged, ten-membered ring anion [(μ-H)B$_2$H$_2$(μ-C$_4$H$_8$)$_2$]$^-$, compound **3** [Reaction (1)].[2b,2c]

$$\text{(1)}$$

(M = Na, K)

1   **3**

Its structure was determined by a single-crystal X-ray analysis (Figure 15-2).[2c]

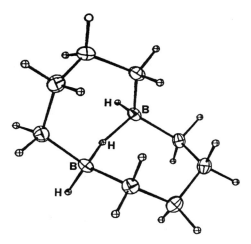

Figure 15-2 An ORTEP drawing of the molecular structure of $[(\mu\text{-}H)B_2H_2(\mu\text{-}C_4H_8)_2]^-$ anion.

Alkali metal hydrides react with **2** to yield the cyclic anion $[H_2BC_5H_{10}]^-$, compound **4** [Reaction (2)].[4c]

$$\text{(structures)} + 2KH \xrightarrow{\text{THF}} 2K\left[\text{structure} \right] \qquad (2)$$

$$\mathbf{2} \qquad\qquad\qquad\qquad \mathbf{4}$$

These reactions can be carried out quantitatively in an atmosphere of nitrogen or under vacuum in THF solution and they are complete within an hour. These hydrides are stable in THF solution for several days. When solvent is removed from the product in Reaction (1), a viscous oil is obtained. However, it can be converted into crystalline $[NEt_4][(\mu\text{-}H)B_2H_2(\mu\text{-}C_4H_8)_2]$ through a metathesis reaction with $[NEt_4]Cl$. On the other hand, a white solid $K[H_2BC_5H_{10}]$ is formed when solvent is pumped away from Reaction (2). The NMR spectra of compound **4** are consistent with its proposed structure.

15.3 HYDRIDE TRANSFER REAGENTS $[(\mu\text{-}H)B_2H_2(\mu\text{-}C_4H_8)_2]^-$, COMPOUND 3, AND $[H_2BC_5H_{10}]^-$, COMPOUND 4

15.3.1 Reduction of Metal Carbonyls by $[(\mu\text{-}H)B_2H_2(\mu\text{-}C_4H_8)_2]^-$, Compound 3

The anion, compound **3**, reacts with metal carbonyls [Reactions (3)–(6)]

$$Cr(CO)_6 + K[(\mu\text{-}H)B_2H_2(\mu\text{-}C_4H_8)_2] \longrightarrow$$
$$K[Cr(CO)_5CHO] + (\mu\text{-}H)_2B_2(\mu\text{-}C_4H_8)_2 \qquad (3)$$

$$Fe(CO)_5 + K[(\mu\text{-}H)B_2H_2(\mu\text{-}C_4H_8)_2] \longrightarrow$$
$$K[Fe(CO)_4CHO] + (\mu\text{-}H)_2B_2(\mu\text{-}C_4H_8)_2 \qquad (4)$$

$$2Cr(CO)_5{\cdot}THF + K[(\mu\text{-}H)B_2H_2(\mu\text{-}C_4H_8)_2] \longrightarrow$$
$$K\{(\mu\text{-}H)[Cr(CO)_5]_2\} + (\mu\text{-}H)_2B_2(\mu\text{-}C_4H_8)_2 \qquad (5)$$

$$Ru_3(CO)_{12} + K[(\mu\text{-}H)B_2H_2(\mu\text{-}C_4H_8)_2] \longrightarrow$$
$$K[HRu_3(CO)_{11}] + (\mu\text{-}H)_2B_2(\mu\text{-}C_4H_8)_2 + CO \qquad (6)$$

in much the same manner as organoborohydride anions, $[HBR_3]^-$ and alkoxy-borohydride anions $[HB(OR)_3]^-$.[7] The advantages of compound **3** over $[HBR_3]^-$ and $[HB(OR)_3]^-$ are that the workup procedure is less demanding since the organodiborane product, compound **1**, is unreactive compared with BR_3 and $B(OR)_3$, which are formed from hydride transfer reactions of $[HBR_3]^-$ and $[HB(OR)_3]^-$. Also, the half-life of the formyl formed in Reaction (3) is longer than that reported when $[HB(OR)_3]^-$ is employed as the hydride transfer agent.[7a]

15.3.2   Reduction of Alkyl Halides, Ketones, and Aldehydes by $[(\mu\text{-}H)B_2H_2(\mu\text{-}C_4H_8)_2]^-$, Compound 3

Several reactions were studied in order to demonstrate the application of compound **3** to the reduction of alkyl halides, ketones, and aldehydes [Reactions (7)–(12)]. Reductive dehalogenation of organic halides through hydride ion transfer from $M[(\mu\text{-}H)B_2H_2(\mu\text{-}C_4H_8)_2]$ (M = Na, K) gave the corresponding alkanes in yields of 95% or higher.

$$(X = Cl, Br; M = K \text{ and } Na)$$

$$+ MX + (\mu\text{-}H)_2B_2(\mu\text{-}C_4H_8)_2 \qquad (7)$$

$$n\text{-}CH_3(CH_2)_6CH_2Br + Na[(\mu\text{-}H)B_2H_2(\mu\text{-}C_4H_8)_2] \xrightarrow{\text{THF}} \qquad (8)$$
$$n\text{-}CH_3(CH_2)_6CH_3 + NaBr + (\mu\text{-}H)_2B_2(\mu\text{-}C_4H_8)_2$$

$$\text{PhCHO} \xrightarrow[0°C, THF]{K[(\mu\text{-}H)B_2H_2(\mu\text{-}C_4H_8)_2]} \xrightarrow{HCl} \text{PhCH}_2\text{OH} \quad (11)$$

$$\text{PhCH=CHCHO} \xrightarrow[0°C, THF]{K[(\mu\text{-}H)B_2H_2(\mu\text{-}C_4H_8)_2]} \xrightarrow{HCl} \text{PhCH=CHCH}_2\text{OH} \quad (12)$$

Ketones and aldehydes were quantitatively reduced to alcohols by compound **3** in THF solution. As indicated in Reaction (9), the reduction product, 4-*tert*-butyl-cyclohexanol, consists of a 90:10 ratio of the axial to equatorial conformation isomers. The stereoselectivity of compound **3** is comparable to other bulky organoborohydride anions.[6b,6c,8] In competition reactions, aldehydes were reduced relatively faster than ketones by compound **3**, probably due to the greater steric bulks of the ketones.

15.3.3 Reduction of alkyl halides by alkali metal hydrides (NaH and KH) Catalyzed by $(\mu\text{-}H)_2B_2(\mu\text{-}C_4H_8)_2$, Compound 1

Although commercially available alkali metal hydrides are strong reducing reagents, they do not reduce alkyl halides under mild conditions,[9] probably due to the limited solubilities of the alkali metal hydrides. In effect, the alkali metal hydride is solubilized by compound **1** in ethers. The fact that $(\mu\text{-}H)_2B_2(\mu\text{-}C_4H_8)_2$ is stable and readily accepts, and then transfers, hydride ion, encouraged us to demonstrate that practical catalytic reductive dehalogenation of organic halides can be achieved by employing compound **1** as a catalyst. Therefore, Reactions (7) and (8) were also carried out under catalytic conditions [Reactions (13) and (14)] using a 1:1 molar ratio of alkali metal hydride to organic halide in the presence of only 0.2 molar equivalent of compound **1**. In the absence of compound **1**, no reaction occurred. But, with the addition of a small amount of compound **1** to the reaction system, quantitative dehalogenation occurred in approximately the same time period as when the reaction was carried out using compound **3** and the respective organic halide in a 1:1 molar ratio [Reactions (7) and (8)]. The rate-determining step is the dehalogenation step.

$$\text{PhCH}_2\text{X} + \text{MH} \xrightarrow[THF]{B{<}^H_H{>}B} \text{PhCH}_3 + \text{MX} \quad (13)$$

$$(X = Cl, Br; M = K \text{ and } Na)$$

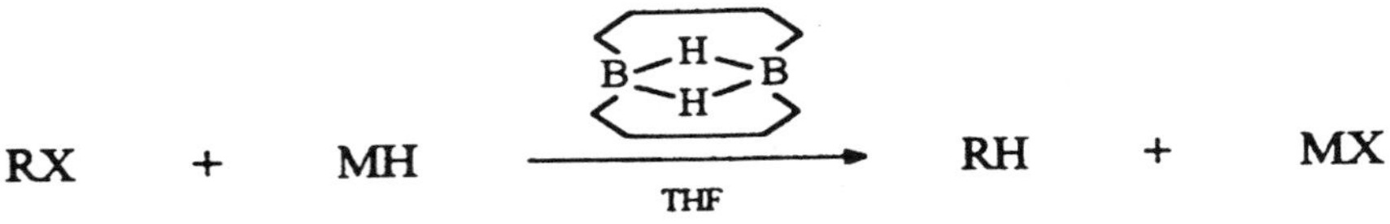
RX + MH
THF
RH + MX

(RX = organic halides, MH = NaH and KH)

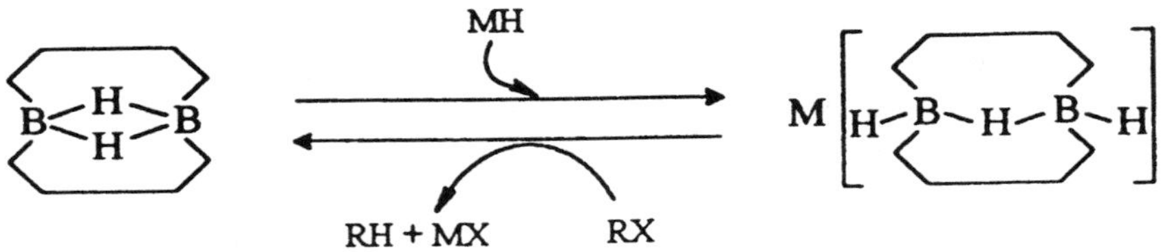
MH
M
RH + MX
RX

Scheme 1 The proposed catalytic pathway for the reduction of alkyl halides by metal hydrides.

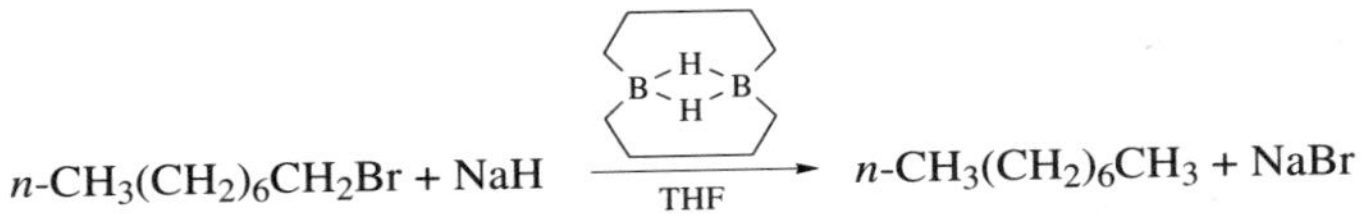
n-CH3(CH2)6CH2Br + NaH
THF
n-CH3(CH2)6CH3 + NaBr

(14)

The proposed catalytic pathway is shown in Scheme 1.

15.3.4 Hydride Transfer by $[H_2BC_5H_{10}]^-$, Compound 4

When one equivalent $K[H_2BC_5H_{10}]$ was added to a BH_3THF solution, hydride ion was immediately transferred to BH_3 to form a precipitate of KBH_4 and the neutral dimer $(\mu\text{-}H)_2B_2(C_5H_{10})_2$, compound **2**, [Reaction (15)].

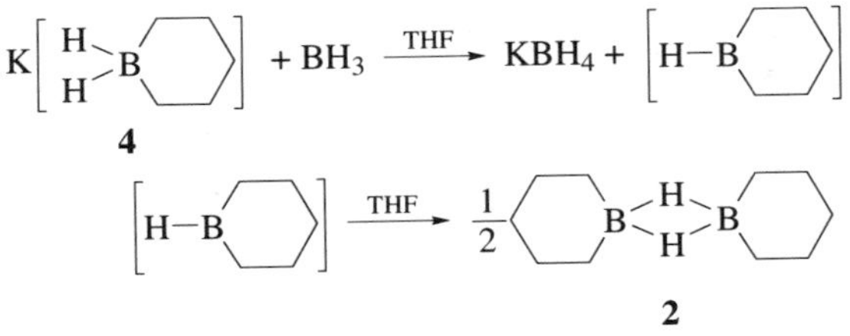
+ BH3
THF
KBH4 +
4
THF
2

(15)

The reaction of $K[H_2BC_5H_{10}]$ with 4-*tert*-butyl-cyclohexanone [Reaction (16)]

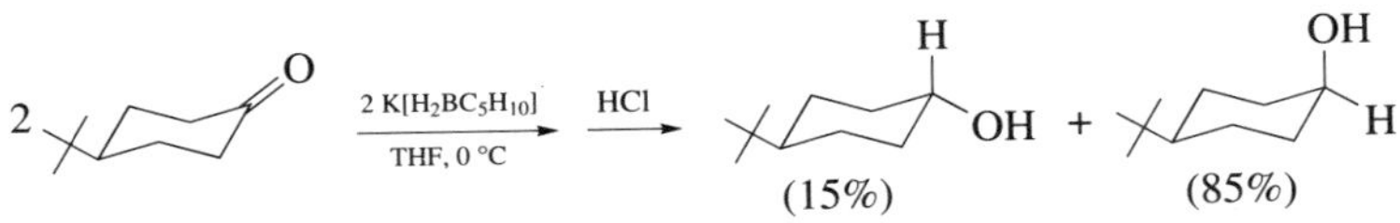
2 K[H2BC5H10]
THF, 0 °C
HCl
(15%)
(85%)

(16)

produced the *trans-* (85% yield) and the *cis-* (15% yield) isomers of 4-*tert*-butyl-cyclohexanol.[4c] These yields of isomers are comparable to those observed in Reaction (9).

15.4   ORGANOBORANE DERIVATIVES OF Cp$_2$MCl$_2$ (M = Zr, Hf)

15.4.1   Reactions of Cp$_2$MCl$_2$ (M = Zr, Hf) with [(μ-H)B$_2$H$_2$(μ-C$_4$H$_8$)$_2$]⁻, Compound 3

In addition to its utility as a hydride transfer agent, as discussed above, it was found that compound **3** undergoes a unique reaction with zirconocene and hafnocene dichlorides. An unanticipated transformation of the anion occurs and the complexes (η^5-C$_5$H$_5$)$_2$ZrCl[(μ-H)$_2$BC$_4$H$_8$], compound **5**, and (η^5-C$_5$H$_5$)$_2$HfCl-[(μ-H)$_2$BC$_4$H$_8$], compound **6** [Reaction (17)], which contain a double hydrogen bridge between the metal and boron are produced.[10] The formation of complexes **5** and **6** are the first reactions in which compound **3** disproportionates to form a five-membered borocyclopentane ring.

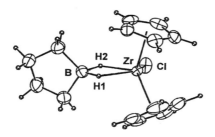

(17)

(M = Zr, compound **5**; Hf, compound **6**)

The molecular structures of these complexes (Figure 15-3) were determined by single-crystal X-ray analyses. Their NMR and IR spectra in solution are consistent with these structures. These structures are analogs of the tetrahydroborate complex (η^5-C$_5$H$_5$)$_2$ZrCl[(μ-H)$_2$BH$_2$].[11] Distances involving metal, hydrogen, and boron are: Zr—H(1) = 2.00(3) Å, Zr—H(2) = 1.97(3) Å, Zr—B = 2.572(3) Å, B—H(1) = 1.24(2) Å, B—H(2) = 1.22 (3) Å, Hf—H(1) = 1.85(7) Å, Hf—H(2) = 2.00(7) Å, Hf—B = 2.537(6) Å, B—H(1) = 1.19(6) Å, and B—H(2) = 1.18 (7) Å. These are consistent with distances observed in other bidentate

Figure 15-3   (*contd.*)

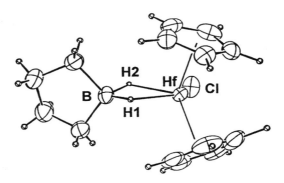

Figure 15-3   (a) An ORTEP drawing of the molecular structure of (η^5-C$_5$H$_5$)$_2$ZrCl-[(μ-H)$_2$BC$_4$H$_8$] and (b) an ORTEP drawing of the molecular structure of (η^5-C$_5$H$_5$)$_2$HfCl-[(μ-H)$_2$BC$_4$H$_8$] with 50% probability thermal ellipsoids.

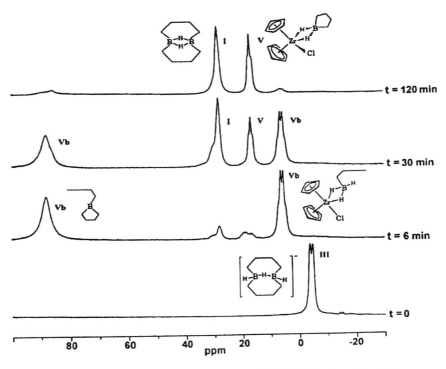

Figure 15-4   Stack-plot of time-elapsed ^{11}B NMR for Reaction (17).

metalloborohydride systems.[11,12] The geometry around the zirconium center is distorted tetrahedral, consisting of a coordination sphere of B, Cl, and the centroids of the two C$_5$H$_5^-$ rings. The Cl—Zr—B angle is 100.47(7)° and the centroid—Zr—Centroid angle is 129.3°. The bridging hydrogens of compounds **5** and **6** are not equivalent on the ^{1}H NMR time scale at room temperature in THF-d_8. Two resonances are observed at room temperature. They collapse to a single signal at around 50°C.

In the formation of complexes **5** and **6**, rearrangement of the anion, compound **3**, is facilitated by the presence of the electron-deficient 16-electron metal center. Time-elapsed ^{11}B NMR spectra, in the formation of compound **5** are provided in a stack-plot (Figure 15-4). They provide insight into the intermediates produced in the formation of compound **5** and most likely compound **6**. Scheme 2 depicts a proposed reaction pathway based upon these spectra. Six

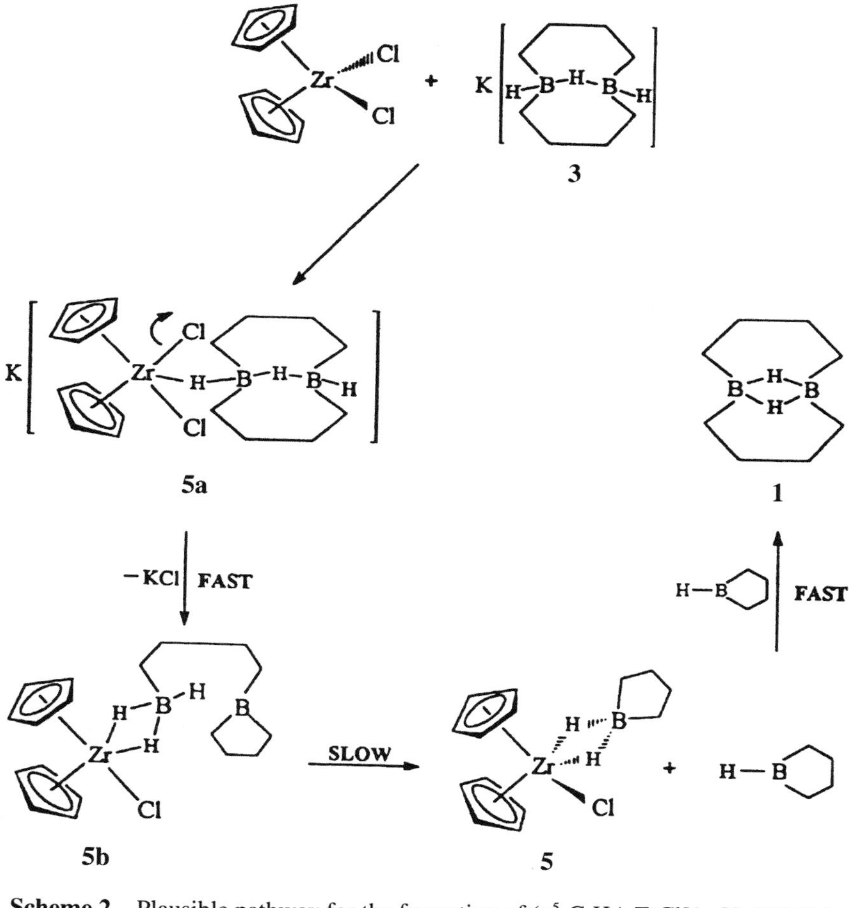

Scheme 2 Plausible pathway for the formation of (η^5-C$_5$H$_5$)$_2$ZrCl[(μ-H)$_2$BC$_4$H$_8$].

minutes after initiating the reaction, all of compound **3** is consumed. A reasonable initial step would involve coordination of compound **3** to the electron-deficient zirconium through a hydrogen-bridge bond to form compound **5a**, an 18-electron complex. Departure of Cl⁻ from compound **5a** and rapid molecular rearrangement result in the formation of the intermediate, compound **5b**. The ^{11}B NMR signals at +88.1 (s) and +5.8 ppm [q, $J(^{11}B,^1H) = 71$ Hz] in a 1:1 ratio support the presence of compound **5b**. The broad downfield resonance is indicative of a trialkylborane.[2c,13] The coupling constant for the upfield quartet is intermediate for terminal and bridge B–H coupling, which is consistent with the bridge hydrogens rapidly exchanging with the terminal hydrogen (rapid exchange of bridging and terminal hydrogens of metallotetrahydroborates at room temperature is well known.[14]) Over the next 2 h, signals for compound **5b** disappear, while those for compounds **1**, (+28.2 ppm)[2a] and **5** appear in a 1:1 ratio that is constant as they increase in intensity with time. Formation of compound **5** results in elimination of the cyclic borane [HBC$_4$H$_8$], which rapidly dimerizes to the transannular hydrogenbridged organoborane, compound **1**.

15.4.2 Reactions of Cp$_2$MCl$_2$ (M = Zr, Hf) with [H$_2$BC$_5$H$_{10}$]⁻, Compound 4

The anion, compound **4**, reacts with zirconocene and hafnocene dichlorides to produce (η^5-C$_5$H$_5$)$_2$ZrCl[(μ-H)$_2$BC$_5$H$_{10}$], compound **7**, and (η^5-C$_5$H$_5$)$_2$HfCl[(μ-H)$_2$BC$_5$H$_{10}$], compound **8** [Reaction (18)].[4c]

$$(18)$$

(M = Zr, compound **7**; Hf, compound **8**)

The formation of compounds **7** and **8** involves a simple metathesis procedure in which the driving force for the reaction is the precipitation of KCl and the formation of a double hydrogen bridge between the metal and boron.

The molecular structure of compound **7** (Figure 15-5) was determined by a single-crystal X-ray analysis. The NMR and IR spectra in solution are consistent with the solid-state structure. Bond lengths for Zr—H1, Zr—H2, and Zr—B are 2.06(2) Å, 2.10(2) Å, and 2.593(2) Å, respectively. These lengths are consistent with those observed in structure **5** [10] and in other reported systems.[11,12] Distorted tetrahedral geometry occurs around the zirconium center formed by B, Cl, and the centroids of the C$_5$H$_5$ rings, as indicated by the angles of Cl—Zr—B [101.12(6)°] and centroid—Zr—centroid (129.97°).

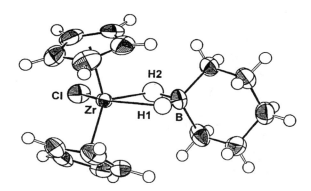

Figure 15-5 An ORTEP drawing of the molecular structure of $(\eta^5$-$C_5H_5)_2$ZrCl-$[(\mu$-H$)_2$BC$_5$H$_{10}]$ with 50% probability thermal ellipsoids.

15.5 REACTIVITIES OF ZIRCONOCENE MONOCHLORO-ORGANOHYDROBORATES $(\eta^5$-$C_5H_5)_2$ZrCl$[(\mu$-H$)_2$BC$_4$H$_8]$, COMPOUND 5, AND $(\eta^5$-$C_5H_5)_2$ZrCl$[(\mu$-H$)_2$BC$_5$H$_{10}]$, COMPOUND 7

15.5.1 Reactions of $(\eta^5$-$C_5H_5)_2$ZrCl$[(\mu$-H$)_2$BC$_4$H$_8]$, Compound 5, and $(\eta^5$-$C_5H_5)_2$ZrCl$[(\mu$-H$)_2$BC$_5$H$_{10}]$, Compound 7, with LiH

Complexes **5** and **7** react with LiH in THF solution to form the complexes $(\eta^5$-$C_5H_5)_2$ZrH$[(\mu$-H$)_2$BC$_4$H$_8]$, compound **9**, and $(\eta^5$-$C_5H_5)_2$ZrH$[(\mu$-H$)_2$BC$_5$H$_{10}]$, compound **10**, [Reaction (19)].

$$\text{Cp}_2\text{ZrCl(H)(H)BR} + \text{LiH} \xrightarrow{\text{THF}} \text{Cp}_2\text{ZrH(H)(H)BR} + \text{LiCl} \qquad (19)$$

$$\left(\text{BR} = \text{B} \bigpentagon , \text{compound 9};\ \text{B} \bighexagon ,\ \text{compound 10} \right)$$

The molecular structures of compounds **9** and **10** were determined by single-crystal X-ray analyses and they are shown in Figure 15-6. Solution NMR and IR spectra are consistent with the solid-state structures. In both structures, there is a terminal hydride atom that is linked to zirconium. There is also a double hydrogen bridge between zirconium and boron. Distances from the zirconium atom to the bridging hydrogen atoms are comparable to those observed in compounds **5** and **7**. In addition; the distances from the zirconium atom to the terminal hydrogen are 1.68(5) (Å) for compound **9** and 1.76(4) Å for compound **10**, which are consistent with previously reported Zr—H(terminal) bond lengths.[15]

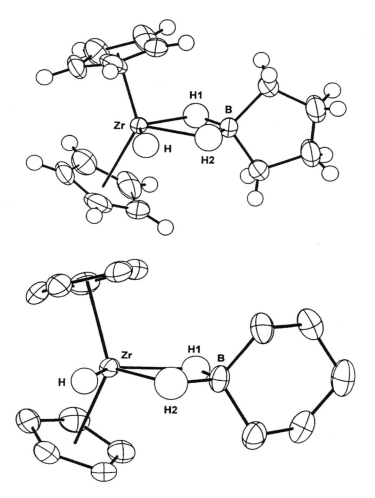

Figure 15-6 (a) An ORTEP drawing of the molecular structure of $(\eta^5\text{-}C_5H_5)_2ZrH$-$[(\mu\text{-}H)_2BC_4H_8]$ and (b) an ORTEP drawing of the molecular structure of $(\eta^5\text{-}C_5H_5)_2ZrH[(\mu\text{-}H)_2BC_5H_{10}]$ with 50% probability thermal ellipsoids.

Complex **9** was also formed when compound **5** reacted with an equivalent of compound **3** [Reaction (20)].

(20)

In this reaction, hydride ion was transferred to compound **5** to replace the chloride ion.

15.5.2 Reactions of $(\eta^5\text{-}C_5H_5)_2ZrCl[(\mu\text{-}H)_2BC_4H_8]$, Compound 5, and $(\eta^5\text{-}C_5H_5)_2ZrCl[(\mu\text{-}H)_2BC_5H_{10}]$, Compound 8, with LiR' (R' = CH$_3$, C$_6$H$_5$)

In addition to hydride displacement of chloride ion in structures **9** and **10** [Reaction (19)], carbanions from organolithium compounds replace the chloride ion [Reaction (21)].

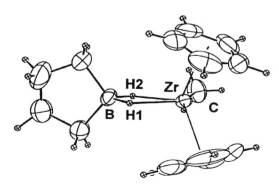

$$\left(BR = B\bigcirc : R' = CH_3,\ \text{compound 11; } C_6H_5,\ \text{compound 12} \right)$$

$$\left(BR = B\bigcirc : R' = CH_3,\ \text{compound 13; } C_6H_5,\ \text{compound 14} \right)$$

The molecular structure of compound **11** was determined by a single-crystal X-ray analysis (Figure 15-7). For the Zr—CH$_3$ group, the zirconium–carbon bond length is 2.286(7) Å. For the double hydrogen bridge between zirconium and boron, the structural parameters are consistent with those found for compounds **5** and **7**. The molecular structures of compounds **12**, **13**, and **14** have not yet been determined by single-crystal X-ray analyses, but can be inferred based upon their NMR spectra. They are most likely closely related to the structure of compound **11**.

Figure 15-7 An ORTEP drawing of the molecular structure of $(\eta^5\text{-}C_5H_5)_2ZrCH_3$-$[(\mu\text{-}H)_2BC_4H_8]$ with 50% probability thermal ellipsoids.

15.5.3 Reactions of $(\eta^5\text{-}C_5H_5)_2ZrCl[(\mu\text{-}H)_2BC_4H_8]$, Compound 5, and $(\eta^5\text{-}C_5H_5)_2ZrCl[(\mu\text{-}H)_2BC_5H_{10}]$, Compound 7, with $C_6H_5CH_2MgCl$

Reactions of compounds **5** and **7** with $C_6H_5CH_2MgCl$ did not result simply in the replacement of chloride atom in the zirconocene complexes; unexpected products were formed. The reaction with complex **7** yielded $(\eta^5\text{-}C_5H_5)_2Zr\{(\mu\text{-}H)_2BC_5H_{10}\}\{(\mu\text{-}H)BC_5H_{10}(CH_2C_6H_5)\}$, compound **15**, and the known complex $(\eta^5\text{-}C_5H_5)_2Zr(CH_2C_6H_5)_2$, compound **16**.[16] The molecular structures of compounds **15** and **16** were determined by single-crystal X-ray analyses and they are shown in Figures 15-8 and 15-9. Complex **15** contains both double and single hydrogen-bridge bonds, between the zirconium and boron atoms. In the single hydrogen-bridge bond, the distance between the zirconium atom and the hydrogen atom is relatively short [1.67(7) Å], significantly less than observed in the double hydrogen bridge [2.08(8) Å and 1.82(8) Å], and significantly less than the sum of the single-bond radii of zirconium and hydrogen (1.77 Å). On the other hand, the hydrogen-boron distance of this single hydrogen-bridged bond [1.61(7) Å] is significantly longer than that observed in the double hydrogen-bridge bond of this molecule [1.32(6) Å and 1.31(6) Å]. It is approximately the same distance as that observed for a terminal Zr—H bond.[15] The "short" Zr—H distance and the "long" B—H distance imply that, in the single hydrogen bridge, the Zr—H is functioning as an electron-pair-donating ligand to the boron atom.

Complex **16** (Figure 15-9) was reported several years ago,[16] but the molecular structure was unreported until the present X-ray analysis. This compound contains two benzyl groups linked to the zirconium atom through methylene groups. The distances between the zirconium atom and the carbons of the methylene groups are 2.314(7) Å and 2.330(8) Å.

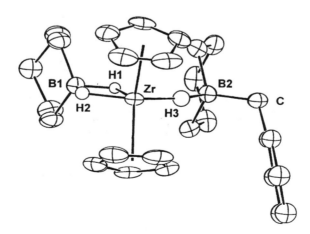

Figure 15-8 An ORTEP drawing of the molecular structure of $(\eta^5\text{-}C_5H_5)_2Zr\{(\mu\text{-}H)_2BC_5H_{10}\}\{(\mu\text{-}H) BC_5H_{10}(CH_2C_6H_5)\}$ with 50% probability thermal ellipsoids.

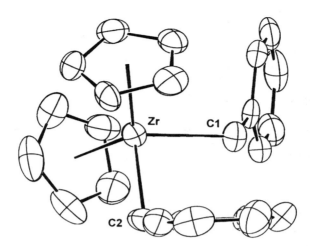

Figure 15-9 An ORTEP drawing of the molecular structure of $(\eta^5\text{-}C_5H_5)_2Zr\text{-}(\mu\text{-}CH_2C_6H_5)_2$ with 50% probability thermal ellipsoids.

From compound **5** reacting with $C_6H_5CH_2MgCl$ in ether, $(\eta^5\text{-}C_5H_5)_2Zr\{(\mu\text{-}H)_2BC_4H_8\}\{(\mu\text{-}H)BC_4H_8(CH_2C_6H_5)\}$, compound **17**, and $(\eta^5\text{-}C_5H_5)_2Zr(CH_2C_6H_5)_2$, $Cp_2Zr(CH_2Ph)[(\mu\text{-}H)_2BC_4H_8]$, compound **18,** were obtained. Molecular structures of these complexes are not yet available, but they can be inferred from their NMR spectra. When a mixed solvent, THF/ether, was employed instead of ether in the reaction between compound **5** and $C_6H_5CH_2MgCl$, complexes **18** and **9** were obtained.

15.6 CONCLUSIONS

Prior to this investigation cyclic organohydroborate anions $[(\mu\text{-}H)B_2H_2(\mu\text{-}C_4H_8)_2]^-$, compound **3**, and $[H_2BC_5H_{10}]^-$, compound **4**, had received virtually no attention with respect to the chemistry that they can undergo. The reactions reported here indicate that these anions have the potential for a rich diversified chemistry.

ACKNOWLEDGMENTS

This work was supported by the National Science Foundation. G.T.J. is grateful to Proctor and Gamble for providing an industrial fellowship.

REFERENCES

1. Brown, H. C., *Boranes in Organic Chemistry*, Part 6, Cornell University Press, Ithaca and London, 1972.

2. (a) Young, D. E. and Shore, S. G., *J. Am. Chem. Soc.*, 1969, **91**, 3497; (b) Clayton, W. R., Saturnino, D. J., Corfield, P. W. R., and Shore, S. G., *J. Chem. Soc., Chem. Commun.*, 1973, 377; (c) Saturnino, D. J., Yamauchi, M., Clayton, W. R., Nelson, R. W., and Shore, S. G., *J. Am. Chem. Soc.*, 1975, **97**, 6063.

3. Zweifel, G., Nagase, K., and Brown, H. C., *J. Am. Chem. Soc.*, 1962, **84**, 183.

4. (a) Köster, R., *Angew. Chem.*, 1960, **72**, 626; (b) Köster, R., *Advan. Organometal. Chem.*, 1964, **2**, 257. (c) Liu, J., Meyers, E. A., Shore, S. G., *Inorg. Chem.*, in press.

5. (a) Hudlicky, M., *Reactions in Organic Chemistry*, Ellis Horwood, Chichester, 1984, and references therein; (b) Walker, E. R. H., *Chem. Soc. Rev.*, 1976, **5**, 23; (c) Brown, H. C. and Krishnamurthy, S., *Tetrahedron*, 1979, **35**, 567.

6. (a) Brown, H. C., Narasimhan, S., and Somayaji, V., *J. Org. Chem.*, 1983, **48**, 3091; (b) Brown, H. C. and Krishnamurthy, S., *J. Am. Chem. Soc.*, 1972, **94**, 7159; (c) Krishnamurthy, S. and Brown, H. C., *J. Am. Chem.*, 1976, **98**, 3383; (d) Kim, S., Moon, Y. C., and Ahn, K. H., *J. Org. Chem.*, 1982, **47**, 3311; (e) Brown, H. C., Singaram, B., and Mathew, C. P., *J. Org. Chem.*, 1981, **46**, 2712.

7. (a) Casey, C. P. and Neumann, S. M., *J. Am. Chem. Soc.*, 1976, **98**, 5395; (b) Collman, J. P. and Winter, S. R., *J. Am. Chem. Soc.*, 1973, **95**, 4089; (c) Darensbourg, M. Y. and Deaton, J. C., *Inorg. Chem.*, 1981, **20**, 1644; (d) Hayter, R. G., *J. Am. Chem. Soc.*, 1966, **88**, 4376; (e) Darensbourg, M. Y. and Slater, S., *J. Am. Chem. Soc.*, 1981, **103**, 5914.

8. Dauben, W. G., Fonken, G. J., and Noyce, D. S., *J. Am. Chem. Soc.*, 1956, **78**, 2579.

9. (a) Caubere, P. and Moreau, J., *Tetrahedron*, 1969, **25**, 2469; (b) Bank, S. and Prislopski, M. C., *Chem. Commun.*, 1970, 1624; (c) Nelson, R. B. and Gribble, G. W., *J. Org. Chem.*, 1974, **39**, 1425.

10. (a) Jordan, G. T. IV and Shore, S. G., *Inorg. Chem.*, 1996, **35**, 1087; (b) Jordan, G. T. IV, Liu, F. C., Shore, S. G., *Inorg. Chem.*, 1997, **36**, 5597.

11. (a) Nanda, R. K. and Wallbridge, M. G. H., *Inorg. Chem.*, 1964, **3**, 1798; (b) Männig, D. and Nöth, H., *J. Organomet. Chem.*, 1984, **275**, 169.

12. (a) Kot, W. K., Edelstein, N. M., and Zalkin, A., *Inorg. Chem.*, 1987, **26**, 1339; (b) Corazza, F., Floriani, C., Chiesi-Villa, A., and Guastini, C., *Inorg. Chem.*, 1991, **30**, 145; (c) Coucouvanis, D., Lester, R. K., Kanatzidis, M. G., and Kessissoglou, D. P., *J. Am. Chem. Soc.*, 1985, **107**, 8279.

13. (a) Good, C. D. and Ritter, D. M., *J. Am. Chem. Soc.*, 1962, **84**, 1162; (b) Nöth, H. and Wrackmeyer, B., in *Nuclear Magnetic Resonance Spectroscopy of Boron Compounds*, Springer-Verlag, Berlin, 1978; Table 9.3, pp. 115–124.

14. (a) Johnson, P. L., Cohen, S. A., Marks, T. J., and Williams, J. M., *J. Am. Chem. Soc.*, **1978**, 100, 2709; (b) Marks, T. J. and Shimp, L. A., *J. Am. Chem. Soc.*, 1972, 94, 1542; (c) James, B. D., Nanda, R. K., and Wallbridge, M. G. H., *Inorg. Chem.*, 1967, **6**, 1979; (d) James, B. D., Nanda, R. K., and Wallbridge, M. G. H., *J. Chem. Soc.(A)*, 1966, 182.

15. (a) Choukroun, R., Dahan, F., Larsoneur, A.-M., Samuel, E., Peterson, J., Meunier, P., and Sornay, C., *Organometallics*, 1992, **10**, 374; (b) Jones, S. B. and Petersen, J. L., *Inorg. Chem.*, 1982, **20**, 2889; (c) Kot, W. K., Edelstein, N. M., and Zalkin, A., *Inorg. Chem.*, 1987, **26**, 1339; (d) Erker, G., Kropp, K., Krüger, C., and Chiang, A.-P., *Chem. Ber.*, 1982, **115**, 2447.

16. Fachinetti, G. and Floriani, C., *J. Chem. Soc., Chem. Commun.*, 1972, 654.

16

IMPORTANCE OF ^{11}B–^{1}H COUPLING CONSTANTS IN ASSIGNING THE ^{11}B-SIGNALS

STANISLAV HERMÁNEK, JAN MACHÁCEK, JIRI FUSEK,

Institute of Inorganic Chemistry, Academy of Sciences of the Czech Republic,
250 68 Rez near Prague, Czech Republic

and

VRATISLAV BLECHTA

Institute of Chemical Process Fundamental, Academy of Sciences of the Czech Republic,
165 02 Prague 6–Suchdol, Czech Republic

16.1   INTRODUCTION

Boron-11 (^{11}B) NMR spectroscopy has proved to be a primary tool in the determining the correct structures of a majority of the known deltahedral boranes and their derivatives, from among the hundreds of thousands of possible candidates. While the values of the B-chemical shifts are well documented [1–8] and often used in structural characterization, the ^{11}B–^{1}H coupling constants are reported infrequently and not systematically. Incorrect and missing values are quite common. Unfortunately, incorrect values are reported in some cases when they should have been easily determined. The reason for the low interest in coupling constants was not only their perceived low information values, but also because of problems in determination, caused by frequent overlaps of the undecoupled signals and by not being able to obtain correct values, even when using apodization. This is the case with substituted dicarbadodecaboranes, in which more than one third of the coupling constants have not been determined accurately, even when measured at 165 MHz for ^{11}B (500 MHz for ^{1}H).

To date, several approaches have been used in attempting to calculate or to estimate $J(^{11}B^1H)$ coupling constants. The main ones have been based upon (1) the CNDO/S-determined element s-orbital electron population,[9] (2) the number of adjacent carbon atoms and cage "umbrella" angle,[10] (both related to hybridization), and (3) more "accurate or precise" calculations at the IGLO-SOS-DFPT/Perdew/III (MPZ(SC)/6-31G) level.[11] In no case (with the exception of Ref. 10) were the results obtained used for a structural determination or the assignments of B-signals. On the contrary, relationships have been found between the ^{11}B–^{1}H coupling constants and electron density on a boron atom.[1] In all cases, the s/p ratio, first mentioned by R. E. Williams,[12] was considered to be the leading factor.[9,10,13]

To obtain correct values for individual coupling constants, a new approach has been elaborated, which is based on measuring proton decoupled and boron undecoupled two-dimensional (2D) spectra. This approach has allowed us to get sufficiently trustworthy values, even with significantly overlapping signals (see Section 16.4).

16.2   THE EFFECT OF HETEROATOM/S ON THE (^{11}B,^{1}H) COUPLING CONSTANTS AT HETEROBORANES

16.2.1   Twelve-Vertex *Closo*-Compounds

During the course of our systematic study of heteroboranes, we found that BH groups that neighbor electronegative heteroatoms, E, show relatively high (^{11}B–^{1}H) coupling constants when compared with other BH vertices [14] (Figure 16-1). We used this observation in assigning a number of signals before the ^{11}B–^{11}B 2D spectra became available.[15]

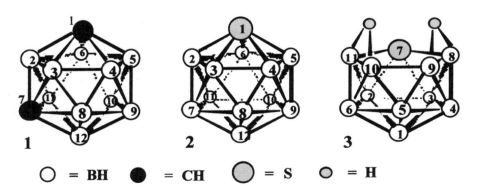

Figure 16-1 Compounds within which increased $J(^{11}$B–^{1}H$)$ coupling constants allowed the assignment of those ^{11}B signals that neighbored the heteroatoms; m-$C_2B_{10}H_{12}$ (compound **1**),[15] $SB_{11}H_{11}$ (compound **2**),[16] and $SB_{10}H_{12}$ (compound **3**).[17]

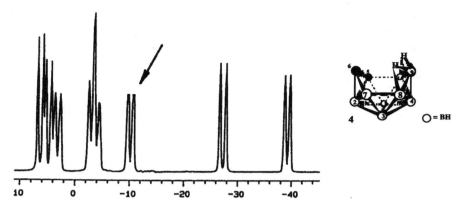

Figure 16-2   An example of a larger-than-average μH splitting at B(10) caused by the neighboring C(5) atom within 5,6-$C_2B_8H_{12}$ (compound **4**).[18]

Not only are the ^{11}B–H coupling constants increased when the boron neighbors electronegative heteroatoms, but coupling to any bridge hydrogens, (^{11}B–H_b–), is also increased [18] (Figure 16-2).

Both types of increased coupling are indicative of a common denominator, i.e., that there is a relatively high δ+ charge on the BH vertices, neighboring the positively charged heteroatom E (when compared with the other BH vertices within the parent deltahedral B-cluster).

According to Lipscomb and colleagues,[19] the inductive effect of the heteroatom E diminishes with increasing distance from the BH vertex under consideration, i.e., the effect diminishes in the order, *ortho* → *meta* → *para* and is additive.[8,19] In order to confirm this hypothesis, we studied the icosahedral $B_{12}H_{12}$ skeletal series within which one or two vertices were substituted with heteroatoms but within which most other factors remained constant—namely, the ($2n + 2$) class, size, and geometry. The $EB_{11}H_{11}$ type (E = BH, CH, NH, S) and *ortho*- (compound **5**), *meta*- (compound **1**) and *para*-$C_2B_{10}H_{12}$ (compound **6**) carboranes were selected as demonstrative examples. Our measurements revealed that the values of the $J(^{11}$B–^{1}H) coupling constants, within the $EB_{11}H_{11}$ examples, decreased significantly from *ortho*- to *meta*- to *para*-positions and allowed us to determinate the values Δ [$J(^{11}$B–^{1}H)$_{subst}$ – $J(^{11}$B–^{1}H)$_{parent}$ in Hz] evoked by single heteroatoms E on the various *ortho*-, *meta*-, and *para*-BH positions (Table 16-1).

With $CB_{11}H_{12}^-$, the Δ increments are 29, 14, and 12 Hz for the *ortho*-, *meta*- and *para*-BH vertices, respectively. Using these increments, we calculated the expected $J(^{11}$B–^{1}H) values for the individual BH vertices at 1,2- (compound **5**), 1,7- (compound **1**), and 1,12-$C_2B_{10}H_{12}$ (compound **6**) carboranes and found that they exhibit an excellent agreement of the calculated with the values found experimentally (Table 16-2). This indicated that the Δ increments have an

TABLE 16-1 The ^{11}B NMR Chemical Shifts δ (ppm), Coupling Constants $J(^{11}$B–^{1}H) (Hz), and Δ (Hz) Increments [J(BH)$_{subst}$ – J(BH)$_{parent}$] Evoked by the Heteroatoms E (E = BH, CH, NH, and S) at the Individual *ortho*-, *meta*-, and *para*-B–H Vertices (BH:E = Mutual BH and E Positions) with the Icosahedral EB$_{11}$H$_{11}$ Compounds

Compound	No.	BH Vertex	δ	$J(^{11}$B–^{1}H)	Δ	BH:E[a]
B$_{12}$H$_{12}^{2-}$	7	1–12	–15.3	122	—	—
CB$_{11}$H$_{12}^{-}$	8	2–6	–17.0	151	29	*o*
		7–11	–14.1	136	14	*m*
		12	–7.8	134	12	*p*
NB$_{11}$H$_{12}$	9 [20]	2–6	–9.8	177	55(?)	*o*
		7–11	–11.9	146	24(?)	*m*
		12	2.7	150	28	*p*
SB$_{11}$H$_{11}$	2	2–6	–5.7	178	56	*o*
		7–11	–3.7	153	32	*m*
		12	18.7	149	27	*p*

[a]*o*, *Ortho*; *m*, *meta*; *p*, *para*.

TABLE 16-2 The ^{11}B Chemical Shifts δ (ppm), Coupling Constants (Hz) Found $J(^{11}$B^{1}H)$_{exp}$] and Calculated $J(^{11}$B^{1}H)$_{cal}$, using Δ Increments (*ortho*-, *meta*-, *para*-) Shown in the BH:E Column for the Individual BH Vertices at *ortho*-, *meta*-, and *para*-carboranes (1,2-C$_2$B$_{10}$H$_{12}$; 1,7-C$_2$B$_{10}$H$_{12}$; 1,12-C$_2$B$_{10}$H$_{12}$)

Compound	No.	BH Vertex	δ	$J(^{11}$B^{1}H)$_{exp}$	$J(^{11}$B^{1}H)$_{cal}$	Δ[a]	BH:E[b]
ortho-Carborane	5	3,6	–15.8	178	180	–2	*o*-, *o*-
		4,5,7,11	–14.7	165	165	0	*o*-, *m*-
		8,10	–10.3	151	150	1	*m*-, *m*-
		9,12	–3.4	150	148	2	*m*-, *p*-
meta-Carborane	1	2,3	–18.0	180	180	0	*o*-, *o*-
		4,6,8,11	–14.3	165	165	0	*o*-, *m*-
		5,12	–7.7	163	163	0	*o*-, *m*-
		9,10	–11.5	150	150	0	*m*-, *m*-
para-Carborane	6	2–11	–15.8	164	165	–1	*o*-, *m*-

[a]Δ = Differences [J (^{11}B^{1}H)$_{exp}$ – J (^{11}B^{1}H)$_{cal}$].
[b]*o*, *Ortho*; *m*, *meta*; *p*, *para*.

additive character and can be used for the calculation of individual BH coupling constants and, consequently, for the assignment of some signals.

Moreover, comparison of $J(^{11}$B–^{1}H)$_{exp}$ values with the electron densities obtained for individual BH vertices by quantum chemical calculations showed that there is an excellent linear dependence with charges on individual atoms at *ortho*-, *meta*-, and *para*-carboranes calculated by the PRDDO method [21]

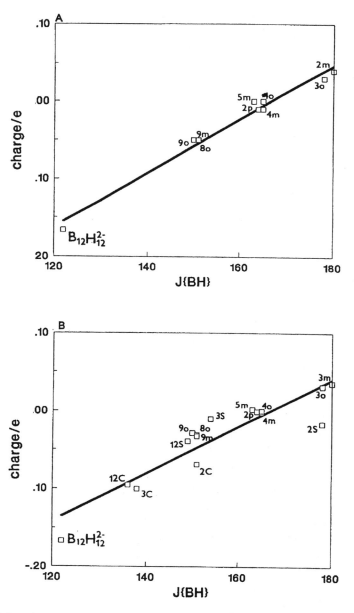

Figure 16-3 The correlation of coupling constants $J(^{11}B^1H)$ with electron densities on individual BH vertices at *ortho*-, *meta*-, and *para*-carboranes ($C_2B_{10}H_{12}$) and $B_{12}H_{12}^{2-}$ calculated (A) by the PRDDO method and (B) by the CNDO-2/STO-3G method at the series in A, enlarged by $CB_{11}H_{12}^{-}$ (compound **8**) and $SB_{11}H_{11}$ (compound **2**).

TABLE 16-3   Coupling Constants $J(^{11}B^1H)$ of Individual BH Vertices at 1, 2, 5, 6, 7, and 8 Compounds and BH Group Charges (Charge in electrons) Calculated using (A) the PRDDO [21] and (B) the CNDO-2/STO-3G Methods

Compound	No.	BH Vertex	$J(^{11}B^1H)$	Charge (e)	
				A	B
$B_{12}H_{12}^{2-}$	7	1–12	122	(−0.1667)	(−0.1667)
$CB_{11}H_{12}^-$	8	2–6	151		−0.0695
		7–11	138		−0.1007
		12	136		−0.0957
$1,2\text{-}C_2B_{10}H_{12}$	5	3,6	178	0.03	0.0313
		4,5,7,11	165	0.00	0.0002
		8,10	151	−0.05	−0.0312
		9,12	150	−0.05	−0.0288
$1,7\text{-}C_2B_{10}H_{12}$	1	2,3	180	0.04	0.0356
		4,6,8,11	165	−0.01	−0.0018
		5,12	163	0.00	0.0023
		9,10	151	−0.05	−0.0329
$1,12\text{-}C_2B_{10}H_{12}$	6	2-11	164	−0.01	−0.0006
$SB_{11}H_{11}$	2	2–6	178		−0.0169
		7–11	154		−0.0104
		12	149		−0.0394

(Table 16-3, Figure 16-3a) and acceptable dependence with charges calculated by the CNDO-2 method using a STO-3G basis, employing GAUSSIAN 94 Program for the **5**, **1**, and **6** compounds (in Table 16-2) in conjunction with the $CB_{11}H_{12}^-$, **8**, and $SB_{11}H_{11}$, **2**, compounds (see Table 16-3, Figure 16-3b).

Figure 16-3 also indicates that B–H interactions can be used to estimate relative electron densities, not only for individual BH vertices, but also over the entire B-skeleton. This follows from the determination of BH coupling constants of bis-(dicarbollide)cobaltate(1-), $[3\text{-}Co\text{-}(1,2\text{-}C_2B_9H_{11})_2]^-$ (compound **10**) and bis-(dicarbollide)nickel, $3\text{-}Ni\text{-}(1,2\text{-}C_2B_9H_{11})_2$ (compound **11**), which are 3-metalla-derivatives of *ortho*-carborane (Figure 16-4). By comparing $J(^{11}B\text{-}^1H)$ of *ortho*-carborane with those of the 3-Co and 3-Ni sandwiches (structures **10** and **11**), we see that there is a clear indication of an increased electron density (e.d.) within the Co-sandwich **10** [a decrease of $J(^{11}B\text{-}^1H)$ by 10 ± 2 Hz] and an almost identical e.d. [$\Delta J(^{11}B\text{-}^1H)$ 0 ± 1 Hz, with the exception of the o-B(8) position] within the Ni-carborane **11** (Table 16-4).

A similar approach can be used for the estimation of electron transfer between the metal and B-cluster ligands in other sandwiches.

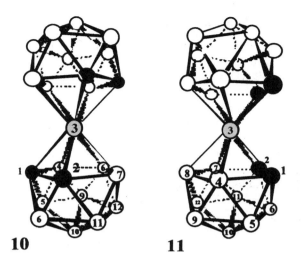

10   **11**

Figure 16-4 Structures of the bis-(dicarbollide)cobaltate(1-), [3-Co-(1,2-C$_2$B$_9$H$_{11}$)$_2$]$^-$ (compound **10**) and bis-(dicarbollide)nickel, 3-Ni(C$_2$B$_9$H$_{11}$)$_2$ (compound **11**)

TABLE 16-4 The ^{11}B NMR Chemical Shifts δ (ppm), Coupling Constants J(^{11}B–^{1}H) (Hz), Δ (Hz) Increments J(^{11}B–^{1}H)$_{sub}$ – J(^{11}B–^{1}H)$_{parent}$ Evoked by the Transition Metal E(3) (E = Co and Ni) at the Individual *ortho-*, *meta-*, and *para*-BH Vertices (BH:E) with [3-Co(1,2-C$_2$B$_9$H$_{11}$)$_2$]$^-$ (Compound **10**) and 3-Ni(C$_2$B$_9$H$_{11}$)$_2$ (Compound **11**)

Compound	No.	BH Vertex	δ	J{BH}	BH:E^a	Δ
ortho-Carborane	**5**	3,6	−15.8	178		
		4,5,7,11	−14.7	165		
		8,10	−10.3	151		
		9,12	−3.4	150		
3-Co	**10**	6	−23.06	168	*m*	−10
		5,11	−17.60	154	*m*	−11
		9,12	−6.39	141	*m*	−9
		4,7	−5.73	153	*o*	−12
		10	1.01	141	*p*	−10
		8	6.15	143	*o*	−8
3-Ni	**11**	6	−16.42	177	*m*	−1
		5,11	−7.08	165	*m*	0
		9,12	2.88	151	*m*	1
		4,7	2.88	165	*o*	0
		10	16.88	149	*p*	−2
		8	21.05	159	*o*	8

^{a}*o, Ortho*; *m, meta*; *p, para*.

16.2.2 Ten-Vertex *Closo*-compounds

A study of 10-vertex *closo*-heteroboranes 1-EB$_9$H$_9$ (E = BH^{2-}, CH$^-$, NH, N$^-$, S) and C$_2$B$_8$H$_8$ showed that, in this skeleton, the influence of the heteroatom E is also similar to the EB$_{11}$H$_{11}$ series, but the *ortho-*, *meta-*, and *para*-effects differ in magnitude if the heteroatom is located in the apex or equatorial positions (Table 16-6). This is understandable if we take a different "umbrella" effect into account, i.e., different coordination and hybridization, as well as different charges at vertices 1 (apex) and 2 (equatorial) (Figure 16-5). In addition, the bond lengths between 2-E and individual B-neighbors (1,3,5,6,9) are different, which makes the surrounding of 2-E more unsymmetrical than it is around 2-B in the parent B$_{10}$H$_{10}^{2-}$ molecule (compound **12**).

In the 10-vertex series, the $J(^{11}$B^{1}H) coupling constants can also be calculated from the increments of the apex C(1) and the equatorial C(2), which must be derived from the experimental $J(^{11}$B^{1}H) of the individual BH(3) to BH(10) vertices obtained from the fully assigned ^{11}B and ^{1}H spectra of the 1,2- or 1,7-C$_2$B$_8$H$_{10}$ carboranes (Table 16-6).

Contemporary experience indicates that other types of BH$^-$ cluster compounds can be treated in a similar fashion. This approach, however, requires the knowledge of $J(^{11}$B^{1}H) of the individual BH vertices of the structure in

TABLE 16-5 The ^{11}B NMR Chemical Shifts δ (ppm), Coupling Constants $J(^{11}$B–^{1}H) (Hz), Increments $\Delta = J(^{11}$B–^{1}H $)_{subst} - J(^{11}$B–^{1}H)$_{parent}$ Evoked by the Heteroatoms 1-E (E = BH, CH, NH, and S) of Connectivity 4 at the Individual *ortho-*, *meta-*, and *para*-BH Vertices with 10-Vertex EB$_9$H$_9$ (E = BH^{2-}, CH$^-$, NH, N$^-$, and S) *Closo*-Compounds 10, and 13–16

Compound	No. [Ref.]	BH Vertex	δ	$J(^{11}$B^{1}H)	Δ	BH:E^a
B$_{10}$H$_{10}^{2-}$	**10**	1,10	−1.96	140	—	—
		2–9	−31.04	128	—	—
CB$_9$H$_{10}^-$	**13**	2–5	−19.45	152	24	*o4*
		6–9	−24.89	138	10	*m4*
		10	29.19	152	12	*p4*
NB$_9$H$_{10}$	**14** [22]	2–5	−6.1	175	47	*o4*
		6–9	−21.5	153	25	*m4*
		10	61.0	165	25	*p4*
NB$_9$H$_9^-$	**15** [22]	2–5	−8.3	161	33	*o4*
		6–9	−18.3	140	12	*m4*
		10	50.0	149	9	*p4*
SB$_9$H$_9$	**16** [23]	2–5	−4.8	170	42	*o4*
		6–9	−17.6	150	22	*m4*
		10	74.5	180	40	*p4*

^{a}*o*, Ortho; *m*, meta; *p*, para.

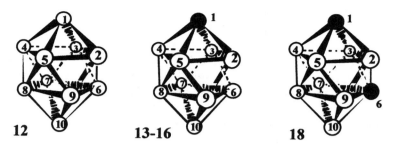

Figure 16-5   Structure and numbering in the parent $B_{10}H_{10}^{2-}$ anion (compound **12**), 1-EB$_9$H$_9$ (compounds **13–16**), and 1,6-C$_2$B$_8$H$_{10}$ (compound **18**).

TABLE 16-6   **The ^{11}B Chemical Shifts δ (ppm), ^{11}B–^{1}H Coupling Constants (Hz) Found (J_{exp}) and Calculated (J_{cal}) using Δ Increments ($o4$, $m4$, and $p4$ Produced by the Apex C-Atom of Connectivity 4, and $o5$, $m5$, and $p5$ Produced by the Equatorial C-Atom of Connectivity 5) in the 10-vertex Dicarbadecaboranes 1,2-C$_2$B$_8$H$_{10}$ (Compound 17); 1,6-C$_2$B$_8$H$_{10}$ (Compound 18) and 1,10-C$_2$B$_8$H$_{10}$ (Compound 19)**

Compound	BH	δ	$J(^{11}B^1H)_{exp}$	$J(^{11}B^1H)_{cal}$	Calculations[a]
17	3,5	−21.06	182	181	$128 + o4 + o5 = 128 + 24 + 29$
	4	+10.08	167	167	$128 + o4 + m5 = 128 + 24 + 15$
	6,9	−27.07	168	168	$128 + m4 + o5 = 128 + 10 + 29$
	7,8	−27.07	154	153	$128 + m4 + m5 = 128 + 10 + 15$
	10	+35.54	170	167	$140 + p4 + m5 = 140 + 12 + 15$
18	2,3	−20.42	179	181	$128 + o4 + o5 = 128 + 24 + 29$
	4,5	−22.56	166	167	$128 + o4 + m5 = 128 + 24 + 15$
	7,9	−27.51	167	167	$128 + m4 + o5 = 128 + 10 + 29$
	8	−18.85	153	153	$128 + m4 + m5 = 128 + 10 + 15$
	10	+22.31	184	181	$140 + p4 + o5 = 140 + 12 + 29$
19	2–9	−13.74	165	162	$128 + o4 + m4 = 128 + 24 + 10$

[a]o, Ortho; m, meta; p, para.

question, from which the individual increments can be deduced. In open skeletons (*nido-, arachno-, hypho-*), only borane structures that do not involve fluxional hydrogens can be used for the determination of the influence of heteroatom(s) substitution. A more detailed study on these effects will be published elsewhere.

16.3   THE EFFECT OF TERMINAL SUBSTITUENTS ON $J(^{11}B^1H)$-SUBSTITUTED HETEROBORANES

Similar effects, but lower by one order than with skeletal heteroatom(s), are produced by terminal substituents, X, replacing terminal hydrogens. Here, also, the inductive effect plays a leading role, i.e., the $J\{BH\}$ values increase in the order: Me < H < Cl < Br < I < COR. The incremental changes, Δ, caused by these substituents are low (1–5 Hz), but also have an additive character (see compounds **22**, **23**, and **25** in Table 16-7.)

TABLE 16-7   The ^{11}B NMR Chemical Shifts δ (ppm), Coupling Constants $J(^{11}B–^1H)$ (Hz), and $\Delta\, J(^{11}B–^1H)_{subst} – J(^{11}B–^1H)_{parent}$ Evoked by the Substituent/s X (X = Cl, Br, and I) at the Individual BH Vertices in *ortho-*, *meta-*, and *para-*Position (BH:E) with Icosahedral $XC_2B_{10}H_{11}$, $X_2C_2B_{10}H_{10}$ *ortho*-Carboranes (Compounds 20–23), and Monocarba Anions $CB_{11}H_{12}^-$ (Compound 8) and $I_6CB_{11}H_6^-$ (Compound 24)

Compound	No.	BH Vertex	δ	$J(^{11}B^1H)$	Δ	BH:E[a]
ortho-Carborane	**5**	3,6	−15.8	178		
		4,5,7,11	−14.7	165		
		8,10	−10.3	151		
		9,12	−3.4	150		
9-Cl	**20**	3,6	−17.25	184	6	*m*
		4,5	−14.77	168	3	*o*
		7,11	−16.08	167	2	*m*
		8,10	−9.68	154	3	*m*
		12	−2.99	152	2	*o*
9-I	**21**	3,6	−15.67	184	6	*m*
		4,5	−13.75	168	3	*o*
		7,11	−14.20	168	3	*m*
		8,10	−8.44	155	4	*o*
		12	−1.90	154	4	*o*
9,12-Cl$_2$	**22**	3,6	−19.4	183	5	*m, m*
		4,5,7,11	−16.3	168	3	*o, m*
		8,10	−9.4	156	5	*o, o*
9,12-Br$_2$	**23**	3,6	−17.8	185	7	*m, m*
		4,5,7,11	−15.4	174	9	*o, m*
		8,10	−8.5	160	9	*o, o*
$CB_{11}H_{12}^-$	**8**	2–6	−17.0	151		
		7–11	−14.1	138		
		12	−7.8	136		
7-12-I$_6$	**25**	2–6	−14.4	166	15	5*o, m*

^{a}*o*, Ortho; *m*, meta; *p*, para.

16.4   METHODS FOR DETERMINING THE ^{11}B–^{1}H COUPLING CONSTANTS

Due to the strong overlap of multiplets, it is usually impossible to estimate the $J(^{11}\text{B}^1\text{H})$ coupling constants from ^{11}B or ^{1}H one-dimensional (1D) spectra. It is necessary to exploit suitable two-dimensional (2D) experiments, which are able to separate overlapping multiplets in the 2D area. The spread of the chemical shift values of protons bonded directly to boron is negligible for the compounds studied, and cannot solve this task. Hence, useful experiments are limited to those producing 2D spectra with the J-splitting along one dimension and the other dimension reflecting the spread of the ^{11}B chemical shift values. Only heteronuclear ^{11}B $\{^1$H$\}$ 2D J-experiments [23] and certain heteronuclear correlation experiments, direct or indirect, fulfill this condition. A second type of experiments appears more suitable because it gives quartets of line intensities 3:1:1:3,[23] which may effectively be considered doublets (if the two weaker inner lines are neglected). Measuring the two outer intense lines of the multiplet in the proton dimension and dividing by three offers about threefold more precise values of $J(^{11}\text{B}^1\text{H})$.

An absolute-value heteronuclear correlation experiment with polarization transfer from ^{11}B to ^{1}H and with ^{1}H detection without ^{11}B decoupling was used routinely. The resulting spectra have crosspeaks ordered according to their ^{11}B chemical shifts along the f1 dimension and show corresponding ^{1}H chemical shift and $J(^{11}\text{B}^1\text{H})$ splitting in the f2 (detection) dimension (see, e.g., spectra of 1,2-Br$_2$-1,2-C$_2$B$_{10}$H$_{10}$, Figure 16-6). Since the initial ^{11}B magnetization has very short T_1 relaxation values, the experiment permits a very fast repetition rate (about 0.1 s) without the appearance of any artifacts on the crosspeaks. A typical experiment with fully sufficient resolution in both dimensions takes a few minutes. Timing of the pulse sequence is as follows:

$$90(11\text{B,fi1}) - (1/2 * J(\text{BH})) - t1/2 - 180(1\text{H,fi2}) - t1/2$$
$$- 90(1\text{H,fi3}), 90(11\text{B,fi4}) - \text{Acq(fi5)}$$

where pulse phases go through the following cycles:

$$\text{fi1} = 0,180; \text{ fi2} = 0,0,0,0,90,90,90,90; \text{ fi3} = 0,0,90,90;$$
$$\text{fi4} = 0, \text{ and fi5} = 0,180,90,270$$

The t_1 value is the evolution time. The pulse sequence is preceded by a proton pulse of about 2 ms duration. This pulse helps primarily to eliminate the magnetization of any slowly relaxing protons not connected with boron. These protons have relatively sharp, intense lines and produce a visible t_1 noise. The main source of the suppression of unwanted peaks is, of course, the phase cycling fi1 of the first boron pulse. The fi3 cycle of the second boron pulse enables quadrature detection in the f1 dimension. The phase cycles of both pulses are accompanied with phase cycling of the receiver, the phase of fi5. The phase fi2 cycle

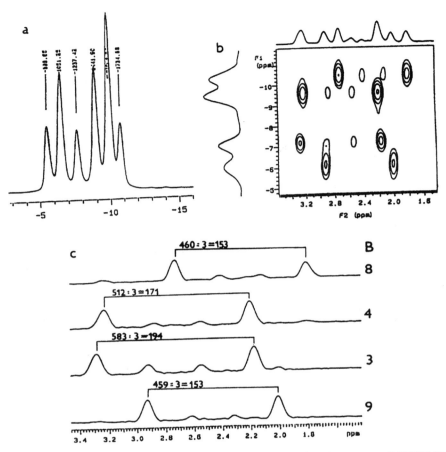

Figure 16-6 Assignment of $J\{BH\}$ to individual BH vertices by means of COSY 2D [^{11}B–^{1}H]$\{^{11}$B$\}$ spectra of 1,12-Br$_2$-1,2-C$_2$B$_{10}$H$_{10}$: (a) ^{1}H undecoupled ^{11}B NMR spectrum at 160.36 MHz; (b) COSY [^{11}B–^{1}H]$\{^{11}$B$\}$ spectrum; (c) cuts showing ^{1}H splitting and $J(^{11}$B^{1}H) values at frequencies of individual ^{11}B signals ordered according to the increasing chemical shielding 9,12 < 3,6 < 4,5,7,11 < 8,10.

cancels coherence connected with ^{1}H transverse magnetization created by the possibly imperfect refocusing proton pulse.

Even in the case of very close ^{11}B chemical shifts when the crosspeaks lie practically in one line (due to almost identical ^{11}B chemical shifts), the pulse sequence described above can be helpful. The key lies in the fact that the inner lines of the quartet are quite suppressed, which simplifies the spectrum and often permits, along with the knowledge of proton chemical shifts obtained from boron decoupled ^{1}H spectra, fitting of the spectrum to its components (see spectra of 1-OH-1,2-C$_2$B$_{10}$H$_{11}$, Figure 16-7).

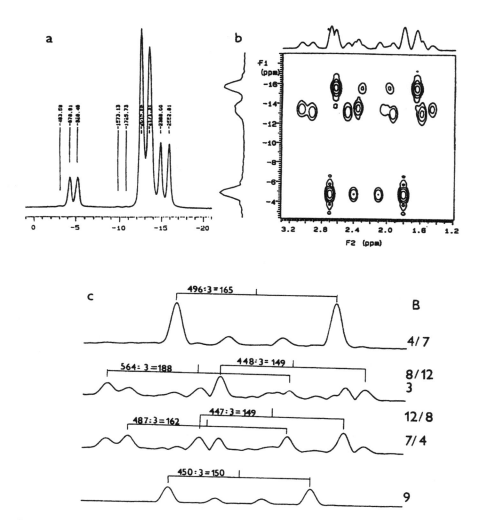

Figure 16-7 Assignment of J(BH) to individual BH vertices by means of COSY 2D [[11]B–[1]H]{[11]B} spectra of 1,–HO–1,2-$C_2B_{10}H_{11}$: (a) ^{1}H undecoupled ^{11}B NMR spectrum at 160.36 MHz; (b) COSY 2D [[11]B–[1]H]{[11]B} 2D spectrum; (c) cuts showing ^{1}H splitting and J(BH) values at frequencies of individual [11]B signals ordered according to the increasing chemical shielding.

REFERENCES

1. Eaton, G. R. and Lipscomb, W. N., *NMR Studies of Boron Hydrides and Related Compounds*, Benjamin, New York, 1969.

2. Nöth, H. and Wrackmeyer, B., *Nuclear Magnetic Resonance of Boron Compounds*, series NMR 14, Basic Principles and Progress, Grundlagen und Fortschritte (P. Diehl, E. Fluck, and R. Kosfeld, eds), Springer Verlag, Berlin, 1978.

3. Todd, L. J. and Siedle, A. R., *Prog. NMR Spectrosc.*, 1979, **13**, 87.

4. Siedle, A. R., *Ann. Rep. NMR Spectrosc.*, 1982, **12**, 177.

5. Kennedy, J. D., in *Multinuclear NMR* (*NMR in Inorganic and Organometallic Chemistry*) (J. Mason, ed.), Plenum Press, New York, 1987, Chap. 8, p. 221.

6. Wrackmeyer, B., *Ann. Rep. NMR Spectrosc.*, 1988, **20**, 61.

7. Siedle, A. R., *Ann. Rep. NMR Spectrosc.*, 1988, **20**, 205.

8. Hermánek, S., *Chem. Revs.*, 1992, **92**, 325.

9. Kroner, J. and Wrackmeyer, B., *J. Chem. Soc. Faraday Trans.* II, 1976, **72**, 2283.

10. Jarvis, W., Abdou, Z. J., and Onak, T., *Polyhedron*, 1981, **2**, 1067.

11. Schleyer, P. v. R. and Hofmann, M., personal communication, 1995.

12. Williams, R. E., Harmon, K. M., and Spielman, J. R., OTS, AD 603782, 1964.

13. Onak, T., Leach, J. B., Anderson, S., and Frisch, M. J., *J. Magn. Reson.*, 1976, **23**, 237.

14. Plešek, J., Štíbr, B., and Hermánek, S., *Chem. Ind.*, 1974, 662.

15. Hermánek, S., Gregor, V., Štíbr, B., Plešek, J., Janoušek Z., and Antonovich, V. A., *Collect. Czech. Chem. Commun.*, 1976, **41**, 1492.

16. Plešek, J., and Hermánek, S., *J. Chem. Soc., Chem Commun.*, 1975, 127.

17. Hermánek, S., Plešek, J., and Štíbr, B., 3rd International Meeting on Boron Chemistry, Munchen, Ettal, 5–9 July, 1976, Abstr. No. 52.

18. Plešek, J., Štíbr, B., and Hermánek, S., *Chem. Ind. (London)*, 1974, 662.

19. Boer, F. P., Hegstrom, R. A., Newton, M. D., Potenza, J. A., and Lipscomb, W. N., *J. Am. Chem. Soc.*, 1966, **88**, 5340.

20. Müller, J., Ph.D. Thesis, Technische Hochschule, Aachen, 1991.

21. Dixon, D. A., Kleier, D. A., Halgren, T. A., Hall, J. H., and Lipscomb, W. N., *J. Am. Chem. Soc.*, 1977, **99**, 6226.

22. Arafat, A., Baer, J., Huffmann, J. C., and Todd, L. J., *Inorg. Chem.*, 1983, **22**, 3721.

23. Finster, D., Hutton, W., and Grimes, R., *J. Am. Chem. Soc.*, 1980, **102**, 400.

17

CLOSO- AND NIDO-CLUSTERS WITH A B₄ OR NB₃ SKELETON

PETER PAETZOLD

Institute of Inorganic Chemistry, Technische Hochschule, Aachen, Aachen, Germany

17.1 INTRODUCTION

Considering the series of tetraboranes B_4H_4, B_4H_6, B_4H_8, and B_4H_{10}, a lot of chemistry has accumulated about *arachno*-B_4H_{10} and its derivatives, particularly those of the type B_4H_8L (L = Lewis base), since A. Stock described B_4H_{10} for the first time in 1912.[1] On the other hand, a substance B_4H_4 is not known, in contrast with two structurally well characterized derivatives, B_4Cl_4 [2] and B_4R_4 (R = *t*-Bu).[3] Up to now, there has been almost no experimental evidence for B_4H_6 or its derivatives, which formally belong to the *closo*-family B_nH_{n+2}. Possible structures of *nido*-tetraborane B_4H_8 were studied by ab initio methods.[4] There are reports on its formation as a short-lived intermediate from the decomposition of B_5H_{11} and other boranes, but it is not an experimentally well-known substance, and the same is true for potential derivatives.

In this chapter, the molecules B_4H_6 and B_4H_8, and also the isoelectronic aza-analogs NB_3H_4 and NB_3H_6, are described by Lipscomb's localized orbital description in a modified manner, as well as by ab initio methods. Experimental results on organic derivatives of the type $B_4H_2R_4$, $B_4H_4R_4$, NB_3R_4, and $NB_3H_2R_4$ are reported.

17.2 PRELUDE: A GENERALIZED TWO- AND THREE-CENTER BOND SCHEME APPLIED TO AZABORANES

The localization of an electron pair at one center (lone pair) or two centers [(2c2e)-bond] is an approximation of theory, which is extremely useful for the simple description of electron-precise and electron-rich molecules. The additional localization at three centers [(3c2e)-bond] gives a tool for the simple description of electron-deficient molecules, as pointed out by W.N. Lipscomb [5]. In the case of molecules of the type $N_aC_bB_cH_d$, a more general formulation of Lipscomb's well-known *styx*-rules may include lone pairs; there are examples of molecules where lone pairs, as well as electron deficiency, are present, e.g., the anions $NB_9H_9^-$ [6] and $NB_{11}H_{11}^-$.[7] If all valence electrons (Σe) are exclusively accommodated in l lone pairs, y (2c2e)-bonds, and t (3c2e)-bonds, and if the sum of valence orbitals (Σo) results from four orbitals for each of the atoms B, C, and N and one orbital for each of the H atoms, we shall arrive at two balances:

$$\Sigma e = 5a + 4b + 3c + d = 2l + 2y + 2t$$

$$\Sigma o = 4(a + b + c) + d = l + 2y + 3t$$

The numbers y and t can definitely be calculated for each reasonable value of l. In most cases, we have $l = 0$ and hence $t = \Sigma o - \Sigma e$ and $y = (\Sigma e - 2t)/2$.

In the special case of a deltahedral cluster molecule, we shall have one *exo*-H atom for each vertex and hence $d - (a + b + c)$ extra-H atoms in either *endo*- or bridging positions. Since y covers all the XH bonds (*exo*- and *endo*-H) and all the skeletal BX or XX bonds (X = B, C, N), and since t includes all the BHB and BXB bonds (open or closed, σ- or π-type, or hybrids of these), one has to consider all possible distributions of the extra-H atoms among *endo*- and bridging positions when going to construct a structural formula. On such construction, one has certainly to establish a binding connection between all neighboring vertices of a given deltahedron or deltahedral fragment; a meeting of (2c2e)- and (3c2e)-bonds at the edge of the same triangle is unfavorable. Moreover, an indispensable condition is an electron octet at each skeletal atom.

The BCB (3c2e)-bonds are considered less favorable than BBB bonds, and certainly BNB (3c2e)-bonds are the least favorable ones, π-bonds excepted. Nevertheless, two BNB (3c2e)-bonds will necessarily be present if the skeletal connectivity of N is $k = 5$ (Figure 17-1a), and N takes part in one such bond in the case of $k = 4$ (Figure 17-1b and c), whereas no (3c2e)-bond is necessary in the case of $k = 3$ (Figure 17-1d). [Note, bold edges and grayish triangles are used to symbolize (2c2e)- and (3c2e)-bonds, respectively!]

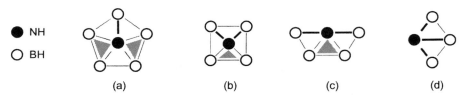

Figure 17-1   The (3c2e)- and (2c2e)-bonds at nitrogen, e.g., in fragments of (a) *closo*-NB₁₁H₁₂, (b) *closo*-NB₉H₁₀, (b) *nido*-NB₁₀H₁₃, and (d).*nido*-NB₉H₁₂.

Going from $k = 5$ via $k = 4$ to $k = 3$, the net positive charge at N decreases from $+5/3$ via $+4/3$ to $+3/3$, and this can be taken as a rough argument to understand the unfavorable situation of the strongly electronegative nitrogen in high coordination. Since the consideration of net charges involves an unrealistic equal distribution of charge density casually between atoms of different elec-

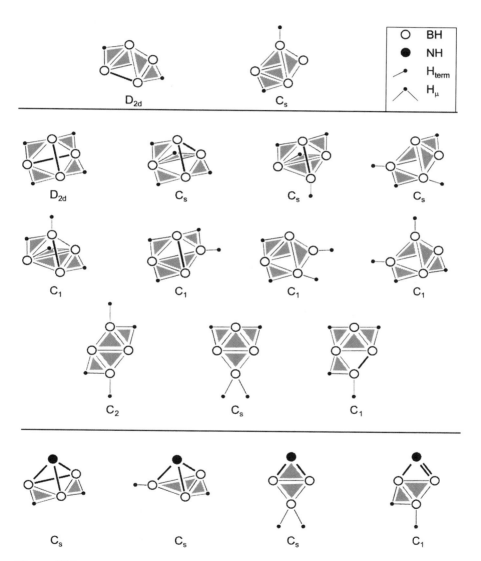

Figure 17-2 Possible structures of B_4H_6, B_4H_8, and NB_3H_6, in accord with the two- and three-center bond scheme [bold edges: (2c2e)-bonds; grayish triangles: (3c2e)-bonds].

TABLE 17-1   Number of (2c2e)- and (3c2e)-Bonds (y and t, Respectively) in the Tetraboranes

Compound	$\Sigma e/\Sigma o$	y/t	Compound	$\Sigma e/\Sigma o$	y/t
B_4H_6	18/22	5/4	NB_3H_4	18/20	7/2
B_4H_8	20/24	6/4	NB_3H_6	20/22	8/2

tronegativities, like B and N, such considerations have by far no quantitative meaning and should not be overemphasized.

The skeletal connectivity does not exceed $k = 3$ in the tetraboranes and azatetraboranes. The number of (2c2e)- and (3c2e)-bonds are shown in Table 17-1. Quite a number of structures can be drawn in accord with the two- and three-center bond scheme. Only tetrahedron-type geometries seem to be possible for B_4H_6, which formally counts as a *closo*-borane. The *nido*-species B_4H_8 and NB_3H_6 can adopt a series of tetrahedron- or bicyclobutane-type structures (Figure 17-2). Interestingly, a cluster molecule NB_3H_4 with formal *closo*-character cannot be drawn in accord with the above-mentioned rules.

The D_{2d} structure looks more reasonable for B_4H_6 than the other one, since we would not expect *endo*-H atoms at a closed polyhedron. Allowing resonance among the four forms that result from assigning the (2c2e)-bond to each of the four equivalent edges of tetrahedral B_4H_6, we arrive at a formal bond order of 8/12 for the bridged BB bonds and 7/12 for the other ones; in this really rough concept of bond order, a (2c2e)-bond corresponds to the bond order 1 and a (3c2e)-bond to the bond order 1/3 for each of the three edges of a (3c2e)-bond. One of the bicylcobutane-type structures is expected for *nido*-B_4H_8 and one for *nido*-NB_3H_6, according to the well-known cluster structure rules; the most probable versions will be the C_1 structures with only one *endo*-H atom.

17.3   TETRABORANES AND AZATETRABORANES BY THEORY

An ab initio calculation of B_4H_6 at the MP2/6-31G* level gives a minimum for a D_{2d} tetrahedron-type structure with opposite BHB bridges,[8] as expected from the localized orbital scheme. The calculated bond lengths are 1.619 (orthogonal to the S_4 axis) and 1.742 Å, in accord with the sequence of formal bond orders mentioned earlier. Bond lengths for the methyl derivative $B_4H_2Me_4$ are 1.661 and 1.766 Å (RHF/6-31G* level); an ^{11}B NMR shift of 13.6 ppm could be derived from the calculated geometry by the IGLO (DZ) procedure.

Four minima and three transition states (Figure 17-3) are found for B_4H_8 at the MP2/6-31G* level.[8] The most stable isomer had also resulted from former theoretical work.[4] It is the same isomer made plausible from the localized orbital scheme described earlier; the isomers of higher energy, one with a bicyclobutane- and two with a tetrahedrane-type structure, are also found in Figure 17-2. The TS1 structure is not in accord with the naive octet rule. Its low energy

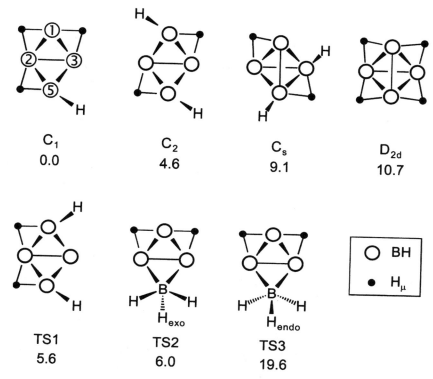

Figure 17-3 Four local minima and three transition states (TS1–TS3) for *nido*-B_4H_8, and their calculated energies (kcal/mol).

makes a dynamic process via TS1 reasonable that transforms the C_1 isomer into its enantiomer, thus making B1 and B5 equivalent by a pseudo-mirror plane. Analogously, a dynamic process via TS2 makes B2 and B3 of the low-energy isomer equivalent, and a combination of both processes would result in C_{2v} pseudo-symmetry. Calculations at the same level had been performed on the methyl derivative $B_4H_4Me_4$. The ^{11}B NMR shifts of $B_4H_4Me_4$ were calculated by the IGLO procedure for an isomer with three *exo*-Me groups and one *endo*-Me group at B5 (considering experimental results, in Section 17-5), but the shifts did not fit the experimental data. When the geometric parameters for the corresponding isomer B_4H_4t-Bu_4 (calculated at the HF/6-31G* level) were used for the shift calculations of $B_4H_4Me_4$, a nice agreement with the experimental values resulted (δ_{calc} = 34.8, 10.3, 8.5, –6.6 ppm for B1, B2, B3, B5, respectively).

Six local minima could be detected on the potential surface of NB_3H_4—again, at the MP2/6-31G* level [9] (Figure 17-4). Diboryl(iminoborane), the low-energy species, can be brought into accord with the localized orbital scheme by assuming a linear BBN and an orthogonal linear BNB (3c2e)-π-

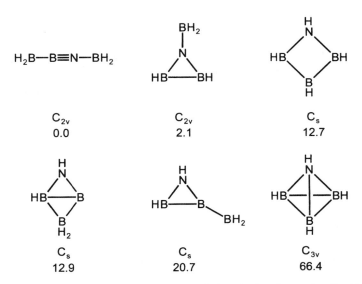

Figure 17-4   Local minima for NB_3H_4 and their calculated energies (kcal/mol).

bond. The azatetrahedrane is a high-energy species with a triplet ground state in C_{3v} symmetry. The Jahn-Teller distortion of the corresponding singlet state leads to the nonplanar four-membered ring of C_s symmetry. The bicyclic C_s isomer looks unconventional. It is isoelectronic, however, with the bicyclic molecule $C_2B_2H_4$, which is energetically favorable [10] and whose derivatives are well characterized.[11]

Finally, the azaborane *nido-*NB_3H_6, represents a minimum of potential energy in its bicyclobutane-type C_s structure (Figure 17-2).[12] The ^{11}B NMR shifts were calculated to be 22.1 (B2,3) and −15.5 ppm (B5) from the optimized geometry by the IGLO (DZ) method.

17.4   TETRABORANE $B_4H_2R_4$

The formation of the tetrahedral tetra-*tert*-butyltetraborane B_4R_4 in a 79% yield from *tert*-butyldifluoroborane and sodium–potassium alloy in pentane is well known.[3] The broad UV absorption of this yellow crystalline material with a maximum at 283 nm (lg ε = 3.70) is comparable to the corresponding absorption of B_4Cl_4 at 245 nm (lg ε = 4.25).[13]

$$12\,RBF_2 + 8\,K \rightarrow B_4R_4 + 8\,K[RBF_3] \quad (R = t\text{-Bu}) \tag{1}$$

Surprisingly, the reduction of *tert*-butyldichloroborane with sodium–potassium alloy gives a mixture of different products from which tetra-*tert*-butyltetraborane(6) can be crystallized in 10% yield. Boiling hexane or toluene at ambient temperature are suitable solvents. The MS data and elemental analysis support the formula $B_4H_2R_4$.[14] It was shown that the two H atoms come from the t-Bu groups by conducting the reaction in deuterated solvents. The dibromoborane $RBBr_2$ gives the same results.[15]

$$4RBCl_2 + 8K + 2 \text{ "H"} \rightarrow B_4H_2R_4 + 8KCl \qquad (2)$$

The ^{11}B NMR doublet at $\delta = 14.6$ ppm corresponds well to the signal calculated for $B_4H_2Me_4$. The broad $^1H\{^{11}B\}$ NMR signal at 0.85 ppm can also be detected by a 2D-$^{11}B/^1H$ NMR experiment. The ^1H and ^{13}C NMR data clearly reveal the presence of one set of symmetrically equivalent t-Bu groups. The asymmetric unit of the crystalline material, space group $Pba2$, comprises three molecules in general positions and two half-molecules on the C_2 axis. Due to a phase transition below 0°C, intensity data had to be collected at room temperature. The only conclusion that can be drawn from the observation of 19,418 reflections is that tetrahedral molecules are present.

All the measured data together leave no doubt that the molecule $B_4H_2R_4$ forms a tetragonally distorted tetrahedron with two H bridges at opposite edges (D_{2d}). It is the first isolated derivative of a neutral borane in the *closo*-series B_nH_{n+2}.

17.5 TETRABORANE B₄H₄R₄

The tetra-*tert*-butyl species $B_4H_4R_4$ is available from $B_4H_2R_4$ by a three-step process: (a) reduction with sodium-potassium alloy in THF, (b) protonation with $HN(SiMe_3)_2$, followed by (c) a protonation with HCl in ether.[15] The experimental [15] and theoretical [8] characterization of the postulated anionic intermediates has not yet been completed.

$$B_4H_2R_4 \xrightarrow{\text{(a)}} [B_4H_2R_4]^{2-} \xrightarrow{\text{(b)}} [B_4H_3R_4]^- \xrightarrow{\text{(c)}} B_4H_4R_4 \qquad (3)$$

The ^1H, ^{11}B, and ^{13}C NMR spectra indicate four nonequivalent B atoms and t-Bu groups at –20°C. The ^{11}B NMR shifts at $\delta = 36.8, 16.1, 11.2, -3.8$ correspond nicely with those calculated for $B_4H_4Me_4$ (see Section 17-3). These data of $B_4H_4R_4$ are in accord with the structure predicted for B_4H_8 by simple rules and by theory: the bicyclobutane-type structure of C_1 symmetry (Figure 17-2). An *endo*-position is assumed for the t-Bu group at B5, since the observed ^{11}B NMR shifts fit well those calculated for that configuration; all three t-Bu groups in *exo*-positions at the triangle B2—B3—B5 would create unbearable steric stress. Such structural assumptions are supported by the observation of all four extra-H atoms in a 2D-$^1H/^{11}B$ NMR spectrum at –20°C at $\delta = 3.75, -0.70, 0.08, 2.14$ ppm, exhibiting crosspeaks with the B atoms B1/2, B1/3, B2/5, B5, respectively. An *endo*-position of the *t*-Bu group at B5 is also indicated by interactions between this group and the *t*-Bu group at B1 in the NOE ^{1}H NMR spectrum.

At rising temperature, a coalescence is observed of the ^{11}B NMR signals of B2 and B3 ($\delta = 13.9$ ppm at 40°C), the ^{1}H NMR signals for the bridges B1–B2, B1–B3 ($\delta = 1.5$ ppm), and for the remaining extra-H atoms ($\delta = 1.2$ ppm). The corresponding dynamic process simulates a pseudo-mirror plane through B1 and B5 by transforming the enantiomers of the C_1 structure one into another. With respect to the transition states TS2 and TS3 (Figure 17-3), the low-energy transformation through TS2 would leave the *t*-Bu group in its *endo*-position at the same side of the molecule, maintaining C_1 symmetry, whereas the transition state TS3 with the *endo-t*-Bu group on the mirror plane allows that isomerization. From the coalescence temperature of 10°C (with respect to the ^{11}B NMR signals), we deduce an activation barrier of $\Delta G^{\neq} = 12$ kcal/mol, which is not too far away from the value $\Delta U^{\neq} = 19.6$ kcal/mol of B$_4$H$_8$ (TS3).

17.6   AZATETRABORANE NB$_3$R$_4$

The B—B bond of tri-*tert*-butylazadiboriridine [16] can be bromoborated by alkyldibromoboranes R′BBr$_2$. The subsequent debromination by lithium gives the azatetraboranes NB$_3$R$_3$R′.[17] The alkyl group R′ can be varied widely, but α-branched groups like R = *t*-Bu cannot be applied. The raw products decompose on attempted purification or storage at ambient temperature. The isobutyl derivative NB$_3$R$_3$*i*-Bu, however, can be gained and stored as a pure substance, but crystals are not available.

The NMR spectra demonstrate the presence of three nonequivalent B atoms and *t*-Bu groups in NB$_3$R$_3$R′ (R′ = *i*-Bu). With reference to the calculated structures of NB$_3$H$_4$ (Figure 17-4), the *B*-boryl cyclopropane-type and the bicyclobutane-type structures are in accord with the NMR data, though the parent molecules are species of rather high energy. The ^{1}H and ^{13}C NMR signals of the *i*-Bu group at low temperature indicate C_1 symmetry, but at a coalescence temperature of 0°C a dynamic process makes the CH$_2$ protons and the Me groups of *i*-Bu equivalent. This process can be easily described as a rotation of the exocyclic BRR′ group around the B—B axis of the one isomer and as the same rotation after a ring opening of the other isomer. The IGLO ^{11}B NMR shifts calculated for the 2-boryl isomer are in accord with the observed values of NB$_3$R$_3$R′, but not the shifts predicted for the bicyclic isomer. This is clear evidence that the azatetraborane NB$_3$R$_3$R′ is neither a cluster molecule nor a four-membered ring nor an unsaturated open-chain molecule, but is a three-membered N—B—B ring with a B-bound boryl group.

17.7 AZATETRABORANE $NB_3H_2R_4$

Azatetraboranes $NB_3H_2R_3R'$ (R = t-Bu) are formed by the addition of $R'BH_2$ to the azadiboriridine NB_2R_3 in THF. Two isomers of $NB_3H_2R_3R'$ are unequivocally identified by NMR methods. The equilibrium ratio of these isomers ranges from 25/75 (R' = t-Bu) to ca. 5/95 (R' = s-Bu).[18] Interestingly, the C_s isomer is the only detectable one in the case of R' = H.[12]

$$\tag{5}$$

With R' = t-Bu, the temperature dependence of the equilibrium was followed and the values $\Delta H = -0.72$ kcal/mol, $\Delta S = -0.24$ cal/(K mol) were deduced. The isomer of C_1 symmetry is the low-temperature species, favored by lower energy. It can be crystallized at –27°C from hexane. The crystals are stable at ambient temperature, but will immediately give the equilibrium mixture if dissolved again in organic media. Nevertheless, the isomerization is slow enough to make both isomers visible by NMR methods. In the case of R' = s-Bu, diastereomers are observable, showing that the enantiomers of the C_1 isomer are not rapidly transformed into one another. Crystal structure data are available for the C_1 isomer $NB_3H_2R_4$ and for the C_s isomer $NB_3H_3R_3$. The B—N distances [1.473(6) and 1.394(6) Å] of the C_1 isomer clearly show B—N bonds of predominantly single- and double-bond character. Both isomers of $NB_3H_2R_4$ represent *nido*-clusters of the expected bicyclobutane-type structure. The bonding situation can qualitatively be well understood by the simple two- and three-center bond scheme, and even the lower energy of the isomer of C_1 symmetry can be predicted by qualitative arguments (see Section 17-2).

The addition of boranes YBH_2 to the azadiboriridine NB_2R_3 can also be achieved with a series of aminoboranes (Y = $R_2'N$) and (organothio)boranes (Y = R'S). The product of C_1 symmetry [Eq. (6)] is the only one in the case of Y = R'_2N, PrS, but (phenylthio)borane gives a 16/84 mixture of both products. For two of the C_1 products, crystal structure analyses are available (Y = Et_2N, PhS) [18,19] that prove the presence of a nonplanar tricyclic N_2B_3 or SNB_3 skeleton. A B—N double bond [1.382(1) and 1.373(2) Å, respectively] is part of both skeletons.

$$(6)$$

From the electron/orbital numbers 26/27 (N$_2$B$_3$H$_7$) and 26/26 (SNB$_3$H$_6$; l = 1), we can easily find out that the skeletons N$_2$B$_3$ and SNB$_3$ of the C_1 isomers need to be built up from one (3c2e)- and five (2c2e)-bonds. The molecules N$_2$B$_3$H$_7$, SNB$_3$H$_6$, and B$_5$H$_{11}$ ($\Sigma e/\Sigma o$ = 26/31, y/t = 8/5) are isoelectronic and have a comparable *arachno*-structure, derived from the pentagonal bipyramid by omitting two vertices.

Let us give a last example of a *nido*-azatetraborane that plays the role of a hypothetical intermediate [Eq. (8a)]. Whereas the azadiboriridine NB$_2$R$_3$ is a stable substance with R = *t*-Bu, the same three-membered ring with only one slightly less bulky ligand undergoes a dimerization that gives either a nonplanar tetracyclic molecule of C_2 symmetry, e.g., N$_2$B$_4$R$_4$(CH$_2$R)$_2$ (R = *t*-Bu),[20] or a *nido*-cluster derived from a pentagonal bipyramid by removing an equatorial vertex, e.g., N$_2$B$_4$R$_4$*i*-Pr$_2$.[16,21]

$$(7)$$

R' = CH$_2$R   R' = *i*Pr

The tetracyclic system contains two B—N double bonds [1.363(8) and 1.373(8) Å] in a nearly orthogonal position, one above the other. A concerted intramolecular (2 + 2) cycloaddition would be allowed as an antarafacial process by orbital symmetry, but should be impossible owing to the steric stress exhibited by the four *t*-Bu groups at the double bonds. Such cycloaddition would give the corresponding *nido*-cluster with two R' groups in equatorial positions (C_{2v} symmetry). In the isolated *nido*-N$_2$B$_4$R$_4$*i*-Pr$_2$, on the other hand, the two *i*-Pr groups are found in an axial and an equatorial position (C_1 symmetry), whereas isomers with a diaxial or diequatorial distribution of *i*-Pr are not detected. How can such stereospecific formation of N$_2$B$_4$R$_4$*i*-Pr$_2$ be explained?

We suppose that one molecule NB_2R_2R' attacks a second one at its basic B—B bond via an acidic B atom, giving a spirocyclic system, one part of which is a *nido*-azatetraborane of the type discussed in this section. The attacking B atom will necessarily be the one with the sterically less demanding group R' [Eq. (8a)]. The subsequent formation of a tetracyclic system proceeds analogously to the equilibria of the Eqs. (5) and (6) [Eq. (8b)]. Two different orientations in the spirocyclic intermediates make two tetracyclic isomers possible. The one with four *t*-Bu groups at the B—N double bonds will be stable, the other one will undergo a (2 + 2) cycloaddition, giving the *nido*-cluster of C_1 symmetry [Eq. (8c)].

In terms of the two- and three-center bond scheme, the numbers $y = 12$ and $t = 2$ can be derived from $\Sigma e/\Sigma o = 28/30$ for a molecule $N_2B_4H_6$. The skeleton contains two (3c2e)- and six (2c2e)-bonds, and a corresponding formula can be drawn without engaging the two N atoms in (3c2e)-bonds. Such unfavorable engagement could not be avoided with the more common alternative six-vertex *nido*-structure, i.e., the pentagonal pyramid.

$$(8)$$

17.8   EPILOG

Some basic one-step reactions, which have been known for decades, can be described in an appropriate manner by applying the two- and three-center bond scheme. Such basic reactions play a role also in the preceding sections. Systematization seems to be desirable. Let us consider two electron pairs, I and II, that may be localized in a (3c2e)- or a (2c2e)-bond or a (1c2e)-lone pair. The pairs I(2c) and II(2c) may bind together the two halves of the molecules AB and CD by (2c2e)-bonds. By reactions according to Figure 17-5, two new molecules ABC and D may be formed. The electron pair I binds the fragments A, B, and C together by a (3c2e)-bond, whereas the electron pair II is now localized at D as

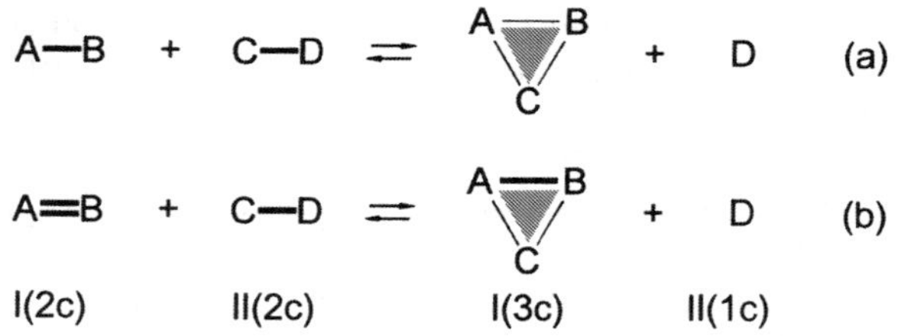

$$A\!-\!B \ + \ C\!-\!D \ \rightleftharpoons \ \overset{A \!=\! B}{\underset{C}{\triangledown}} \ + \ D \qquad (a)$$

$$A\!=\!B \ + \ C\!-\!D \ \rightleftharpoons \ \overset{A \!=\! B}{\underset{C}{\triangledown}} \ + \ D \qquad (b)$$

I(2c)   II(2c)   I(3c)   II(1c)

Figure 17-5 The [2c,2c]-dislocation (from left to right) and the [3c,1c]-collocation (from right to left) in (a) the σ- and (b) the π-version.

a lone pair. We call this reaction a *[2c,2c]-dislocation* and the back reaction a *[3c,1c]-collocation*. The pair I(2c) may be a σ- or a π-bond.

A lot of examples have been worked out either experimentally or theoretically, or by both methods.[22] Several elements can be bound together in the molecule AB by the bond I(2c) that functions as a Lewis base towards the moiety C of CD. In Scheme 1, the H—H bond of H$_2$ and the C—C π-bond of

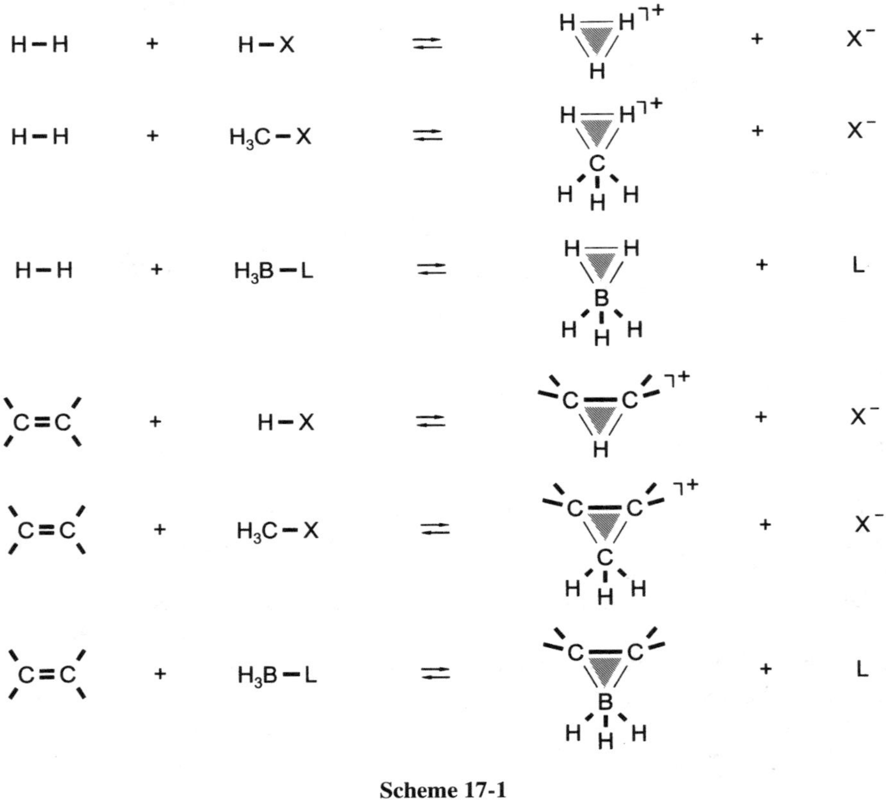

Scheme 17-1

alkenes are taken as examples for AB, and Brönsted acids HX (e.g., in superacidic medium), methylation reagents H_3CX (e.g., in combination with Lewis acidic acceptors for X^-), or borane adducts H_3BL (where L is a neutral ligand) play the role of CD.

The contribution of this chapter to [2c,2c]-dislocation reactions is the addition of boranes XBH_2 to the basic B—B bond of the three-membered ring NB_2R_3 [Eqs (5), (6), and (8)]. The basicity of the hypothetical molecule NB_2H_3 towards BH_3 is greater than that of ammonia towards BH_3, according to the ab initio calculated reaction energies of -45.7 and -28.5 kcal/mol, respectively [12]; the same is true for the basicity of the real molecule NB_2R_3, which accepts BH_3 from the amine adducts $R_3'N—BH_3$. The formation of the *arachno*-cluster $N_2B_3H_2R_3R_2'$ and $SNB_3H_2R_3R'$ [Eq. (7)] apparently proceeds via a [2c,2c]-dislocation [Eq. (9a)] and subsequent intramolecular [3c,1c]-collocation (π-type) [Eq. (9b)].

$$(9)$$

An analogous sequence needs be taken into account for the protonation of NB_2R_3 by acids HX, like triflic acid [23] [Eq. (10)].

$$(10)$$

A second basic one-step reaction that involves (3c2e)-bonds is well known, especially in the chemistry of carbocations [22]: i.e., the simultaneous intramolecular transformation of an electron pair I from a (3c2e)- to a (2c2e)-bond pair, and vice versa for an electron pair II. We call this type of rearrangement a *[3c,2c]-translocation.* A σ- and a π-version can be distinguished (Figure 17-6).

The first example that is given in Scheme 17-2 plays an important role in carbocation chemistry, the second example deals with the dynamic process observed for $B_4H_4R_4$ (Section 17-5); the third example is the equilibrium presented in Eq. (5) (Section 17-7).

All three of these basic processes can be nicely exemplified in a reaction sequence that describes a most probable mechanism of the hydroboration of alkenes: i.e., [2c,2c]-dislocation [Eq. (11a)], [3c,2c]-translocation [Eq. (11b)], and [3c,2c]-collocation [Eq. (11c)].

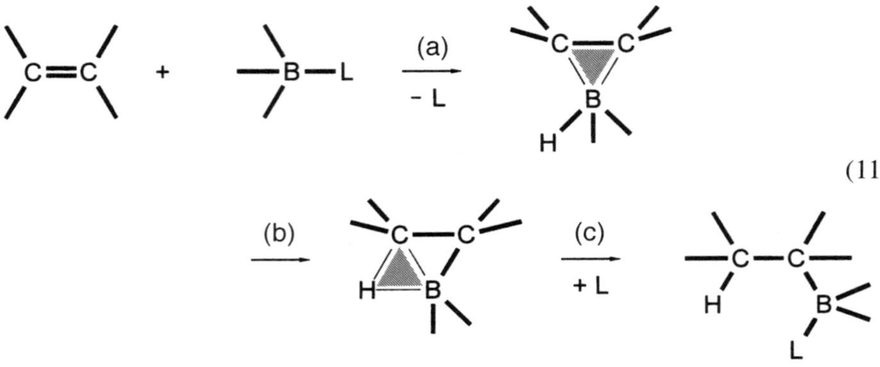

(11)

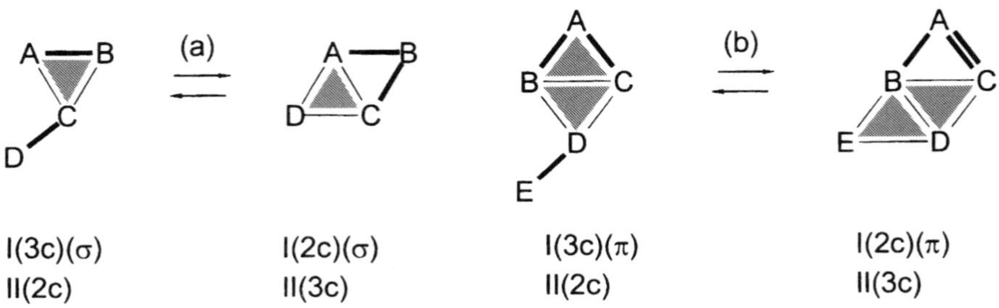

I(3c)(σ)   I(2c)(σ)   I(3c)(π)   I(2c)(π)
II(2c)   II(3c)   II(2c)   II(3c)

Figure 17-6   The [3c,2c]-translocation in (a) the σ- and (b) the π-version.

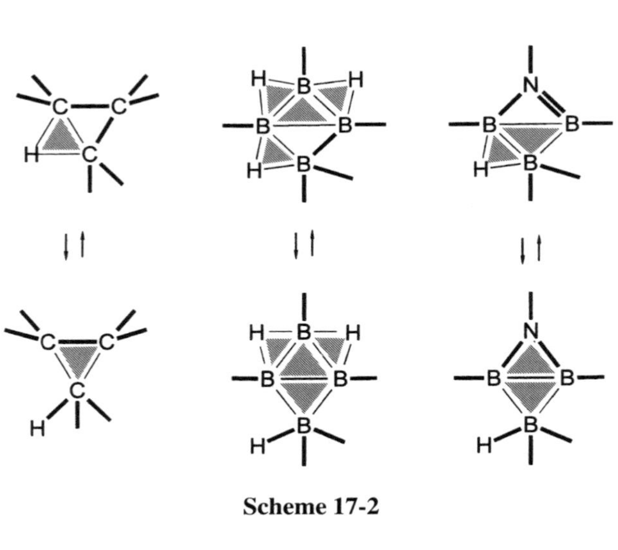

Scheme 17-2

REFERENCES

1. Stock, A. and Massenez, C. *Berichte.*, 1912, **45**, 3539.
2. Urry, G., Wartik, T., and Schlesinger, H. I., *J. Am. Chem. Soc.*, 1952, **74**, 5809.
3. Mennekes, T., Paetzold, P., Boese, R., and Bläser, D., *Angew. Chem. Int. Ed. Engl.*, 1991, **30**, 173.
4. McKee, M. L. and Lipscomb, W. N., *Inorg. Chem.*, 1982, **21**, 2846.
5. Lipscomb, W. N., *Boron Hydrides*, Benjamin, New York, 1963.
6. Arafat, A., Baer, J., Huffmann, J. C., and Todd, L. J., *Inorg. Chem.*, 1986, **25**, 3757.
7. Müller, J., Runsink, J., and Paetzold, P., *Angew. Chem. Int. Ed. Engl.*, 1991, **30**, 175.
8. von Ragué Schleyer, P., Bühl, M., and Hofmann, M., Universität Erlangen, private communication.
9. Müller, J., Technische Hochschule Aachen, private communication.
10. Budzelaar, P. H. M., von Ragué Schleyer, P., and Krogh-Jespersen, K., *Angew. Chem. Int. Ed. Engl.*, 1984, **23**, 825.
11. Wieczorek, C., Allwohn, J., Schmidt-Lukasch, G., Hunold, R., Massa, W., and Berndt, A., *Angew. Chem. Int. Ed. Engl.*, 1990, **29**, 398.
12. Paetzold, P., Redenz-Stormanns, B., Boese, R., Bühl, M., and von Ragué Schleyer, P., *Angew. Chem. Int. Ed. Engl.*, 1990, **29**, 1059.
13. Massey, A. G., and Urch, D. S., *J. Chem. Soc.*, 1965, 6180.
14. Mennekes, T., Dissertation, Technische Hochschule Aachen, 1992.
15. Neu, A., Paetzold, P., Technische Hochschule Aachen, unpublished results.
16. Boese, R., Kröckert, B., and Paetzold, P., *Chem. Ber.*, 1987, **120**, 1913.
17. Müller, M., Dissertation, Technische Hochschule Aachen, 1995.
18. Müller, M., Wagner, T., Englert, U., and Paetzold, P., *Chem. Ber.*, 1995, **128**, 1.
19. Müller, M., Englert, U., and Paetzold, P., *Chem. Ber.*, 1995, **128**, 1105.
20. Luckert, S., Dissertation, Technische Hochschule Aachen, 1996.
21. Paetzold, P., Redenz-Stormanns, B., and Boese, R., *Chem. Ber.*, 1991, **124**, 2435.
22. (a) Olah, G. A., Prakash, G. K. S., Williams, R. E., Field, L. D., and Wade, K., *Hypercarbon Chemistry*, Wiley, New York, 1987; (b) Olah, G. A., Wade, K., and Williams, R. E., (eds), *Electron Deficient Boron and Carbon Clusters*, Wiley, New York, 1991.
23. Müller, M., Eversheim, E., Englert, U., Boese, R., and Paetzold, P., *Chem. Ber.*, 1995, **128**, 99.

18

NEW PERSPECTIVES IN THE CHEMISTRY OF C₂B₄- AND C₄B₈-CARBORANES

NARAYAN S. HOSMANE* AND JOHN A. MAGUIRE

Department of Chemistry, Southern Methodist University, Dallas, TX 75275

18.1   INTRODUCTION

Carbaboranes, commonly known as "carboranes," can formally be considered as arising from the substitution of one or more $\{BH\}^-$ vertices in a polyhedral borane framework by isolobal $\{CR\}$ (where R = H or a cage carbon substituent) units without altering the gross geometry of the cage molecules. A carborane cage can be expanded by the incorporation of other atoms into the polyhedral framework to give the heterocarboranes. When the heteroatom is a metal or a metal moiety, the term metallacarborane is used to describe the compound. Much of the early, and present, interest in such compounds has derived from the recognition by Hawthorne that the frontier orbitals of the *nido*-carborane dianion, $[R_2C_2B_9H_9]^{2-}$, should be similar to those of the cyclopentadienide ion, $[C_5R_5]^-$ (Cp).[1] This similarity has been experimentally verified in a number of studies that are the subjects of reviews and monographs.[2,3] The heterocarborane geometries can range from completely closed polyhedra (*closo*), to more open ones derived by removing one (*nido*), two (*arachno*), or more vertices from the closed structures. The structures of many of these carboranes and their derivatives can be predicted from the number of electron pairs involved in cage bonding using the electron-counting/structure-relationship rules of Wade and Williams.[4] However, there are numerous examples where the rules cannot be easily applied or do not give the correct geometries.[2,3] In order to avoid confusion, we shall use an alternative nomenclature, derived from cyclopentadi-

---

*Author to whom the correspondence should be made.

enyl–metal chemistry, in describing the particular geometry of a metallacarborane polyhedron. In most of the metallacarboranes, the metal is bonded to an open $\{C_2B_3\}$ or $\{C_2B_4\}$ carborane face. When the metal occupies the apical position above this open face, the resulting compound is said to be a half-sandwich complex, while the full-sandwich (sometimes termed *commo*) complexes are those in which a centrally located metal atom is bonded to the open faces of two carboranes. The metallacarborane complexes can have idealized geometries, i.e., the metal can be bonded equally to the facial atoms. Alternately, the metal atom can be dislocated toward one side of the bonded face, to give slipped full- or half-sandwich complexes. In the full-sandwich complexes, the two carborane cages could be either directly opposite one another to give a linear cage(1)—M—cage(2) geometry with parallel bonding faces, or the complex could be bent such that the bonding faces make acute dihedral angles: such compounds are called bent-sandwich complexes. The terms half, full, slipped, and bent sandwich are strictly geometric descriptors, and hence have some advantages over the alternative *nido*, *closo*, and *commo* notations, which carry both structural and electronic connotations. We will use both sets of terminology in this account.

Most of the metallacarboranes discussed here are those derived from the *nido*-$\{C_2B_4\}$ carborane system, in which the carboranes have a C_2B_3 bonding face. There are two different possible arrangements of the facial atoms in such carboranes: the two cage carbons can be next to one another in the "carbons-adjacent" carboranes or they can be separated by a boron atom in the "carbons-apart" isomers. Most of the structural data on these metallacarboranes have been collected on the "carbons-adjacent" isomers. This is dictated by the increased ease of preparation of this isomer rather than by any enhanced stability or inherent interest.

18.2   LARGE-SCALE PREPARATIONS OF AIR-STABLE (C—SiMe₃)-SUBSTITUTED "CARBONS ADJACENT" *NIDO*-$\{C_2B_4\}$ AND *NIDO*-$\{C_4B_8\}$ CARBORANES

The reactions that led to the formation of tris[(trimethylsilyl)-1-alkenyl]borane, and also to several air-stable C-trimethylsilyl-substituted *nido*-C_2B_4 carboranes, were described in the mid-1980s [Eqs (1) and (2)].[5] Since *nido*-$[(Me_3Si)(R)C_2B_4H_6]$ derivatives (where R = $SiMe_3$, Me, n-C_4H_9, t-C_4H_9, and H) can be produced in high yields (12–13 g per batch when R = $SiMe_3$ or n-C_4H_9), these have been the major subject for most of our investigations.[3(b)–(d)]

$$nido\text{-}B_5H_9 + 4\ Me_3SiC\equiv CR \xrightarrow[\text{no solvent}]{25°C} (Me_3SiC\!=\!C(H)R)_3B\ +$$

(crystalline solid)

$$[2\text{-}(SiMe_3)\text{-}3\text{-}(R)\text{-}nido\text{-}2,3\text{-}C_2B_4H_6] \qquad (1)$$

(volatile colorless liquid)

$$nido\text{-}B_5H_9 + 4\ Me_3SiC{\equiv}CR \xrightarrow[\substack{\text{no solvent} \\ \text{in stainless-steel} \\ \text{reactor}}]{140°C}$$

$$Me_3SiH + [2\text{-}(SiMe_3)\text{-}3\text{-}(R)\text{-}nido\text{-}2,3\text{-}C_2B_4H_6] + \text{polymer} \qquad (2)$$
$$\text{(as the most volatile reaction products)} \qquad \text{(nonvolatile)}$$

The trimethylsilyl-substituted carboranes offer a number of advantages over their hydrogen-substituted analogs: (1) all are air-stable liquids that do not require special handling procedures; (2) they can be routinely prepared in multi-gram quantities as single pure isomers; (3) due to two B—H—B bridge hydrogens, they are fairly reactive compounds even in the absence of solvents; (4) the C—$SiMe_3$ moiety is conducive to single-crystal growth; (5) in general, they are converted to fairly stable products; (6) the C—$SiMe_3$ bond can be readily cleaved selectively; (7) the products of their reactions tend to be more soluble in organic solvents than their C—H, C-alkyl, or C-aryl-substituted analogs; (8) their reactions generally proceed quantitatively to yield carborane products; and (9) their chemistry has been developed more systematically than that of their alkyl or aryl analogs.

The controlled reaction of liquid $nido$-$[(Me_3Si)_2C_2B_4H_6]$, either with solid $NaHF_2$ or gaseous HCl (in a pyrex-glass reactor or stainless-steel reactor) at 140°C, produces $nido$-$[(Me_3Si)(H)C_2B_4H_6]$ by elimination of one of the cage-$\{SiMe_3\}$ groups, while not affecting the terminal B—H and B—H—B bridge bonds.[6] The 2-(trimethylsilyl)-$nido$-2,3-dicarba-hexaborane(8) product has been obtained in very high purity by this method, so that its molecular structure could be determined by gas-phase electron diffraction.[6] The structure is shown in Figure 18-1. Even though it is somewhat indirect, this synthesis avoids the use of expensive starting materials, such as $Me_3SiC{\equiv}CH$, that were needed in the more direct preparations that were used previously.[5,7] At higher temperatures, the $[2,3\text{-}(SiMe_3)_2\text{-}nido\text{-}2,3\text{-}C_2B_4H_6]$ reacts with different molar ratios of dry HCl gas to produce a number of carborane products [see Eq. (3)].[8] For example, when the carborane to HCl molar ratio was 1:3.3 and the mixture was heated at 160–170°C for 4 days, the carborane products $[2\text{-}(SiMe_3)\text{-}nido\text{-}2,3\text{-}C_2B_4H_7]$ and $nido$-2,3-$C_2B_4H_8$ (the parent carborane) were obtained in 31% and 27% yields, respectively.

$$[2,3\text{-}(SiMe_3)_2\text{-}nido\text{-}2,3\text{-}C_2B_4H_6] + 3.3\ HCl \xrightarrow[\substack{\text{no solvent} \\ \text{in stainless steel} \\ \text{cylinder}}]{\substack{160\text{--}170°C \\ 4\ \text{days}}}$$

$$Me_3SiCl + [2\text{-}(SiMe_3)\text{-}nido\text{-}2,3\text{-}C_2B_4H_7]$$
$$+ nido\text{-}2,3\text{-}C_2B_4H_8 + BCl_3 + B_2Cl_6 + H_2$$
$$+ \text{chlorinated products} \qquad (3)$$

When the molar ratio of mono(trimethylsilyl)-carborane [produced as in Eq. (3)] to HCl was 1:1.5, the yield of the parent carborane increased slightly at the

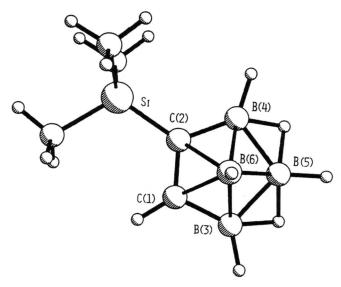

Figure 18-1 Structure of [2-(SiMe$_3$)-*nido*-2,3-C$_2$B$_4$H$_7$] obtained by gas-phase electron diffraction analysis.

expense of other carboranes. These last were evidently converted to BCl$_3$, B$_2$Cl$_6$, and other chlorinated volatile products. For a high-yield preparation of the parent {C$_2$B$_4$} carborane, the ideal molar ratio between the bis(C—SiMe$_3$)-substituted carborane and HCl was found to be 1:3.3.[8]

The *nido*-carborane [2,3-(SiMe$_3$)$_2$-2,3-C$_2$B$_4$H$_6$] undergoes direct fusion at 210°C to give *nido*-[(SiMe$_3$)$_2$C$_4$B$_8$H$_{10}$].[9] This reaction occurs without the need of a metal catalyst [see Eq. (4)].

$$2 \text{ [2,3-(SiMe}_3)_2\text{-}nido\text{-2,3-C}_2\text{B}_4\text{H}_6] \xrightarrow[\substack{\text{neat} \\ \text{in vacuum sealed} \\ \text{glass tube}}]{210°C}$$

$$2 \text{ Me}_3\text{SiH} + [(\text{SiMe}_3)_2\text{-}nido\text{-2,3,7,8-C}_4\text{B}_8\text{H}_{10}] \qquad (4)$$
$$+ \text{ nonvolatile residue}$$

Although the exact mechanism of the fusion process is not known, it could well involve an initial high-temperature formation of a trimethylsilyl radical, which could then extract one of the carborane bridging hydrogens to form Me$_3$SiH and a reactive carborane fragment that could condense with another such fragment to give the final C$_4$B$_8$ product.[9] Irrespective of the exact mechanism, the presence of the trimethylsilyl moiety is directly involved in the synthesis of the tetracarbon carborane.

18.3   REACTIVITIES OF (C—SiMe₃)-SUBSTITUTED "CARBONS-ADJACENT" *NIDO*-{C₂B₄} CARBORANES TOWARD REAGENTS OF GROUP 1 METALS

The Group 1 metallacarboranes are the most commonly prepared of all the metallacarboranes in that they are normally the precursors for other heterocarborane compounds; however, until recently, they have been scarcely examined. The most general synthetic route to the metallacarboranes is through the reactions of the *nido*-carborane anions with metal reagents, usually metal halides. This procedure was used by Hawthorne and coworkers in the syntheses of the initially reported metallacarboranes having the general formulae [3,1,2-M(CR)$_2$B$_9$H$_9$] (where M = metal moiety; R = H or an alkyl group).[1] The carborane anions are formed by an initial degradation of *closo*-1,2-(CR)$_2$B$_{10}$H$_{10}$ by alcoholic KOH to produce the monoanions [7,8-(CR)$_2$B$_9$H$_{10}$]⁻, which are assumed to be *nido*-carboranes containing a single bridging hydrogen. The bridging hydrogen is then removed as a proton by the reaction with NaH to give the dianionic ligands *nido*-[7,8-(CR)$_2$B$_9$H$_{11}$]²⁻. Since equivalent results are obtained with different alkali metals, and also when tetralkylammonium cations replaced the alkali metals,[1(c)] the Group 1 metals were assumed to be just spectator ions and their interactions with the carborane dianions were not probed. On the other hand, the situation in the smaller, {C₂B₄}-cage system was not so straightforward. Since the *nido*-[2,3-(CR)$_2$B$_4$H$_6$] carboranes have two bridging hydrogens on adjacent boron atoms on the open pentagonal face (see Figure 18-1), the dianion should, in principle, be easily obtainable by removal of the bridging hydrogens as protons by a suitable base. However, Onak and coworkers have found that the reaction of *nido*-2,3-C₂B₄H₈ with NaH produced only the monosodium compound, as shown in Eq. (5).[10]

$$nido\text{-}C_2B_4H_8 + \text{excess NaH} \xrightarrow[\text{0-150°C}]{\text{THF}} \text{Na}[nido\text{-}C_2B_4H_7] + H_2 \qquad (5)$$

The most interesting facet of this reaction is its stoichiometry: the monoanion was found to be the exclusive product, even when the reaction was carried out in the presence of excess NaH and at elevated temperatures. Grimes and coworkers have previously studied the kinetics of this deprotonation reaction using various substituted carboranes of the type 2,3-R,R′-*nido*-2,3-C₂B₄H₆ (where R = alkyl, arylmethyl and phenyl; R′ = R, H), and using both NaH and KH in tetrahydrofuran (THF).[11] Their results were consistent with a mechanism involving a direct reaction of the carborane with a hydride site on the solid metal hydride matrix. As was found for the unsubstituted carboranes, only a single bridging hydrogen was removed in these heterogeneous reactions. This lack of further reactivity of the monoanion toward the metal hydride bases was somewhat surprising because the monoanions react readily with bases such as BuLi to give mixed natralithacarborane compounds (Scheme 1), which have proven to be useful synthons in the preparation of a number of heterocarboranes (see

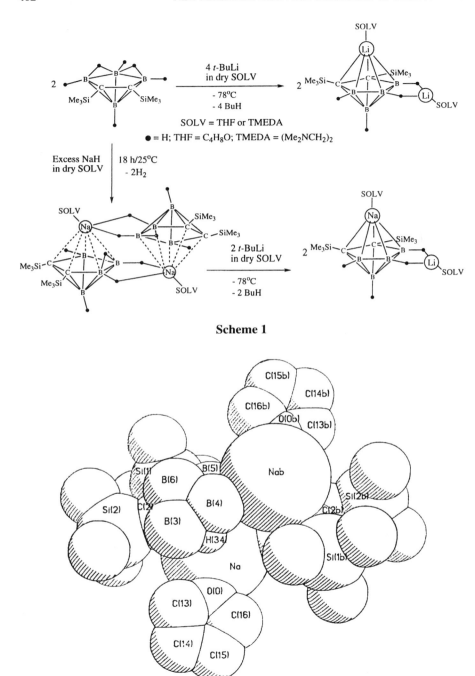

Scheme 1

Figure 18-2   Space-filling model of the dimeric structure of [1-Na(THF)-2,3-(SiMe$_3$)$_2$-2,3-C$_2$B$_4$H$_5$]$_2$.

below).[12] A possible explanation for the exclusive monodeprotonation has been provided by the crystal structure of the sodium compound of the trimethylsilyl-substituted carborane monoanion $[2,3-(SiMe_3)_2-2,3-C_2B_4H_5]^-$.[13] The solid-state structure of the THF-solvated species is that of an extended network of dimeric $[1-Na(THF)-2,3-(SiMe_3)_2-2,3-C_2B_4H_5]_2$ clusters that are layered symmetrically on top of one another. Each remaining bridging hydrogen on the bonding face in the dimer is well shielded sterically by the surrounding groups, as can be seen in the space-filling model of this compound shown in Figure 18-2. The reaction of $[2,3-(SiMe_3)_2-nido-2,3-C_2B_4H_6]$ with NaH in the alternative solvent tetramethylethylenediamine (TMEDA) has resulted in the exclusive formation of a monosodium compound that crystallized as a $[1-Na(TMEDA)-2,3-(SiMe_3)_2-2,3-C_2B_4H_5]_2$ dimer. The basic crystal structure, shown in Figure 18-3, is similar to that of the THF-solvated one, except that the THFs in Figure 18-2 are replaced by TMEDA molecules. The major effect of this solvent substitution is that the larger TMEDA molecules tend to break up the chain structure found in THF, leading to isolated dimeric clusters.[14] However, steric shielding of the remaining bridging hydrogen is apparent in both structures. It has been argued that, in THF or TMEDA solutions, the natracarboranes would exist as intimate ion-pair clusters, similar to those shown in Figures 18-2 and 18-3, and that the steric shielding of the bridging hydrogens in such

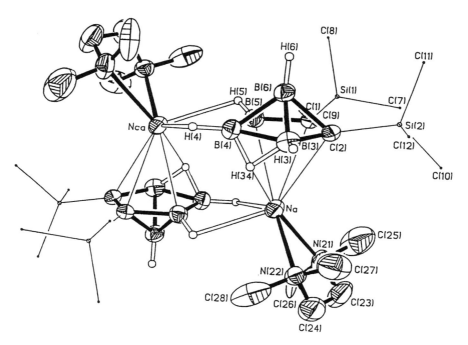

Figure 18-3 Perspective view of a discrete dimeric unit of $[1-Na(TMEDA)-2,3-(SiMe_3)_2-2,3-C_2B_4H_5]_2$.

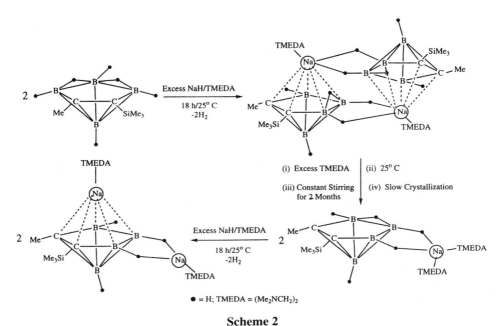

Scheme 2

Figure 18-4 Crystal structure of the zwitterionic, TMEDA-separated, monomeric monosodium salt, [exo-4,5-{(μ–H)$_2$Na(TMEDA)$_2$}-2-(SiMe$_3$)-3-Me-nido-2,3-C$_2$B$_4$H$_5$].

clusters would effectively prevent their reaction with a surface H⁻ site on the solid NaH to give the dianion.[14,15] Thus, the stoichiometry of Eq. (5) seems to be dictated more by the heterogeneous nature of the reaction and steric factors than by an inherent stability of the monoanion. Support for this argument has been furnished by the synthesis of the more solvated species *exo*-4,5-[(μ-H)$_2$Na(TMEDA)$_2$]-2-(SiMe$_3$)-3-Me-*nido*-2,3-C$_2$B$_4$H$_5$, shown in Scheme 2 and Figure 18-4.[15] It was found that the less sterically protected bridging hydrogen H(56) could be easily removed by NaH to give the dinatracarborane.[15] Furthermore, the highly solvated species [Li(TMEDA)$_2^+$][2,3-(SiMe$_3$)$_2$-*nido*-2,3-C$_2$B$_4$H$_5^-$] has also been synthesized, structurally characterized (Figure 18-5) and found to react with NaH to give the natralithacarborane compound.[16] The direct synthesis of the dilithacarborane [*exo*-4,5-{(μ-H)$_2$Li(TMEDA)}-1-Li(TMEDA)-2,3-(SiMe$_3$)$_2$-*closo*-2,3-C$_2$B$_4$H$_4$] is described in Scheme 1, and its solid-state geometry, which is probably typical for these dimetalated carbo-

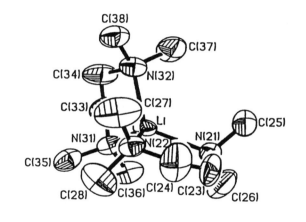

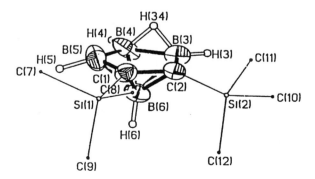

Figure 18-5 Perspective view of the solid-state structure of [Li(TMEDA)$_2^+$][2,3-(SiMe$_3$)$_2$-*nido*-2,3-C$_2$B$_4$H$_5^-$].

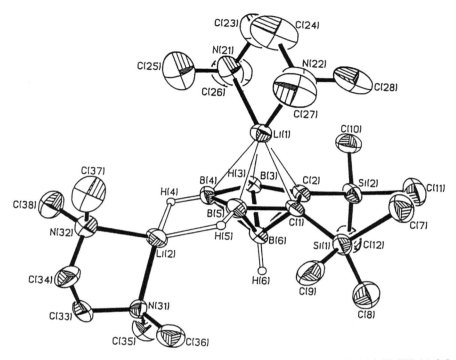

Figure 18-6   Crystal structure of [*exo*-4,5-{(μ-H)₂Li(TMEDA)}-1-Li(TMEDA)-2,3-(SiMe₃)₂-*closo*-2,3-C₂B₄H₄].

ranes, is shown in Figure 18-6.[14] The two lithium atoms occupy nonequivalent positions in the cluster: one is *exo*-polyhedrally bound to two adjacent borons on the {C₂B₃} face of the carborane, while the other occupies the apical position above the bonding face. The ^{7}Li NMR spectrum of this complex in C₆D₆ shows two resonances: one at δ −1.60 ppm (relative to external aqueous LiNO₃), due to the *exo*-polyhedral lithium, and the more shielded apical lithium with a peak at δ −6.03 ppm.[14] The presence of two distinct ^{7}Li NMR resonances indicates that the ion cluster is quite stable and that any ion exchange that takes place in solution is slow; this would be expected in the nonpolar solvents used in these studies. The ^{7}Li NMR spectrum of the corresponding mixed natralithacarborane compound showed a single resonance at δ −1.82 ppm, indicating that the Li occupied the *exo*-polyhedral position with the Na in an apical position.[14] Thus, the order of reactivity establishes the location of the Group 1 metal: the metal that is introduced first occupies the apical position, with the second metal being the *exo*-polyhedral one.

Slow sublimation of the TMEDA-solvated monolithium carborane complex [16,17] at 160–170°C over a period of 6–7 h in vacuo produced the full-sandwich lithacarborane complex, [Li(TMEDA)₂][*commo*-1,1′-Li{2,3-(SiMe₃)₂-

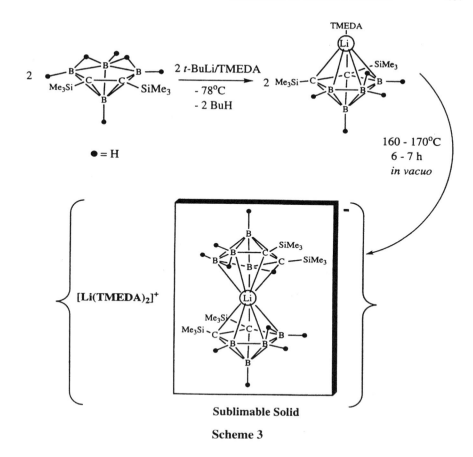

Sublimable Solid

Scheme 3

2,3-$C_2B_4H_5$}$_2$], as a transparent crystalline solid (see Scheme 3).[18] The mechanism of the formation of this complex is not known. Since alkyllithiums have been shown to be sublimable,[19] the monolithium compound could be the subliming species, which then disproportionates to give the ionic full-sandwich lithacarborane complex. The spectroscopic data for the fully sandwiched lithacarborane are consistent with its solid-state structure, shown in Figure 18-7. The distances from lithium to the ring centroids in the complex (2.047 and 2.071 Å) are longer than the value of 1.906 Å found in the half-sandwich dilithacarborane,[14] but are comparable to the corresponding metal–centroid distance of 2.008 Å found in the [Cp_2Li]$^-$ sandwich complex.[20] The sensitivity of the metal-to-ligand distance to the ligand charge is consistent with a predominantly ionic interaction between the Group 1 metal and the carborane cages. Careful inspection of Figure 18-7 shows that the lithium atom is displaced toward the cage carbons and one of the basal borons. This slip distortion of the lithium is most probably due to the presence of B—H—B bridging hydrogens on the bonding faces. The high yield for this reaction indicates that it may prove to be

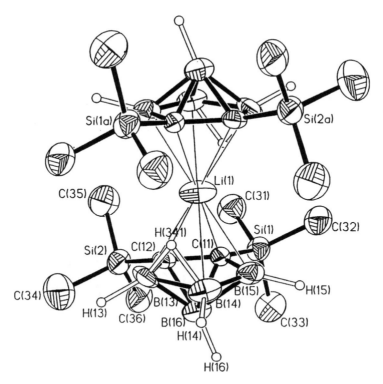

Figure 18-7   Crystal structure of the full-sandwich lithacarborane anion $[1,1'\text{-Li}\{2,3\text{-}(SiMe_3)_2\text{-}2,3\text{-}C_2B_4H_5\}_2]^-$.

a general method for the selective synthesis of a number of hitherto unknown full-sandwich metallacarboranes of Group 1 and other metals.

18.4   METAL-PROMOTED OXIDATIVE CAGE CLOSURE AND CAGE FUSION: SYNTHESES OF (C—SiMe₃)-SUBSTITUTED "CARBONS-ADJACENT" *CLOSO*-{C₂B₄} CARBORANES AND "CARBONS-APART" *NIDO*-{C₄B₈} CARBORANES

The reaction of anhydrous $NiCl_2$ with the dilithacarborane $[exo\text{-}4,5\text{-}\{(\mu\text{-}H)_2Li(TMEDA)\}\text{-}1\text{-}Li(TMEDA)\text{-}2,3\text{-}(SiMe_3)_2\text{-}closo\text{-}2,3\text{-}C_2B_4H_4]$ (Figure 18-6) [14] in 1:1 stoichiometry affords a mixture of three products. These are $1,2\text{-}(SiMe_3)_2\text{-}closo\text{-}1,2\text{-}C_2B_4H_4$ (see Figure 18-8),[14,21] isolated as a colorless liquid, and a white crystalline solid that was found to be a 50:50 mixture of two isomeric "carbons apart" *nido*-{C₄B₈} carboranes (Scheme 4).[22] Fractional crystallization of the solid from a 1:1 solution of benzene and hexane facilitated the separation of the two isomers.

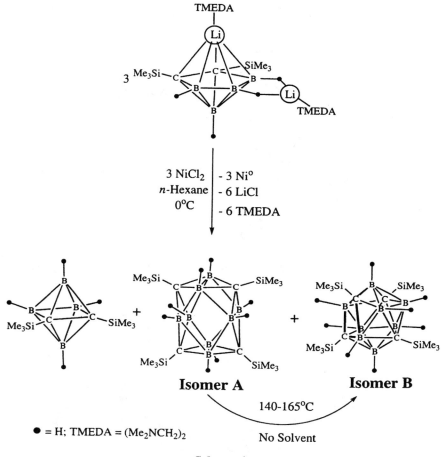

Isomer A   **Isomer B**

140-165°C

● = H; TMEDA = (Me$_2$NCH$_2$)$_2$ No Solvent

Scheme 4

The structures of the two isomers, called isomer A and B, are shown in Figures 18-9 and 18-10, respectively. A neat sample of isomer A was found to undergo a thermal isomerization at 140–165°C to yield, exclusively, isomer B. The isomerization showed no tendency for reversibility, indicating that B is the more stable isomer. This conclusion is supported by ab initio molecular-orbital calculations on the model isomers of C$_4$B$_8$H$_{12}$, which showed that the {C$_4$B$_8$} cage with C_{2v} symmetry is more stable than that with D_{2h} symmetry by 124.2 kJ/mol^{-1}.[22] The solid-state structure of isomer A (Figure 18-9) is best described as a distorted cuboctahedron. This isomer has the same general geometry as that of the intermediate proposed by Lipscomb in his diamond-square-diamond (DSD) mechanism for the thermal rearrangement of *ortho*-carborane (*closo*-1,2-C$_2$B$_{10}$H$_{12}$) to the corresponding *meta*-isomer (*closo*-1,7-

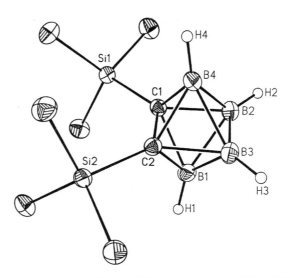

Figure 18-8   Low-temperature X-ray solid-state structure of $[1,2\text{-}(SiMe_3)_2\text{-}closo\text{-}1,2\text{-}C_2B_4H_4]$.

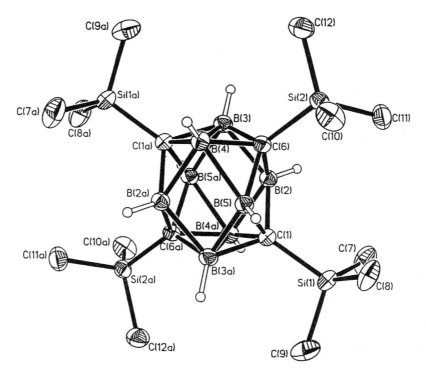

Figure 18-9   Solid-state structure of isomer A of the "carbons-apart" tetracarbadode-caborane $[(CSiMe_3)_4B_8H_8]$.

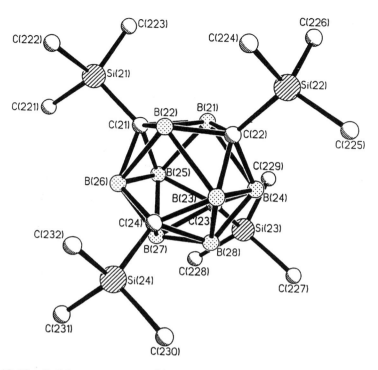

Figure 18-10   Solid-state structure of isomer B of the "carbons-apart" tetracarbadode-caborane [(CSiMe$_3$)$_4$B$_8$H$_8$].

$C_2B_{10}H_{12}$).[23] The distortion from D_{4h} symmetry results in a structure where the $\{C_2B_2\}$ quadrilateral faces are not planar and the B—B distances are not all equal; much of this distortion is expected from the heteronuclear nature of the faces. The structures of isomer A and isomer B differ mainly in the relative positions of the cage carbons (see Figure 18-10).[22] Inspection of Figures 18-9 and 18-10 shows that, whereas the cage carbons might be equivalent in isomer A, or nearly equivalent in isomer B, there are two sets of borons in both isomers. Therefore, the appearances of only single resonances in the ^{11}B NMR spectra of both compounds indicate that the $\{C_4B_8\}$ cages are fluxional on an NMR time scale. These isomers are quite different from those of the "carbons-adjacent" (CR)$_4$B$_8$H$_8$ isomers, reported by Grimes and coworkers, in which the cage carbons were localized on the same side of the cage.[24] It is of interest to note that neither isomer A nor isomer B could be converted to the "carbons-adjacent" isomers. Moreover, the use of Grimes' method for oxidative ligand fusion,[24] using the reaction of FeCl$_2$ with the monosodium salt of 2,3-(SiMe$_3$)$_2$-$nido$-2,3-C$_2$B$_4$H$_6$, failed to produce the "carbons-adjacent" [(CSiMe$_3$)$_4$B$_8$H$_8$] isomers: the ^{11}B NMR spectrum of the product was the same as that reported for isomer B. Thus, the SiMe$_3$ moieties that are *exo*-polyhedrally C$_{(cage)}$-bound to the cage play a direct role in determining the course of the cage-fusion reaction.[22] Even though the mechanism for the formation of either of isomers A or B is not known, the isolation and structural characterization of isomer A (Scheme 4 and Figure 18-9) demonstrate that the cuboctahedral geometry in carboranes can be stabilized by using carbon atoms to increase the total number of skeletal electrons in a 12-vertex cage to 28, and by introducing a steric bulk on the cluster faces.

18.5   CAGE-OPENING REACTIONS OF *CLOSO*-{C$_2$B$_4$} CARBORANES AND SUBSEQUENT PROTONATIONS OF "CARBONS-APART" METALLACARBORANES OF GROUP 1 ELEMENTS

The *closo*-carborane 1-(SiMe$_3$)-2-(R)-1,2-C$_2$B$_4$H$_4$ (where R = SiMe$_3$, Me, *n*-C$_4$H$_9$, *t*-C$_4$H$_9$, and H) (see Figure 18-8 as an example)[14,21] can be reductively opened to give a Group 1 metal-complexed "carbons-apart" dianion, [2-(SiMe$_3$)-4-(R)-2,4-C$_2$B$_4$H$_4$]$^{2-}$, as outlined in Scheme 5.[25,26] The X-ray

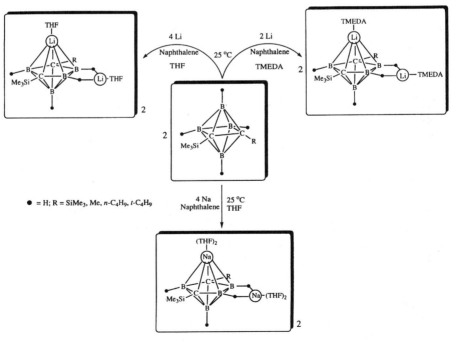

Scheme 5

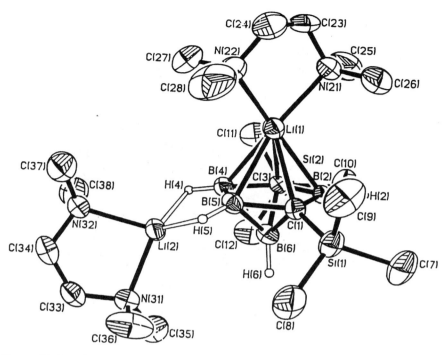

Figure 18-11 Crystal structure of the TMEDA-solvated "carbons-apart" dilithacarborane [*exo*-5,6-{(μ-H)$_2$-Li(TMEDA)}-1-Li(TMEDA)-2,4-(SiMe$_3$)$_2$-*closo*-2,4-C$_2$B$_4$H$_4$].

analysis of the dilithacarborane shows that the solid-state structure of the TMEDA-solvated species (Figure 18-11) is almost identical to its corresponding "carbons-adjacent" isomer (Figure 18-6).[14] On the other hand, the THF-solvated dilithacarborane exists as a fairly tight ion cluster consisting of two {Li$_2$C$_2$B$_4$} units (Figure 18-12).[25] The structure of the corresponding disodium compound shows that this compound also exists as a discrete dimer in which each carborane dianion is associated with an *endo*-sodium, which adopts an essentially η^5-bonding posture with respect to the {C$_2$B$_3$} face, and an *exo*-polyhedral sodium that is situated over the B$_3$ trigonal face formed by the apical boron and the two basal boron atoms (see Figure 18-13).[26] The sodium atoms are also coordinated to a sufficient number of THF molecules to give irregular tetrahedral arrangements about the metals.[26] In contrast to the results from the 2,3-{C$_2$B$_4$} carborane system (see above), the "carbons-apart" carborane ligands react with NiCl$_2$ to give either the full-sandwich or the half-sandwich nickel complex, depending on the solvent used in the reaction.[25] It is of interest to note that, while the neutral and monoanionic compounds of the "carbons-

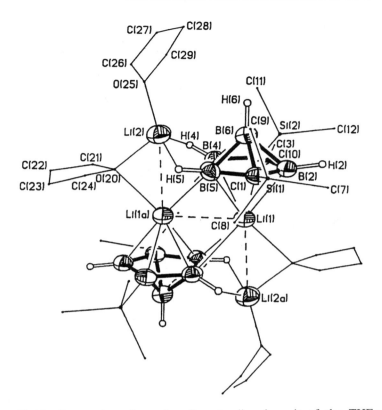

Figure 18-12 Perspective view of a discrete dimeric unit of the THF-solvated "carbons-apart" dilithacarborane [exo-5,6-{(μ-H)$_2$Li(THF)}(μ-THF)-1-Li-2,4 (SiMe$_3$)$_2$-$closo$-2,4-C$_2$B$_4$H$_4$]$_2$.

adjacent" {C$_2$B$_4$} carboranes were the first ones to be synthesized, with dianion formation not being reported until some 20 years later, the opposite is true for the "carbons-apart" {C$_2$B$_4$} carboranes. The $nido$-{2,4-C$_2$B$_4$} carborane ligands are prepared by the two-electron reduction of the corresponding $closo$-{C$_2$B$_4$} carborane precursors to give directly the dianion, as described above.[14,21] The "carbons-apart" monoanionic species can only be made by the careful reaction of the corresponding disodium or dilithium compound with anhydrous HCl; this reaction results in the protonation of the two adjacent borons, B(4) and B(5), as shown in Figure 18-14, to give the corresponding Group 1 monoanionic complex, similar in structure to those shown in Figure 18-3 and Scheme 3. Attempts at more extensive protonation led to the decomposition of the carborane.[27]

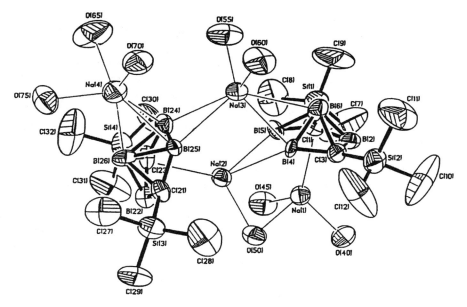

Figure 18-13   Solid-state structure of a discrete dimeric cluster unit of the THF-solvated "carbons-apart" dinatracarborane [*exo*-5,6-{Na(THF)$_2$}-1-{Na(THF)(μ-THF)}-2,4-(SiMe$_3$)$_2$-*closo*-2,4-C$_2$B$_4$H$_4$]$_2$.

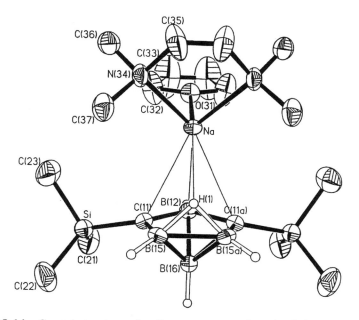

Figure 18-14   Crystal structure of a discrete monomeric unit of the "carbons-apart" monosodium compound [1-(THF)-1-(TMEDA)-1-Na-2,4-(SiMe$_3$)$_2$-*nido*-2,4-C$_2$B$_4$H$_5$].

18.6   REACTIVITIES OF METALLACARBORANES OF GROUP 1 METALS TOWARD METAL ALKYLS OF GROUP 2 ELEMENTS

Although the organometallic chemistry of Group 2 elements has received special attention in recent years and a large number of novel cyclopentadienyl complexes have been synthesized and crystallographically characterized, [28–31] the literature on the analogous metallacarborane complexes is severely limited. Over 25 years ago, Hawthorne and coworkers reported the first Group 2 metallacarborane, the beryllacarborane $[(CH_3)_3NBeC_2B_9H_{11}]$, which was characterized spectroscopically but not by X-ray diffraction.[32,33] It was only recently that the studies of the Group 2 metallacarboranes have been extended to the syntheses and structural determinations of the calca- and strontacarbo-ranes, $[1,1,1-(MeCN)_4-closo-1,2,4-CaC_2B_{10}H_{12}]$ and $[1,1,1-(MeCN)_3-closo-1,2,4-SrC_2B_{10}H_{12}]_n$.[34,35] Although magnesocene has been known for some time,[36] and an *exo*-polyhedrally bound magnesium has been reported in the metallaborane $[(THF)_2Mg(B_6H_8)_2]$,[37] until recently there were no reported π-complexes of magnesium in any polyhedral borane or carborane system. The first report of such complexes was in 1995. This described the syntheses and solid-state structures of both the half-sandwich and full-sandwich magnesa-

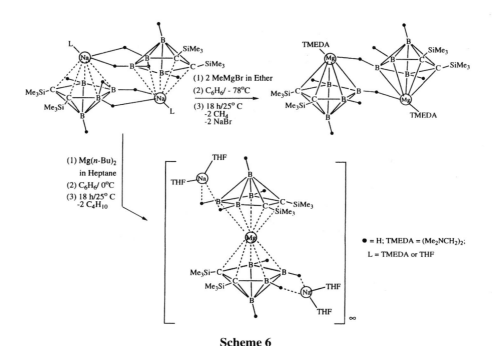

Scheme 6

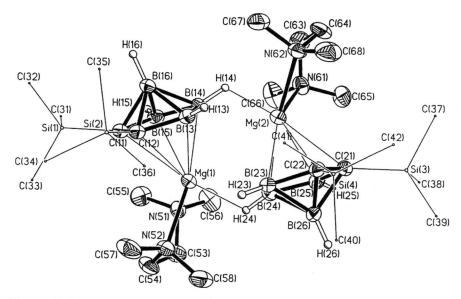

Figure 18-15 Crystal structure of the half-sandwich magnesacarborane dimer [1-Mg(TMEDA)-2,3-$(SiMe_3)_2$-2,3-$C_2B_4H_4]_2$.

carboranes in the pentagonal bipyramidal $\{C_2B_4\}$ cage system.[38] Scheme 6 outlines the high-yield reactions between the TMEDA-solvated monosodium derivative or salt of the [2,3-$(SiMe_3)_2$-2,3-$C_2B_4H_5]^-$ anion (Figure 18-3) and the alkylating agents [MeMgBr] and [(n-$C_4H_9)_2$Mg] in molar ratios of 1:1 and 2:1, respectively, to produce the novel half-sandwich magnesium complex, [1-$\{Mg(TMEDA)\}$-2,3-$(SiMe_3)_2$-2,3-$C_2B_4H_4$], and the full-sandwich magnesacarborane, the [$commo$-1,1'-Mg$\{2,3$-$(SiMe_3)_2$-2,3-$C_2B_4H_4\}_2]^{2-}$ dianion.[38] No alkylations of the B—H bonds of the carborane cage were evident in the reaction. The structure of the half-sandwich compound (Figure 18-15) is that of a dimer in which the solvated magnesium center is situated above the $\{C_2B_3\}$ face of the carborane and is also bonded to the unique boron B(14) of an adjacent carborane unit by a single Mg—H—B bridge.[38] The magnesium atom is not symmetrically bonded to the $\{C_2B_3\}$ face but is slipped toward the unique boron B(14,24) in Figure 18-15. Since the carboranes are η^5-bonded to the magnesium atom in the full-sandwich complex (Figure 18-16), it is doubtful that the slippage found in the half-sandwich complex reflects inherent metal–carborane bonding preferences. If the interaction between the metal and the carborane were essentially ionic, then the observed dimer formation could give rise to the distortion shown in Figure 18-15.

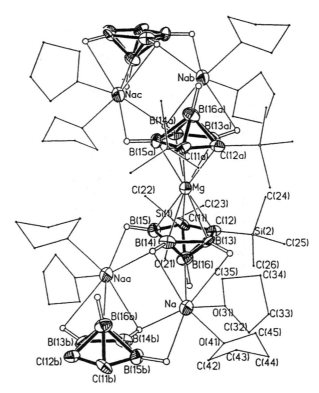

Figure 18-16 Crystal structure of the full-sandwich dianionic magnesacarborane [*commo*-1,1′-Mg{2,3-(SiMe$_3$)$_2$-2,3-C$_2$B$_4$H$_4$}$_2$]$^{2-}$

18.7   REACTIVITIES OF (C—SiMe$_3$)-SUBSTITUTED "CARBONS-APART" *NIDO*-{C$_4$B$_8$} CARBORANES TOWARD GROUP 1 AND GROUP 2 METALLIC ELEMENTS

Grimes and co-workers have studied the two-electron reduction of the C-alkyl-substituted "carbons-adjacent" species R$_4$C$_4$B$_8$H$_8$ with excess Na/naphthalene in THF.[39] Although the structure of the resulting dianion was not determined, the solid-state structures of several of its metalla- and bimetallacarboranes suggested an *arachno*-{C$_4$B$_8$} cage that can be envisaged as arising from the removal of two vertices from a 14-vertex hexagonal antiprismatic *closo*-carbo-rane.[39] Since essentially all of the cited examples involve the use of metal naphthalenes as reducing agents,[40] the extent to which the exact nature of the reducing agent determines the course of these reactions has not been considered. However, a recent report on the reactivity of the (C—SiMe$_3$)-substituted "carbons-apart" species [2,4,7,9-(SiMe$_3$)$_4$-*nido*-2,4,7,9-C$_4$B$_8$H$_8$] toward lithium

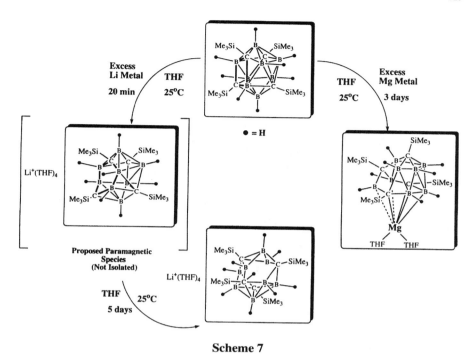

Scheme 7

and magnesium metals in the absence of naphthalene and aromatic solvents indicates that a significant dependence may exist.[41] As outlined in Scheme 7, treatment of a THF solution of the "carbons-apart" [(SiMe$_3$)$_4$C$_4$B$_8$H$_8$] isomer (Figure 18-10)[22] with finely cut excess lithium metal at 25°C resulted in the formation of a red/orange-colored heterogeneous mixture, without any gas evolution. The initial red/orange compound was found to be EPR-active, with a g value of 2.0030 that suggested a π(pi)-type radical formation. After stirring the mixture for 5 days, a monolithium compound, {[(THF)$_4$Li][(SiMe$_3$)$_4$C$_4$B$_8$H$_9$]}, was produced as an EPR-silent transparent crystalline solid.[41] A direct reaction of [(SiMe$_3$)$_4$C$_4$B$_8$H$_8$] with pure magnesium metal over a period of 3 days resulted in the formation of a novel magnesacarborane, [(THF)$_2$Mg(SiMe$_3$)$_4$C$_4$B$_8$H$_8$] (Scheme 7).[41] The driving forces for these reactions are not known, nor is it apparent why, even in the presence of excess lithium metal, the monolithium compound is formed instead of the expected two-electron reduction product. Nevertheless, these reactions suggest that the "carbons-apart" {C$_4$B$_8$} carborane could perhaps be effectively used to oxidize a single metal-atom species, thus facilitating the formation of the corresponding 1:1 ionic products, without the loss of any metal- or cage-bound moieties. If this is proven correct, this method could then constitute a general approach to the synthesis of a series of ionic (noncoordinating) or predominantly ionic (weakly coordinating) metal–{C$_4$B$_8$} species. Such substances would be of particular

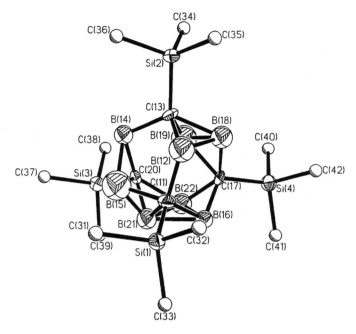

Figure 18-17 Solid-state X-ray structure of the EPR-silent monolithium "carbons-apart" $\{C_4B_8\}$ compound $[\{Li(THF)_4\}\{(SiMe_3)_4C_4B_8H_9\}]$.

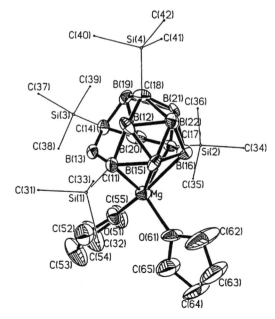

Figure 18-18 Crystal structure of the novel "carbons-apart" $\{C_4B_8\}$ magnesacarborane $[(THF)_2Mg(SiMe_3)_4C_4B_8H_8]$.

interest as noninteracting anions in the production of cationic Group 4 metallocenes used in the Ziegler-Natta catalyzed olefin polymerization reactions. Irrespective of the potential uses of the products, to our knowledge these are the first examples of reactions in which a neutral carborane acts as a restricted electron-acceptor (electron sponge) that removes only the valence electrons of a single Group 1 or Group 2 metal, even when a large excess of the particular metal is available. Whereas the solid-state structure of the lithium compound (Figure 18-17) consists of a single anionic $[(SiMe_3)_4C_4B_8H_9]^-$ unit that is well separated from a discrete $[Li(THF)_4]^+$ cation, the structure of the magnesacarborane (Figure 18-18) constitutes the first example of a cluster that combines both electron-precise atoms [three-coordinate B(13), four-coordinate C(14), and the four-coordinate $(THF)_2Mg$ moiety] and an electron-deficient molecular unit (the *arachno*-$(CR)_3B_6H_9$ fragment) in a single cage framework. At present, it is not known whether this compound is an isolated example of this behavior, or the first of a series of structurally new metallacarboranes.

18.8 CURRENT AND FUTURE DIRECTIONS

The above discussion clearly demonstrates that metallacarboranes of the *s*-block elements possess unique reactivity and structural properties. With the discovery of a convenient synthetic route to the "carbons-apart" $\{C_2B_4\}$ carborane anions, and the "carbons-apart" neutral $\{C_4B_8\}$ carboranes, it is expected that the metallacarboranes of these new ligands will display even more fascinating structural and reactivity patterns than those found in the "carbons-adjacent" system. Also, they may provide new opportunities in their utilization as least coordinating or noncoordinating 1:1 "ionic" species.

ACKNOWLEDGMENTS

Dedicated, with all best wishes, to Dr. Robert E. Williams of the Loker Hydrocarbon Research Institute of the University of Southern California on the occasion of his 70th birthday. This work was supported, in part, by grants from the National Science Foundation, the Robert A. Welch Foundation (N-1016 and N-1322), and the donors of the Petroleum Research Fund, administered by the American Chemical Society. N.S.H. thanks the Camille and Henry Dreyfus Foundation for a Scholar/Fellow Award. The perseverance of numerous undergraduate students, postdoctoral associates, and other coworkers in many of these studies is gratefully acknowledged.

REFERENCES

1. (a) Hawthorne, M. F., Young, D. C., and Wegner, P. A., *J. Am. Chem. Soc.* 1965, **87**, 1818; (b) Hawthorne, M. F., Young, D. C., Garrett, P. M., Owen, D. A., Schwerin,

S. G., Tebbe, F. N., and Wegner, P. A., *J. Am. Chem. Soc.*, 1968, **90**, 862; (c) Hawthorne, M. F., Young, D. C., Andrews, T. D., Howe, D. V., Pilling, R. L., Pitts, A. D., Reintjes, M., Warren, L. F., Jr., and Wegner, P. A., *J. Am. Chem. Soc.*, 1968, **90**, 879.

2. For general discussions see: *Comprehensive Organometallic Chemistry*, Vol. 1. (Abel, E. W., Stone, F. G. A., and Wilkinson, G., eds), Pergamon, Oxford, 1995, Chaps. 6–8.

3. (a) Grimes, R. N., *Carboranes*, Academic Press, New York, 1970; (b) Hosmane, N. S. and Maguire, J. A., in *Electron-Deficient Boron and Carbon Clusters* (Olah, G. A., Wade, K., and Williams, R. E., eds), Wiley, New York, 1991, Chap. 9; (c) Hosmane, N. S. and Maguire, J. A., *Adv. Organomet. Chem.*, 1990, **30**, 99; (d) Hosmane, N. S. and Maguire, J. A., *J. Cluster Science*, 1993, **4**, 297.

4. (a) O'Neill, M. E. and Wade, K., in *Comprehensive Organometallic Chemistry*, Vol. 1, (Wilkinson, G. and Stone, F. G. A., eds), Pergamon Press, Oxford, 1982, Chap. 1, and references therein; (b) Wade, K., *Adv. Inorg. Chem. Radiochem.*, 1976, **18**, 1; (c) Williams, R. E., *Adv. Inorg. Chem. Radiochem.*, 1976, **18**, 67; (d) Williams, R. E., *Chem. Rev.*, 1992, **92**, 177.

5. Hosmane, N. S., Sirmokadam, N. N., and Mollenhauer, M. N., *J. Organomet. Chem.*, 1985, **279**, 359; Hosmane, N. S., Mollenhauer, M. N., Cowley, A. H., Norman, N. C., *Organometallics*, 1985, **4**, 1194; Barreto, R. D. and Hosmane, N. S., *Inorg. Synth.*, 1992, **29**, 89.

6. Hosmane, N. S., Maldar, N. N., Potts, S. B., Rankin, D. W. H., and Robertson, H. E., *Inorg. Chem.*, 1986, **25**, 1561.

7. Ledoux, W. A. and Grimes, R. N., *J. Organomet. Chem.*, 1971, **28**, 37.

8. Hosmane, N. S., Islam, M. S., and Burns, E. G., *Inorg. Chem.*, 1987, **26**, 3236.

9. Hosmane, N. S., Dehghan, M., and Davies, S., *J. Am. Chem. Soc.*, 1984, **106**, 6435.

10. Onak, T. and Dunks, G. B., *Inorg. Chem.*, 1966, **5**, 439.

11. Fessler, M. E., Whelan, T., Spencer, J. T., and Grimes, R. N., *J. Am. Chem. Soc.*, 1987, **109**, 7416.

12. Siriwardane, U., Islam, M. S., West, T. A., Hosmane, N. S., Maguire, J. A., and Cowley, A. H., *J. Am. Chem. Soc.*, 1987, **109**, 4600.

13. Hosmane, N. S., Siriwardane, U., Zhang, G., Zhu, H., and Maguire, J. A., *J. Chem. Soc., Chem. Commun.*, 1989, 1128.

14. Hosmane, N. S., Saxena, A. K., Barreto, R. D., Zhang, H., Maguire, J. A., Jia, L., Wang, Y., Oki, A. R., Grover, K. V., Whitten, S. J., Dawson, K., Tolle, M. A., Siriwardane, U., Demissie, T., and Fagner, J. S., *Organometallics*, 1993, **12**, 3001.

15. Hosmane, N. S., Jia, L., Wang, Y., Saxena, A. K., Zhang, H. and Maguire, J. A., *Organometallics*, 1994, **13**, 4113.

16. Wang, Y., Zhang, H., Maguire, J. A., and Hosmane, N. S., *Organometallics*, 1993, **12**, 3781.

17. Hosmane, N. S., de Meester, P., Siriwardane, U., Islam, M. S., and Chu, S. S. C., *J. Am. Chem. Soc.*, 1986, **108**, 6050.

18. Hosmane, N. S., Yang, J., Zhang, H., and Maguire, J. A., *J. Am. Chem. Soc.*, 1996, **118**, 5150.

19. Cotton, F. A. and Wilkinson, G., *Advanced Inorganic Chemistry*, 5th Edition, Wiley, New York, 1988.

20. Harder, S. and Prosenc, M. H., *Angew. Chem. Int. Ed. Engl.*, 1994, **33**, 1744.

21. Hosmane, N. S., Barreto, R. D., Tolle, M. A., Alexander, J. J., Quintana, W., Siriwardane, U., Shore, S. G., and Williams, R. E., *Inorg. Chem.*, 1990, **29**, 2698.

22. Hosmane, N. S., Zhang, H., Maguire, J. A., Wang, Y., Thomas, C. J., and Gray, T. G., *Angew. Chem. Int. Ed. Engl.*, 1996, **35**, 1000.

23. Lipscomb, W. N. and Britton, D., *J. Chem. Phys.*, 1960, **33**, 275; Lipscomb, W. N., *Science*, 1966, **153**, 373.

24. Maxwell, W. M., Miller, V. R., and Grimes, R. N., *Inorg. Chem.*, 1976, **15**, 1343; Maxwell, W. M., Miller, V. R., and Grimes, R. N., *J. Am. Chem. Soc.*, 1976, **98**, 4818; Maynard, R. B., Grimes, R. N., *J. Am. Chem. Soc.*, 1982, **104**, 5983; Maynard, R. B. and Grimes, R. N., *Inorg. Synth.*, 1983, **22**, 215; Grimes, R. N., *Adv. Inorg. Chem. Radiochem.*, 1983, **26**, 55.

25. Zhang, H., Wang, Y., Saxena, A. K., Oki, A. R., Maguire, J. A., and Hosmane, N. S., *Organometallics*, 1993, **12**, 3933.

26. Hosmane, N. S., Jia, L., Zhang, H., Bausch, J. W., Prakash, G. K. S., Williams, R. E., and Onak, T. P., *Inorg. Chem.*, 1991, **30**, 3793.

27. Ezhova, M. B., Zhang, H., Maguire, J. A., and Hosmane, N. S., *J. Organomet. Chem.*, 1998.

28. Lindsell, W. E., in *Comprehensive Organometallic Chemistry*, Vol. 1 (Wilkinson, G., Stone, F. G. A., and Abel, E. W., eds), Pergamon, Oxford, 1982, Chap. 4.2.

29. Hanusa, T. P., *Polyhedron*, 1990, **9**, 1345.

30. Raston, C. L. and Salem. G., in *Chemistry of the Metal Carbon Bond*, Vol. IV (Hartley, F. R., ed.), Wiley, Chichester, 1987, Chap. 2.

31. Hanusa, T. P., *Chem. Rev.*, 1993, **93**, 1023.

32. Popp, G. and Hawthorne, M. F., *J. Am. Chem. Soc.*, 1968, **90**, 6553.

33. Popp, G. and Hawthorne, M. F., *Inorg. Chem.*, 1971, **10**, 391.

34. Khattar, R., Knobler, C. B., and Hawthorne, M. F., *J. Am. Chem. Soc.*, 1990, **112**, 4962.

35. Khattar, R., Knobler, C. B., and Hawthorne, M. F., *Inorg. Chem.*, 1990, **29**, 2191.

36. Bunder, W. and Weiss, E., *J. Organomet. Chem.*, 1975, **92**, 1.

37. Denton, D. L., Clayton, W. R., Mangion, M., Shore, S. G., and Meyers, E. A., *Inorg. Chem.*, 1976, **15**, 541.

38. Hosmane, N. S., Zhu, D., McDonald, J. E., Zhang, H., Maguire, J. A., Gray, T. G., and Helfert, S. C., *J. Am. Chem. Soc.*, 1995, **117**, 12362.

39. Maxwell, W. M., Bryan, R. F., and Grimes, R. N., *J. Am. Chem. Soc.*, 1977, **99**, 4008; Grimes, R. N., Pipal, J. R., and Sinn, E., *J. Am. Chem. Soc.*, 1979, **101**, 4172; Maxwell, W. M., Weiss, R., Sinn, E., and Grimes, R. N., *J. Am. Chem. Soc.*, 1977, **99**, 4016.

40. Grimes, R. N., in *Comprehensive Organometallic Chemistry*, Vol. 1 (Wilkinson, G., Stone, F. G. A., and Abel, E. W., eds.), Pergamon, Oxford, 1982; Chap. 5.5, p. 459.

41. Hosmane, N. S., Zhang, H., Wang, Y., Lu, K-J., Thomas, C. J., Ezhova, M. B., Helfert, S. C., Collins, J. D., Maguire, J. A., Gray, T. G., Baumann, F., and Kaim, W., *Organometallics*, 1996, **15**, 2425.

INDEX